The New Naturalist Library

A SURVEY OF BRITISH NATURAL HISTORY

THE NEW NATURALISTS

HALF A CENTURY OF BRITISH NATURAL HISTORY

The aim of this series is to interest the general reader in the wildlife of Britain by recapturing the enquiring spirit of the old naturalists. The editors believe that the natural pride of the British public in the native flora and fauna, to which must be added concern for their conservation, is best fostered by maintaining a high standard of accuracy combined with clarity of exposition in presenting the results of modern scientific research.

'The New Religion', the unofficial title of an oil painting of four naturalists by L.J. Watson, 1946. The four are (from top left, clockwise): A.G. Tansley, A.S. Watt, Cyril Diver and E.B. Ford. A bystander, observing the quartet, asked what they were doing. 'Don't you know?' replied Watson. 'It's the new religion!' (*Photo: Peter Wakely/English Nature*)

The New Naturalist

THE NEW NATURALISTS

Peter Marren

With sixteen colour plates and over 100 black and white photographs and drawings

COLLINS

HarperCollins*Publishers*
London Glasgow Sydney Auckland
Toronto Johannesburg

For Christopher

And this our life, exempt from public haunt,
Finds tongues in trees, books in the running brooks,
Sermons in stones, and good in everything.

William Shakespeare
As You Like It, act 2, scene 1, 15-17.

Discovery consists of seeing what everybody has seen and thinking what nobody has thought.

Albert von Szent-Györgyi 1893-1986, Hungarian biochemist

First published 1995
Second edition published 2005

ISBN 000 719716 0 (Hardback)
ISBN 000 719715 2 (Paperback)

Designed and typeset by British Wildlife Publishing, Rotherwick, Hampshire
Printed and bound in Thailand by IMAGO

Contents

Foreword

Stefan Buczacki

I've always thought of myself as a New Naturalist – a hard-covered version of course – and with rather good reason. The series and I were born in the same year, we grew up together and I have read them all. It's my great pleasure to own a complete set of first editions, together with all the Monographs, Country Naturalists and Countryside series. They are, however, not only among the most prized possessions in my library, they are also among the most used; no brown paper wrappers in dark places for my set. I refer to one or other almost daily and that in itself, to use an appropriate expression, speaks volumes. Whilst New Naturalists are now bought, collected and hoarded by some like stamps or coins, they are far more importantly still practical, for the most part highly readable, and certainly highly relevant compilations, and accounts of the numerous facets of British natural history. And as I pondered my response to Peter Marren's most kind invitation to write this Foreword to the new edition of his utterly excellent and beautifully-written series tribute, two things struck me. First, that I must be reasonably typical of a whole generation of naturalists in that by profession and training I embrace one discipline (in my case botany) but in interest and activity, I take in many others; and second, that the scope of the New Naturalist series remarkably mirrors both my own career and the evolution of British natural history in my lifetime. It is all a startling testament to the vision and foresight of its creators.

I have always lived in a *Country Parish*, I began my natural history like so many others, by collecting *Butterflies* and *Moths*, and became excited whenever one of our *British Mammals* or better still, one of the few *British Amphibians and Reptiles* made an appearance in the garden. I became passionate about *British Plant Life* and started my own herbarium in order to identify *Wild Flowers*, especially the *Wild Flowers of Chalk and Limestone* and the *Mountain Flowers* in the *Peak District* close to where I grew up. We lived alongside a small stream and I was entranced by *Life in Lakes and Rivers* and, as a trout fisherman, inevitably took an interest in *An Angler's Entomology*. Then, in the nineteen fifties and early sixties, I travelled to more distant areas of the country, fell in love with *Mountains and Moorlands*, saw the *Highlands and Islands, The Weald, Dartmoor, Snowdonia* and began to discover *The World of Spiders, Insect Migration* and *Woodland Birds*.

As a native of Derbyshire, arguably the most inland of British counties, venturing to *The Sea Coast* opened up a new world to me, the world of *Sea Birds*, *The Open Sea* and the *Flowers of the Coast.* As I matured as a naturalist, I began to appreciate an understanding of the earth on which we lived and became fascinated by *Britain's Structure and Scenery*, by *Fossils*, indeed by the whole complex partnership of *Man and the Land*. This saw fruit at university where as an undergraduate, I read both botany and geology and then became hooked on *Mushrooms and Toadstools* – to the extent that years later I was to become one of John Ramsbottom's successors as President of the British Mycological Society – and then during my research years in forestry, admired the complex interrelationship between *Trees, Woods and*

Man. Later still, when I was engaged in horticultural research and immersed in *The World of the Soil* in the nineteen seventies, it was the post-Rachel Carson era, when *Pesticides and Pollution* were beginning to be understood and the uneasy co-existence of *Farming and Wildlife* appreciated. I learned about *Grass and Grasslands, Weeds and Aliens* and *Hedges*. And more recently, of course, the importance of *Nature Conservation in Britain* has dawned on everyone. How unerringly and accurately the New Naturalists have charted the course of my life and career, and of post-war British natural history. How cleverly the Collins founders foresaw such changes in legislation and attitude as those embodied in the 1981 Wildlife and Countryside Act which created the SSSIs and the 2000 Countryside and Rights of Way Act – the 'right to roam' law.

For a young naturalist like me in the nineteen fifties and sixties, the New Naturalist authors were iconic figures, and it's a great sadness that by the time I became old and bold enough to approach such people, the earlier ones had largely passed away. The first author I met was the extraordinary E. B. Ford at Oxford, but the one I came to know best and was the most influential by far was a truly quintessential naturalist. She was however one of the forgotten few, the 'contributing' authors whom Peter glossed over in his first edition and who were sadly never accorded a proper biography because they didn't appear on a title page. Joyce Lambert contributed chapters on broadland vegetation and on her own seminal studies of broadland origin to the original version of *The Broads* and in tribute to the unsung heroes and heroines behind the New Naturalist series, I shall take just a few lines to amend Peter's omission.

Joyce taught me ecology and instilled in me her quite perfect and utterly simple definition of what is now considered so complex a discipline. 'Ecology' she said, 'is what lives where, and why'. She had learned her ecology at Cambridge under Sir Harry Godwin and once recounted to me a story of her own early undergraduate days when Godwin gave her a huge ecological tome and told her to read it and distil its critical message into a 3000 word essay by the following Tuesday. 'But Professor Godwin', she complained, 'it's by Braun-Blanquet and I don't speak German'. 'Then I suggest you learn it' was Godwin's reply. She did.

What an essential naturalist Joyce Lambert was, in the very finest New Naturalist tradition. Although a professional botanist, she was well versed in all branches of the living world and I have two enduring undergraduate memories of her. The first was on a university field course visit to Ted Ellis at his home on Wheatfen Broad. We all sat around Ted's house, having a Sunday lunch picnic among the reeds on a sweltering August day, with most of the insect life of the Norfolk Broads taking a serious interest in our meal, while Ted and Joyce, utterly oblivious to being eaten alive, argued the identity of some of the largest and most vicious species of Diptera I have ever seen. The second was watching with dismay in the New Forest as our teacher, this short, stout woman, missed her footing while lecturing her charges on the structure of a Schwingmoor or floating bog and began to descend downwards through the raft of peat with slow and graceful inevitability towards the cold watery depths below. Joyce's proportions however meant she became wedged at the waistline and later, having been hauled free from the filthy black hole by a small army of students, expressed equal anxiety at having possibly disturbed the breeding site of a rare dragonfly and at the soggy state of her Woodbines. As may be said so accurately of so many early New Naturalist authors – they don't make them like that any more.

Since that meeting with E. B. Ford, I have been privileged to meet, know or correspond with other series authors and most recently, was delighted to receive much help from the only one to have three bars to his distinction. Eric Simms, who penned four bird titles, was a fount of information and anecdote when I was writing my *Fauna Britannica*. It has been suggested however that Eric Simms' contributions, marvellous as they are, in fact highlighted a series flaw – that if the Monographs are included, there have been disproportionately too many bird books. Possibly, but there are after all plenty of birds – far more than mammals, reptiles and amphibians put together and here too, I think the editors – under James Fisher's tutelage – spotted a trend early on. If membership of the RSPB charts the progress of a national interest in birds, the figures speak for themselves. When the series began in 1945, there were but a handful of members; they only passed 10,000 in 1960. Today there are over a million, ample justification I would have thought for the New Naturalists being generous with ornithology. I must add that whilst sadly I never met James Fisher, I treasure my all too brief friendship with his son and editorial successor Crispin, a marvellously personable and gentle man although one who I don't think ever forgave me my fondness for fuchsias, plants that he abhorred beyond all reason.

In mainstream botany and zoology, there is a constant argument between the protagonists of regional floras and faunas on the one hand and the protagonists of monographs on the other. The New Naturalist series has very gingerly trodden a median path. The Monographs themselves were an erratic adjunct to the main series, were published for only 23 years; and sad to relate, included only one botanical title. But I do hope the present editors will continue to weave a little monography into the main series and not perceive the monograph as a purely scientific publication rather than genuine natural history. Although in saying this, I confess to not really knowing where science ends and natural history begins and that perhaps is the secret of the New Naturalists' success. They are neither popular natural history nor hard-core science. They are quite simply unique, and that really is their enduring merit and appeal.

As they created the yardsticks by which the progress of natural history in Britain can be charted, the huge talent of the founding New Naturalist authors was that although they were of an earlier and older generation, most of them instinctively understood how to communicate with a modern and younger one. Not many were teachers in a professional sense, but it was in the skill of the editors to have found men and women of science who bucked the trend – because, let's be honest, most scientists have traditionally been abysmal communicators. Perhaps things are becoming a little easier as the modern author generation has grown up with more immediate access to a wider range of communication media. I do hope so, as I look forward with some anxiety to joining this distinguished throng with my forthcoming *Garden Natural History*. I reassure myself that when people have asked me what are my driving passions, I have always replied, with genuine honesty, 'Living things – and telling people about them'. And that does seem after all to be the very ethos of The New Naturalists, the books and the people.

Author's Foreword

Genuine surprises are rare, or so I have usually found. Pleasure or disappointment at the unexpected tend to be tempered by the fact that you half-expected it all along. But a telephone conversation two years ago was something else – as real a knock-me-down-flat, well I'm expletive-deleted surprise that one could possibly wish for. Knowing that it would soon be 50 years since the publication of E.B. Ford's famous book, *Butterflies*, I had written a couple of trifling pieces to draw attention to the New Naturalist series. I had also, in a desultory way, accumulated a certain amount of background material on the authors of the series, partly to gratify my own curiosity but also with the vague intention of turning it into something a bit meatier than a short article. One day, while tramping across the withered heath of Thorne Waste with Derek Ratcliffe, I happened to mention these biographical rummagings. Derek responded with some first-hand stories of his own about Pearsall, Ford *et al*, and there, I thought, the matter had ended.

Then came the surprise. Some time in June 1993 I had telephoned HarperCollins to ask if any internal archive on the series existed, and if so whether I could have a look at it. The cheerful female voice on the other end of the line thought that there was, and, yes, that I probably could.

'Oh, and by the way,' she added, 'we are planning a special volume for the 50th anniversary of the series. I thought you sounded familiar. Your name has been put forward as a possible author.'

At which point, I said either 'who?' or 'what!' or possibly both.

'Come to lunch next week, and we'll talk about it then.'

It transpired that the idea of celebrating the series had been on the editorial table for some time, so that what for me was a lightning bolt from an empty sky was for those more closely involved an increasingly urgent problem. As it happened, I was coming to the end of a long book and was rather wondering what to do next. So I ate my lunch, and took what was offered, and said, yes please.

No sooner had I started to research the series in earnest than I realised it was necessary, as the American writer Shelby Foote put it, 'to go spread-eagle whole hog on the thing'. There are 102 New Naturalist books, if we include the monographs (as, indeed, we must), and at least that number of authors. The main series covers almost every imaginable facet of British natural history as it has developed during these past 50 years and the authors themselves form a display of biodiversity as kaleidoscopic as any latter-day environmentalist might wish for. I did some more delving, in London, Oxford, Edinburgh and other places; I wrote a lot of letters and read an awful lot of files, papers, memoirs and other biographical odds and ends; and renewed some old acquaintances and made some new ones.

From the resultant heap emerged a story of New Naturalist life and times. I say 'a' rather than 'the' because my story is a personal interpretation, a matter of selection and parings, of deductions, opinions and no doubt prejudices. It is a story that has not yet ended, thank goodness, though it has a very definite beginning: in the publishing ambitions of Billy Collins, in the popular science of Julian Huxley and James Fisher, and in the cultural impact of the Second World War. The books were an innovation on several levels: as the first serious attempt to bridge the divide between popular natural histories and the professional journal; as the first large-scale use of outdoor colour photographs; and, as the most wide-ranging natural survey of these islands ever made. The series could probably not have happened at any other time. Its success was a happy combination of technological advance, the right authors, a novel way of commissioning books and, above all, of perfect timing. And it was therefore not only a scientific sensation and a *succes d'estime*, but a commercial success as well. The first titles sold by the tens of thousands: to ex-servicemen, starved of the sights and sounds of the countryside, to evacuees who had just experienced their first taste of them, to schools and colleges with a new interest in field study, and to anyone with a more than passing interest in nature. The series was a product of the age, and an upbeat age it was, despite the tragedies and austerities of the war. The people had been promised a new Britain, and National Parks, nature reserves and better access to the countryside were all part of that vision. Nature was the perfect antidote to war memories.

It follows that this is a story with roots and tentacles. It begins at a luncheon for four in a Soho restaurant, followed by the establishment of an editorial board, the search for authors and for colour photographs, and for a distinctive image exemplified by the distinctive Ellis dust jackets. Since we are celebrating the half centenary not only of the series but of a particular book, I have devoted a chapter to *Butterflies* and its singular author, E.B. Ford. We then move on to a review of the series as a whole, of writers and their writings and what they have to tell us about the nature of field study in Britain. After a short digression on promising titles that were never made into books, we reach the more recent years of the series and its survival in more uncertain times. Finally we take a briefer look at the role of the New Naturalist series and authors in shaping the development of nature conservation in Britain. The chapters are all to some extent watertight, and can be read in any order, though they are intended to form a gradually unrolling story. To round off the tale, I have commissioned two short appreciations of the New Naturalist library told from a more personal perspective by Peter Schofield and Peter Laslett, both keen collectors and readers of the series. Two of the three appendices are factual: short biographies of all but a handful of the authors of the series, and a full bibliography based on the internal records of HarperCollins. The third appendix is about collecting New Naturalist books, a healthy pursuit which has grown in popularity during the past ten years.

I have not interrupted the main text with source references since much of this book is based on unpublished material, though I have indicated my main sources at the end of the book. If anyone finds that I have quoted them out of context, or without permission, I apologise in advance. The responsibility for any mistakes or distortions is, of course, entirely my own.

Here is the book. I hope you read it with as much enjoyment as I have experienced in writing it. And let us hope that there will be another like it, 50 years onwards, to mark the centenary of the New Naturalist library and numberless classics yet unwritten.

Acknowledgements

I have many people to thank. I could not have asked for a more perceptive and constructive editor than Derek Ratcliffe, nor a more cheerfully efficient and helpful one than Isobel Smales. Both have read and commented on the full text, and working with them has been, to put it bluntly, good fun. HarperCollins gave me the full run of their New Naturalist archive and current files, and were admirably broad-minded in allowing me to include material, such as print-runs, which is normally a commercial secret. I thank Myles Archibald for sharing some of his memories of the series and reading some of the text with a friendly eye. To the New Naturalist authors and photographers with whom I have corresponded, or interviewed, I owe a special debt of thanks, since the book could not have been written in this way without their help. They are: Mr Robert Atkinson, Mr Sam Beaufoy, Professor R.J. Berry, Dr J. Morton Boyd CBE, Dr Colin G. Butler FRS, Mr Niall Campbell, Mr William Condry, Professor Philip Corbet, Dr Brian Davis, Mr Richard Fitter, Professor John Free, Mr Fred Goldring, Professor W.G. Hale, Mr Laughton Johnston, Dr Peter Maitland, Dr Michael Majerus, Dr Norman Moore, Dr Ernest Neal MBE, Dr Ian Newton FRS, Mr Max Nicholson CB, Dr Christopher Page, Dr Christopher Perrins, Dr Ernest Pollard, Dr Michael Proctor, Dr Miriam Rothschild FRS, Mr Eric Simms DFC, Dr Max Walters, Dr E.B. Worthington CBE and Dr Peter Yeo. The following have kindly helped me with information on deceased authors: Dr Douglas Bassett (F.J. North), Dr J. Morton Boyd (F. Fraser Darling), Dr Cedric Collingwood (D. Wragge-Morley), Mrs Phyllis Ellis (E.A. Ellis), Dr J. Elmes (M.V. Brian), Dr Clemency Fisher (James Fisher), Dr Tegwyn Harris (L.A. Harvey), Mr David Hosking (Eric Hosking), Desmond and Maimie Nethersole-Thompson, Mr Ian Pettman (T.T. Macan), Mr Steven Simpson (Malcolm Smith), Mr Ioan Thomas (Ian Hepburn), Mr A.F. Vizoso (Monica Shorten), Mr Tom Wall (A.W. Boyd), Dr Charles Watkins (K.C. Edwards) and Dr Max Walters (John Gilmour and John Raven). For memories of E.B. Ford, I thank Professor Berry, Professor Bryan C. Clarke FRS, Mr John S. Haywood, Mr Thomas Huxley, Dr John Pusey and Miriam Rothschild. I thank Miriam Rothschild, too, for allowing me to reprint her article *My First Book*, first published in *The Author*, the journal of the Society of Authors.

I am particularly indebted to Rosemary Ellis and her daughters, Penelope and Charlotte, for their hospitality, for providing valuable information about the life and work of Clifford and Rosemary Ellis, and for commenting so constructively on that part of the book. I also spent an enjoyable day in the company of Robert Gillmor, and am grateful to him for the insights into his work for the series and for reading the relevant chapter.

I thank Dr Peter Laslett and Mr Peter Schofield for their respective contributions reproduced at the end of this book.

The artwork reproduced in this book was all photographed under my supervision with great skill and patience by Mr Mike Amphlett. Many people have kindly entrusted photographs to me, and I am indebted to the following for their help in tracking down suitable material: Mrs Mary Briggs, Mrs Sheila Edwards, Dr Clemency Fisher, Mr John Haywood, Mr David Hosking, Dr A.F. Millidge, Mr Ian Pettman, Dr Derek Ratcliffe, Mr John Thackray, Mr Ioan Thomas, Mr Peter Wakely, Dr Max Walters and Dr Charles Watkins. I am grateful too for the assistance of archivists and librarians at The Bodleian Library, The Natural History Museum, The Royal Botanic Gardens at Kew and Edinburgh, English

Nature, the British Trust for Ornithology, the Royal Society, the Linnaean Society and the Freshwater Biological Association. Bloomsbury Book Auctions kindly allowed me access to the private papers of James Fisher.

I thank the following for permission to reproduce copyright material: Eric and David Hosking (pp. 14, 20, 22, 30, 37, 43, 51, 53, 121, 129, 149, 169, 191, 196, 198), the Natural History Museum picture library (pp. 126, 128, 167, 211), English Nature (frontispiece, pp. 47, 110, 151, 230, 234, 236, 237), Mr John S. Haywood (pp. 93, 95, 100, 103, 114), Royal Botanic Gardens, Kew (pp. 150, 160, 161), Freshwater Biological Association (pp.125, 133, 201), Rothamsted Experimental Station (pp. 58, 166), Botanical Society of the British Isles (pp. 27, 138), John Innes Institute (p. 189), Mr Sam Beaufoy (pp. 53, 106), Institute of Terrestrial Ecology (p.144), Dept. of Zoology, Cambridge (p. 152), St Catherine's College, Cambridge (p. 235), Dept. of Geography, Royal Holloway (p. 168), Dept. of Geography, Nottingham University (p. 158), Paul Sterry/Nature Photographers (p. 16), Kate Collinson/Norman Parkinson Studio (p. 66), *The Old Oundelian* (p. 159), Dr Nigel Westwood (p. 203), Mrs Faith Raven and King's College, Cambridge (p. 136), Mr D.R. Nellist (p. 140), Ms Anna Fitter (p. 120), Mr Euan Dunn (p. 196), Mr Dick Balharry (Plate 16).

Finally I have, as ever, to thank three faithful friends: Dick Seamons for his help with library searches and his roving eye for the more bizarre items in current journals, and Maureen Symons for deciphering my heavily edited manuscripts and turning them with great speed into faultless type with a cheerfulness and humour that never showed a sign of the long, ungodly hours she spent on it. Also to Nicky Salfranc for reading and commenting on several chapters with the gleeful zeal of a linguist. Thanks, chums.

That I was able to spend a year with the New Naturalists was in large part due to my brother Christopher, who provided a roof over my head while I scribbled. To him this book is dedicated, with fraternal affection.

P.R.M.

Newtown Lodge,
Ramsbury.
September 1994

Author's Foreword to the Second Edition

The 60th anniversary of the New Naturalist Library has provided a welcome opportunity to update *The New Naturalists*. The book was first published in 1995 and went out of print within two years. Like most recent out-of-print titles, it is now scarce and expensive to obtain secondhand. For the new edition I have made a substantial number of changes and additions. There is a new chapter devoted mainly to the events of the past ten years, which include 13 new titles and many more in preparation. Wherever possible, I have corrected such minor mistakes that have been pointed out over the years (bibliography is as exact a trade as any science), and have also amplified the original text with a series of notes. The appendices have all been substantially updated and in part rewritten, and they include potted biographies of all the recent authors in the series. As a refresher, the original postscript has been replaced by a witty and fantastical short story of the-book-that-never-was by John Sykes, first published in the New Naturalists Book Club newsletter, and reprinted here by permission of its editor, Robert Burrow. The original colour plates have been revised to include more of Robert Gillmor's artwork for the magnificent jackets he has designed for the hardbacks since 1986. Revising the book was a great pleasure, and I can only hope you enjoy reading it as much.

I have many people to thank. First to the new boys of the series for their willing cooperation in distilling their distinguished careers into a pithy line or two, and for sharing their experiences in writing for the series: John Altringham, Trevor Beebee, David Cabot, Oliver Gilbert, Richard Griffiths, Peter Haywood, David Ingram, Andrew Lack, John Mitchell, Brian Moss and Derek Ratcliffe. I also thank the following for their kind help over earlier authors omitted from the original book: Mrs Jane Bright (for Deryk Frazer), Dr Jimmy Chubb (J.W. Jones) and Dr Carolyn King (Harry Thompson). Bob Burrow, who runs the New Naturalist Bookshop in Jersey, advised me on current secondhand prices. John Sykes allowed me to include a slightly revised version of his very funny and well-imagined short story. Simon Appleton of Saxon Ltd patiently explained the intricacies of colour printing. Michael Majerus generously took time out of a busy schedule to pen a defence of the great E.B. Ford, whose reputation has been maligned recently in a much-reviewed book. Robert Gillmor contributed to this book in all sorts of ways, not least in designing another beautiful jacket for it. And Stefan Buczacki, as a forthcoming New Naturalist writer himself, and a long-standing fan of the series, nobly agreed to pen a foreword. Finally, I am grateful to Myles Archibald for commissioning a new edition and agreeing to be interviewed for it, and to my editor, Helen Brocklehurst, for seeing it through to publication. Against all the odds, the series is in remarkably good heart – better indeed than in 1995. Perhaps a good idea is immortal after all.

Peter Marren

Ramsbury
September 2004.

1

A New Natural History

The first New Naturalist book I can remember was a battered copy of *Butterflies* in the school library when I was about 14. I remember being struck by the dust jacket, torn and grubby as it was, which was not like that of any other natural history book I had ever seen. The contents were even more curious: a magnificent set of colour plates, and a readable text which explained the significance of all those varieties and 'aberrations' in Nature's grand scheme. But mixed in with the interesting stuff were incomprehensible chapters about breeding, involving words like homozygote, allelomorph and polymorphism. I gave those chapters a miss.

A year or two later, I acquired an ex-library copy of Summerhayes' *Wild Orchids of Britain*, which seemed to me wholly excellent. When I saved up enough money to buy a single-lens reflex camera, it was Robert Atkinson's pictures of orchids that I wanted to emulate. Better still was J.E. Lousley's *Wild Flowers of Chalk and Limestone*, back in print again after a long absence. Following Lousley's lead, I took my camera to the White Cliffs of Dover, Painswick Beacon, Avon Gorge and Berry Head, and for a year or two became rather hooked on wild flower hunting. I realise now that I should have been drinking beer and chasing girls, but I don't blame Summerhayes and Lousley for that. Their books seemed to me to be inspirational in a way that others were not. They were about the things I liked: the English countryside and its seasons; the 'localities' hidden away behind the fields and lanes; and nature's quirky and sometimes comic ways: the sex-starved wasps which, for lack of a mate, make do with an orchid flower, the yellow centre of a white rock-rose which spreads out like a fan when touched, and the mysterious *Epipogium* which may flower only once in a lifetime.

The list on the back of the books showed them to be part of an already lengthy and expanding series. By the time I graduated (in botany: thank you, Lousley and Summerhayes) I owned about a third of them. A few I could never get into, but most of them were such a refreshing change from university textbooks and identification guides that I wondered why on earth there were not more like them.

A quarter of a century later, I still wonder that, but I can appreciate why there is only one New Naturalist library. No ordinary mortal sets out on such a venture, nor would he do so in normal times. The New Naturalist library exists because one man wanted to produce a series of books so different to what had gone before that it would set a new standard of natural history publishing. It developed in the way it did for two main reasons. The first was 'the coming of age' of the relatively new science of ecology, which not only united the hitherto disparate strands of field study but promoted new ways of observing and understanding nature. Ecology was the science of *relationships*, between living things and their different environments – food, hiding places, burrows and nests, air, water and rock. It was field study transformed into a respectable (because quantifiable) science. The second reason was the new developments in photography, pioneered by Kodak in America, that enabled skilled naturalists for the first time to record wild nature in colour. That in itself was a major development which changed the habits of naturalists and publishers alike. The butterfly, bird or flower need no longer be depicted pinned, stuffed or pressed but as it occurs in life, and in its natural habitat.

The circumstances might have been propitious, but without the right kind of experts, willing to organise and write for the series or take the photographs, it would not have lasted five years, let alone (the so far) fifty. Here lay the greatest good fortune of all. In 1945, there was a considerable number of full-time and amateur naturalists who were not only expert on a particular subject but could think in the round and put their thoughts down on paper in a way that any reasonably well-educated person could understand. Some of these people had already written successful books, or were known to the public through wireless broadcasts and newspaper nature diaries. One, in particular – Julian Huxley – was not only a gifted scientific populariser, but a man of influence. It was the commercial resources of a publisher, W.A.R. Collins, allied to the intellectual resources of Huxley, that provided the key to the success of the series. Collins could print the books, Huxley could find the authors. And at this time, authors of the highest calibre were willing and able to write the sort of books that both of them wanted.

But was there a market for such books? The New Naturalist library charted unknown publishing waters. Other publishers seemed convinced that it would be a flop, and Collins must have been only too aware of the commercial risk he was taking, especially since the investment in the series, in terms of photography and editorial time, had been enormous. In the event, the subscription sales of the first books more than fulfilled his expectations. There must have been something special about the readers of 1945. They had not been starved of books during the war. But there had been very little colour printing (with the significant exception of the 'Britain in Pictures' series, published by Collins), and, of course, for most people the opportunities for walking in the country and looking at birds and wild flowers had been very limited. The New Naturalist books must have made a colourful impact in the bookshops of postwar Britain. There was a great deal of popular interest in nature, and, unlike today, no colour television to cater for it. Instead people read books. Their attention spans seem to have been longer, and there was a great deal of interest in adult education, including special grants to turn ex-servicemen into schoolmasters. Popular science was at its peak (who could not have been stirred by the advancing machines of war, culminating in the atomic bomb?). So was a desire for a more peaceful future, in which there would be green spaces around towns and national parks in the hills. These must have been factors that contributed to the success of the series. It was no coincidence that the New Naturalist library and the institutionalisation of field study and nature conservation after the war ran on parallel tracks. Both were responses to a similar and widely felt need. The series was a triumph of good timing. And since these books were marketed as a collectable series, the success of the first half dozen titles ensured a similar success for the second.

What sort of person was W.A.R., usually called 'Billy', Collins? Physically, he was a tall, athletic man, with a quizzical expression, the combination of a gimlet gaze and attractively feathered eyebrows. His manner was gruff and direct, and he could seem impatient. He had three qualities that made him a great publisher: he was an enthusiast; he had an attention for detail that was almost obsessive; and he drove both himself and his colleagues hard. Even his weekends were, we are told, 'vigorously occupied in farming, hunting, gardening, and above all reading

W.A.R. Collins (1900-1976), photographed in 1966 by Eric Hosking.

manuscripts'. He was the old-fashioned sort of publisher who actually read books.

He was Billy Collins the Fifth. Collins the First was a Glasgow schoolmaster who had set up as a printer of religious and educational books in 1819. When his great-great-grandson took over the reins of the family firm in the late 1930s, Collins & Co was still more of a manufacturer than a publisher, printing stationery, diaries and bibles. Its reputation as a publisher of natural history books and novels was largely the work of our Billy Collins. He had entered the family firm in the mid-1920s, and his interests had always been less in the printing side of things than in the more glamorous world of London publishing. He was a good businessman, and knew his way around the international book world. The Dictionary of National Biography notes that 'he knew what the leading booksellers throughout the world had ordered, when they should reorder, when they were overstocked'. He and his wife visited the United States regularly, and on one such trip returned with the manuscript of James Jones' *From Here to Eternity*. He built up a loyal and devoted staff, including Ronald Politzer, the greatest book promoter of his day, and F.T. Smith, the inventor of the Collins Crime Club, and who was, for many years, his right-hand man on the New Naturalist series. It was also our Billy Collins who introduced the clear and beautiful 'Fontana' typeface used in Collins publications from the mid-1930s onwards, and based on eighteenth century lettering. Despite conscription and bombs, Collins had a good war, recruiting works from a number of significant novelists and historians to his list, including Arthur Bryant's patriotic *English Saga*.

In 1944, the Collins London offices were blitzed, and the firm moved to new premises at 13-14 St James's Place, its home up to the 1980s. It was here, in the two Georgian houses, one of brick, the other painted cream, that the New Naturalist Board used to meet around the big polished boardroom table. It was a fitting office for what was still a close-knit family firm: a graceful winding staircase and panelled rooms in which publisher and author could talk books over a glass of madeira. The noise and bustle of the printing factory lay far away in Glasgow, where 2,500 people were employed – 'compositors, machinemen, electro- and stereo-typers, lithographers, binders, marblers, cutters, sewers, pagers, bundlers, chemists and engineers' – over some 13 acres of floor space. It was here that the New Naturalist pages were printed – an edition of 15,000 copies could, on these machines, be printed in as little as an hour – and also their colour plates, rolling round the flashing cylinders 'like a whirling rainbow'.

For most of the war, Billy Collins was thinking about the peace. A countryman and amateur naturalist himself, he wanted more than anything else to develop the firm into a major natural history publisher. *The House of Collins* (1952) relates how 'for a long time he had pondered a new series of illustrated nature books, which would be not merely a popular addition to the literature of natural history, but a series of definitive texts judged by the strictest scientific standards'. This, the New Naturalist library, was intended to be the Collins flagship. Compared with the New Naturalists, all the other well-known natural history books published later by Collins – the Pocket guides, the Field Guides, the New Generation Guides – were secondary in prestige. Billy Collins is said to have regarded the New Naturalist library as his most significant achievement, and when one looks at what else he achieved, that is no small tribute.

The immediate steps taken by Billy Collins to prepare and launch the New Naturalist series are the subject of the next two chapters. What he himself wanted from it was what he called 'a new survey of Britain's natural history .... popular in price, presentation and appeal, that would make available to the general public all the wealth of new scientific knowledge that had been acquired during the present century'. He reckoned it to be 'one of the biggest ventures on which any publishing house has ever embarked', and that was at a time when the library was conceived of as a programme of 'only' 50 titles, produced over six years or so. Today it is said to be the most significant natural history library ever published. It was, above all else, a product of its time, though, like all the best stories, it has grown in the telling. It captures a moment when the great field traditions of British natural history merged with the cutting edge of science and the founding of institutions that remain with us to this day. It was a fleeting moment, and it did not last. But the spirit of the New Naturalist lives on in the new titles being added to the library, and even newer naturalists still refresh their minds by reading the timeless classics of Pearsall, Ford and Yonge. This book is about that spirit, and about the people who embodied it best.

Table 1. The New Naturalist Library 1945-1995

1. BUTTERFLIES (1945) by E.B. Ford
2. BRITISH GAME (1946) by Brian Vesey-Fitzgerald
3. LONDON'S NATURAL HISTORY (1945) by R.S.R. Fitter
4. BRITAIN'S STRUCTURE AND SCENERY (1946) by L. Dudley Stamp
5. WILD FLOWERS Botanising in Britain (1954) by John Gilmour and Max Walters
6. THE HIGHLANDS AND ISLANDS (1964) by F. Fraser Darling and J. Morton Boyd
7. MUSHROOMS AND TOADSTOOLS A Study of the Activities of Fungi (1953) by John Ramsbottom
8. INSECT NATURAL HISTORY (1947) by A.D. Imms
9. A COUNTRY PARISH Great Budworth in the County of Chester (1951) by A.W. Boyd
10. BRITISH PLANT LIFE (1948) by W.B. Turrill
11. MOUNTAINS AND MOORLAND (1950) by W.H. Pearsall
12. THE SEA SHORE (1949) by C.M. Yonge
13. SNOWDONIA The National Park of North Wales (1949) by F.J. North, Bruce Campbell and Richenda Scott
14. THE ART OF BOTANICAL ILLUSTRATION (1950) by Wilfred Blunt
15. LIFE IN LAKES AND RIVERS (1951) by T.T. Macan and E.B. Worthington
16. WILD FLOWERS OF CHALK AND LIMESTONE (1950) by J.E. Lousley
17. BIRDS AND MEN The Bird Life of British Towns, Villages, Gardens & Farmland (1951) by E.M. Nicholson
18. A NATURAL HISTORY OF MAN IN BRITAIN Conceived as a study of changing relations between Men and Environments (1951) by H.J. Fleure
19. WILD ORCHIDS OF BRITAIN, with a key to the species (1951) by V.S. Summerhayes
20. THE BRITISH AMPHIBIANS AND REPTILES (1951) by Malcolm Smith

The New Naturalists library up to No. 58. (*Photo: Paul Sterry/Nature Photographers*)

21. BRITISH MAMMALS (1952) by L. Harrison Matthews
22. CLIMATE AND THE BRITISH SCENE (1952) by Gordon Manley
23. AN ANGLER'S ENTOMOLOGY (1952) by J.R. Harris
24. FLOWERS OF THE COAST (1952) by Ian Hepburn
25. THE SEA COAST (1953) by J.A. Steers
26. THE WEALD (1953) by S.W. Wooldridge and Frederick Goldring
27. DARTMOOR (1953) by L.A. Harvey and D. St Leger Gordon
28. SEA-BIRDS An Introduction to the Natural History of the Sea-Birds of the North Atlantic (1954) by James Fisher and R.M. Lockley
29. THE WORLD OF THE HONEYBEE (1954) by Colin G. Butler
30. MOTHS (1955) by E.B. Ford
31. MAN AND THE LAND (1955) by L. Dudley Stamp
32. TREES, WOODS AND MAN (1956) by H.L. Edlin
33. MOUNTAIN FLOWERS (1956) by John Raven and Max Walters
34. THE OPEN SEA I THE WORLD OF PLANKTON (1956) by Alister Hardy
35. THE WORLD OF THE SOIL (1957) by Sir E. John Russell
36. INSECT MIGRATION (1958) by C.B. Williams
37. THE OPEN SEA II FISH AND FISHERIES With chapters on Whales, Turtles and Animals of the Sea Floor (1959) by Sir Alister Hardy
38. THE WORLD OF SPIDERS (1958) by W.S. Bristowe
39. THE FOLKLORE OF BIRDS An Enquiry into the Origin & Distribution of some Magico-Religious Traditions (1958) by Edward A. Armstrong
40. BUMBLEBEES (1959) by John B. Free and Colin G. Butler
41. DRAGONFLIES (1960) by Philip S. Corbet, Cynthia Longfield and N.W. Moore
42. FOSSILS (1960) by H.H. Swinnerton
43. WEEDS AND ALIENS (1961) by Sir Edward Salisbury
44. THE PEAK DISTRICT (1962) by K.C. Edwards
45. THE COMMON LANDS OF ENGLAND AND WALES (1963) by W.G. Hoskins and L. Dudley Stamp
46. THE BROADS (1965) by A.E. Ellis
47. THE SNOWDONIA NATIONAL PARK (1966) by W.M. Condry
48. GRASS AND GRASSLANDS (1966) by Ian Moore
49. NATURE CONSERVATION IN BRITAIN (1969) by Sir Dudley Stamp
50. PESTICIDES AND POLLUTION (1967) by Kenneth Mellanby
51. MAN AND BIRDS (1971) by R.K. Murton
52. WOODLAND BIRDS (1971) by Eric Simms
53. THE LAKE DISTRICT A Landscape History (1973) by W.H. Pearsall and W. Pennington
54. THE POLLINATION OF FLOWERS (1973) by Michael Proctor and Peter Yeo
55. FINCHES (1972) by Ian Newton
56. PEDIGREE Essays on the Etymology of WORDS FROM NATURE (1973) by Stephen Potter and Laurens Sargent
57. BRITISH SEALS (1974) by H.R. Hewer
58. HEDGES (1974) by E. Pollard, M.D. Hooper and N.W. Moore
59. ANTS (1977) by M.V. Brian
60. BRITISH BIRDS OF PREY (1976) by Leslie Brown
61. INHERITANCE AND NATURAL HISTORY (1977) by R.J. Berry
62. BRITISH TITS (1979) by Christopher Perrins
63. BRITISH THRUSHES (1978) by Eric Simms

64. THE NATURAL HISTORY OF SHETLAND (1980) by R.J. Berry and J.L. Johnston
65. WADERS (1980) by W.G. Hale
66. THE NATURAL HISTORY OF WALES (1981) by William M. Condry
67. FARMING AND WILDLIFE (1981) by Kenneth Mellanby
68. MAMMALS IN THE BRITISH ISLES (1982) by L. Harrison Matthews
69. REPTILES AND AMPHIBIANS IN BRITAIN (1983) by Deryk Frazer
70. THE NATURAL HISTORY OF ORKNEY (1985) by R.J. Berry
71. BRITISH WARBLERS (1985) by Eric Simms
72. HEATHLANDS (1986) by Nigel Webb
73. THE NEW FOREST (1986) by Colin R. Tubbs
74. FERNS Their Habitats in the British and Irish Landscape (1988) by C.N. Page
75. FRESHWATER FISHES OF THE BRITISH ISLES (1992) by P.S. Maitland and R.N. Campbell
76. THE HEBRIDES A Natural History (1990) by J.M. Boyd and I. L. Boyd
77. THE SOIL (1992) by B.N.K. Davis, N. Walker, D.F. Ball and A.H. Fitter
78. BRITISH LARKS, PIPITS AND WAGTAILS (1992) by Eric Simms
79. CAVES AND CAVE LIFE (1993) by Philip Chapman
80. WILD AND GARDEN PLANTS (1993) by Max Walters
81. LADYBIRDS (1994) by Michael E.N. Majerus
82. THE NEW NATURALISTS Half a century of British natural history (1995) by Peter Marren
83. THE NATURAL HISTORY OF POLLINATION (1996) by Michael Proctor, Peter Yeo and Andrew Lack.
84. IRELAND. A natural history (1999) by David Cabot.
85. PLANT DISEASE. A natural history (1999) by David Ingram and Noel Robertson.
86. LICHENS (2000) by Oliver Gilbert.
87. AMPHIBIANS AND REPTILES. A natural history of the British herpetofauna (2000) by Trevor J.C. Beebee and Richard A. Griffiths.
88. LOCH LOMONDSIDE. Gateway to the Western Highlands of Scotland (2001) by John Mitchell.
89. THE BROADS. The People's Wetland (2001) by Brian Moss.
90. MOTHS (2002) by Michael E.N. Majerus.
91. NATURE CONSERVATION. A review of the conservation of wildlife in Britain 1950-2001 (2001) by Peter Marren.
92. LAKELAND. The wildlife of Cumbria (2002) by Derek Ratcliffe.
93. BRITISH BATS (2003) by John D. Altringham.
94. SEASHORE A natural history of the seashore (2004) by Peter J. Hayward.
95. NORTHUMBERLAND (2004) by Angus Lunn.
96. FUNGI (2005) by Brian Spooner and Peter Roberts.

THE MONOGRAPHS

M1. THE BADGER (1948) by Ernest Neal
M2. THE REDSTART (1950) by John Buxton
M3. THE WREN (1955) by Edward A. Armstrong
M4. THE YELLOW WAGTAIL (1950) by Stuart Smith
M5. THE GREENSHANK (1951) by Desmond Nethersole-Thompson
M6. THE FULMAR (1952) by James Fisher
M7. FLEAS, FLUKES AND CUCKOOS A study of bird parasites (1952) by Miriam Rothschild and Theresa Clay

M8. ANTS (1953) by Derek Wragge Morley
M9. THE HERRING GULL'S WORLD A study of the social behaviour of birds (1953) by Niko Tinbergen
M10. MUMPS, MEASLES AND MOSAICS A study of animal and plant viruses (1954) by Kenneth M. Smith and Roy Markham
M11. THE HERON (1954) by Frank A. Lowe
M12. SQUIRRELS (1954) by Monica Shorten
M13. THE RABBIT (1956) by Harry V. Thompson and Alastair N. Worden
M14. THE BIRDS OF THE LONDON AREA, since 1900 (1957) by the London Natural History Society
M15. THE HAWFINCH (1957) by Guy Mountfort
M16. THE SALMON (1959) by J.W. Jones
M17. LORDS AND LADIES (1960) by Cecil T. Prime
M18. OYSTERS (1960) by C.M. Yonge
M19. THE HOUSE SPARROW (1963) by D. Summers-Smith
M20. THE WOOD-PIGEON (1965) by R.K. Murton
M21. THE TROUT (1967) by W.E. Frost and M.E. Brown
M22. THE MOLE (1971) by Kenneth Mellanby

2

A Lunch and its Consequences

> 'Most of us were old enough to remember the disaster of the early twenties, following the First World War, when there was a dismal failure to produce the promised world fit for heroes. All during the years of the Second World War was the determination to be ready for peace.....
> 'A bright new series [of books] was a godsend to harassed seekers after presents as well as to a public hungry for peace and forgetfulness of war. What better than natural history?'
> *L. Dudley Stamp.* From original Author's Preface of *Nature Conservation in Britain* (1969)

The first informal meeting between the publisher, the producer and the future editors of the New Naturalist series took place in June 1942, at a point, as Billy Collins expressed it, when 'this country's fortunes were at their most perilous hazard'. Rommel had taken Tobruk and driven the 8th Army back into Egypt, Britain was losing the Battle of the Atlantic, and three German capital ships had steamed through the English Channel unharmed and in the process had shot down a lot of British torpedo planes. It was not the most immediately threatening moment of the war, but militarily it was one of the low points. The earliest New Naturalist meetings were in that sense an assertion of optimism – Dudley Stamp referred to them as 'a form of escapism'. If Germany were to win the war, the British people would presumably have more pressing matters on their mind than natural history.

For the planned series of books, Billy Collins turned to Wolfgang Foges, the director of Adprint, with whom he had collaborated recently on the 'Britain in Pictures' series (*see* Chapter 3). Foges, an Austrian refugee from Hitler's Third Reich, specialised in colour reproduction and believed that the camera would shortly revolutionise publishing in the natural sciences. Foges was to be responsible for commissioning the colour pictures and also for the production of maps and diagrams, based on the authors' sketches. It is to Foges and Adprint that we owe the exquisitely lettered maps in the earlier New Naturalist titles.[1]

Some time before the first meeting, but certainly after April 1942, Collins and Wolfgang Foges had approached Julian Huxley to try to gain his active support for the proposed series. The idea of blending the fresh insights of ecology and ethology (the study of animal behaviour) with traditional British natural history and presenting the results to the public was, they knew, very much Huxley's own forte. The 54-year-old Julian Huxley had by degrees abandoned an academic career to take a more active part in promoting science. He was in many ways the prototype of the new naturalist, combining, in his earlier career, laboratory research on genetics and the development of embryos with his better-known field studies of grebes and divers, and above all in his writings on evolution, animal language and social insects. By successfully introducing a wide public to the latest advances of field-based science, he, more than anyone else, had literally, as well as

Sir Julian Huxley (1887-1975), photographed in 1966 by Eric Hosking.

metaphorically, brought nature study back to life. It has been said of Huxley that, almost single-handedly, he made field natural history scientifically respectable again. In the opinion of his friend the late W.H. Thorpe, he was almost the only senior British zoologist in the inter-war years who thought ecology and behaviour important. In combining the two great arms of field study, he was perhaps the most influential naturalist since Darwin.

Huxley's achievement was helped by his own genetics: like other members of his family, he wrote well, with great elegance and *savoir faire.* Which is precisely what one would expect from his ancestry. His grandfather, the great Darwinian T.H. Huxley, had whetted the young Julian's appetite for evolution studies. His father and younger brother distinguished themselves respectively as a biographer and magazine editor and as a novelist and essayist. On Julian's maternal-side he inherited the organisational gifts of the famous Rugby School headmaster and educational reformist, Thomas Arnold; Matthew Arnold was a great-uncle, and for an aunt he had another successful novelist, Mrs Humphrey Ward. He had less cause for gratitude for another facet of the Arnold genes: a tendency for uncontrollable bursts of depression. This drove Julian's younger brother Trevenan to suicide, and was responsible for the troughs in Julian's own life, when he felt incapable of work and, shunning all company except that of his wife, wandered through the countryside or shut himself away in his study.

In 1925 Huxley left Oxford to chair the Department of Zoology at King's College, London. Presumably Professorship was not his *métier*, for only two years later he resigned to work with H.G.Wells on a fortnightly popular science publication, *The Science of Life*, which Wells intended to be 'a summary of contemporary knowledge about life and its possibilities'. In 1935 Huxley was appointed Secretary of the Zoological Society of London, that is to say he was placed in day-to-day charge of London Zoo. Over the next few years he edited two important books, *The New Systematics* (1940) and *Evolution, the modern synthesis* (1942), and was elected a Fellow of the Royal Society. When Billy Collins contacted him, he had just returned from a lecture tour of the United States and was about to resign his position with the Zoological Society to concentrate once again on writing and journalism.

Huxley's reputation was at its height in the early 1940s. Billy Collins had not published any of his books, but the two evidently knew one another. In the circumstances, Collins' approach to Huxley was exceedingly well-timed. Huxley showed interest in Collins' proposal to publish an important series of nature books, and so the latter invited him to lunch 'to discuss the scheme as a practical proposition'.

The 'exploratory lunch', at the Jardin des Gourmets in Soho, was a foursome. Collins brought with him Wolfgang Foges of Adprint, and Huxley was accompanied by his young protégé, James Fisher. With the appearance of Fisher, we meet the most dynamic member of the New Naturalist team. We also at once enter the land of legends: many readers will, I am sure, know the story of the chance meeting of Billy Collins and James Fisher in an air-raid shelter during a bombing raid on the capital. Fisher is supposed to have remarked, 'What this country needs is a good series of books on natural history to take people's minds off this carnage'. Collins agreed: 'Quite right. We will need an editorial board. You see to it, and I'll provide them with tea and cream buns'. It is a nice story. But is it true? In a letter written more than forty years later, James' son Crispin referred to the story as if it were fact. But in his *Bookseller* article, written in 1946, Billy Collins says only that 'together publisher and photographer [i.e. Collins and Foges] approached Dr Julian Huxley, who was interested, and he in his turn *introduced* [my italics] Mr James Fisher, the ornithologist'. It reads as though Collins had not in fact met Fisher until that time, but the sentence may be only a manifestation of Billy Collins' rather formal style. Quite possibly the exchange did occur, but, if so, it did not at the time lead anywhere.

James Fisher's alleged remark in the shelter was at least in character. He was a 'Julian Huxley' for the masses; the New Naturalist incarnate. At the time of the Soho lunch, he was only 29, but already one of the best-known British ornithologists through his position as Secretary of the British Trust for Ornithology (BTO), his authorship of a best-selling book *Watching Birds*, and most of all through his

James Fisher (1912-1970), photographed in 1970 by Eric Hosking.

unusual energy and strength of purpose. As with Huxley, brains were in no short supply among the Fishers. James' father was Kenneth Fisher, headmaster of Oundle School, whose pupils included Peter Scott and Leslie Brown. He was a keen birdwatcher, though the greatest influence on the young James' natural history was not his father but his maternal uncle Arnold Boyd, the Cheshire naturalist (of whom we shall hear more). James was bright enough to be sent to Eton as a King's Scholar, and from thence went up to Magdalen College, Oxford in 1931 to read medicine. He soon changed to zoology: birds were by now his life's passion. He joined the Oxford Arctic expedition in 1933 as ornithologist. This resulted in his first scientific paper and a twenty-year obsession with the fulmar. In 1936, he married Margery Lilian Edith Turner, later a writer and critic of children's books and with whom Fisher wrote a biography of the explorer Ernest Shackleton. Their honeymoon was spent watching sea-birds on Fair Isle. The quest for sea-birds matched Fisher's love of travel and his fondness for boating and climbing for which his tall, broad-shouldered physique was ideal. Reading about him, one has the impression that he threw himself into every project with the greatest drive and enthusiasm. Even at that time he seemed to know everyone and everything in ornithological circles. If some people found him rather exhausting, he was nonetheless a popular man. Unlike some learned people, he was out-going and affable, with a sense of humour (and hence a sense of proportion). He was by all accounts a good mixer, equally at home in the Savile Club or at meetings of the Northamptonshire Naturalists' Trust. He was a great *bon viveur.* Niall Campbell recalled an evening with James Fisher on a cruise ship anchored off St Kilda, around 1950: 'At the end of our cheerful evening I had to be lowered like a sack of potatoes into an inflatable and taken ashore to my billet. I wish that I had met him more often.'

Having gone down from Oxford with a disappointing second-class degree (having spent most of his time there birdwatching), Fisher taught briefly at Bishops Stortford before joining the staff at London Zoo as assistant curator. There he saw a great deal of Julian Huxley, both on the Association for the Study of Animal Behaviour and in their joint work for the BTO, organising popular participation in bird study. James Fisher was to dedicate his New Naturalist book, *Sea-Birds*, to Julian Huxley 'for his guidance and encouragement, and in recollection of the many happy days we have spent together watching sea-birds'. In 1939, the two of them had taken part in an expedition to St Kilda to survey gannets and other sea-birds. St Kilda was thereafter Fisher's spiritual home. Had it not been for the war, he would have been back there year after year, and perhaps written a *Gannet* as well as a *Fulmar.*

Instead, Fisher spent the war studying rooks for the Ministry of Agriculture. But first he had found time to write a natural history classic, *Watching Birds*, which, more than any other book, established birdwatching as a popular scientific hobby. *Watching Birds* was a more pertinent title then than it would be now, for in 1940 more people shot birds than watched them. The book had many of the hallmarks of Huxley's influence, with chapters devoted to topics still relatively little known to the ordinary naturalist, like migration, bird habitats, bird language and behaviour. The style also was novel in its presentation of the latest discoveries in easy, everyday language, and its confident, not to say exuberant, manner. *Watching Birds* was published as a cheap Pelican 'science for the masses' paperback, and proved a bestseller. The subscription form on the back helped to double the membership of the BTO between 1940 and 1944. Of special interest is the original preface which gives us a good idea why James Fisher became so interested in the New Naturalist books:

> 'Some people might consider an apology necessary for the appearance of a book about birds at a time when Britain is fighting for its own and many other lives. I make no such apology. Birds are part of the heritage we are fighting for. After this war ordinary people are going to have a better time than they have had; they are going to get about more; they will have time to rest from their tremendous tasks; many will get the opportunity, hitherto sought in vain, of watching wild creatures and making discoveries about them.'
>
> *James Fisher.* Preface in *Watching Birds* (1940)

Fisher had little sympathy with scientific elites. Nor had he anything but scorn for the type of 'popular' natural history that prevailed before the war, which he considered simplistic rather than simple and sometimes gruesomely sentimental, or, at the other pole, rather arid and concerned mainly with identification. The clearest statement of James Fisher's views on natural history are contained in the first three pages of his ill-fated New Naturalist Journal, compiled in 1947. He probably held similar views in 1940. Essentially, it was that the amateur field naturalist had come of age; the natural history tradition of England – 'this great stream' – was being broadened by popular participation and deepened by the new science of ecology. It was time for the naturalist to leave his skins, slides and cabinets, and step outside to observe how wild animals, birds and plants live. Now, above all, was the time for the expert to speak to the amateur through journals, books and lectures, and for the amateur to contribute to solving the problems of distribution, ecology and life history. In a letter to Alister Hardy, Fisher expressed confidence that 'the regrettable confusion between popular scientific writing and vulgar journalism is being swept away .... In another twenty years every field of natural history will be invaded by amateurs, who will outnumber trained professionals in almost every sphere. We seem [with the New Naturalist library] to be getting prepared for this.'

Perhaps Fisher underestimated the tenacity of vulgar journalism and the malign influence of television. But it was views like these that he brought to bear on the development of the New Naturalist series, so much so that it became the written embodiment of Fisher's new natural history. It is said that James Fisher was the only person to have been unsurprised at the astounding sales of the first New Naturalist books. In his eyes, their success was no more than a validation of the beliefs he had expressed so confidently in *Watching Birds*, five years earlier.

No one took notes on what was said at the Jardin des Gourmets, but an article written a few years later by Billy Collins for *The Bookseller* summarises his own intentions and Huxley's response to them. His idea was for 'a new survey of Britain's natural history'. As Collins expressed it, this was to take the form of a complete library of nature books which would combine low price and presentation, and yet make available to the public 'all the wealth of scientific knowledge that had been acquired during the present century'. Though no doubt commercially courageous, this was not a revolutionary idea. On the contrary, it corresponded with the mood of the time, with such projects as the Huxley-Wells *Science of Life* series, and popular interest in the latest discoveries of science had, if anything, increased during the war. Billy Collins believed his trump card lay in his ability to illustrate these books lavishly with colour photographs 'to show nature in her true colours'. This was in fact Huxley's main doubt about the enterprise. He believed that colour reproduction had not yet advanced that far, and needed a great deal of convincing that the miraculous qualities of American quarter plate Kodachrome, in which Collins and Foges had great faith and had acquired large stocks of (see Chapter 3), would overcome hitherto insuperable problems in nature photography. Fisher's contribution to the lunch might have been to emphasise the importance of modern ornithology in the new natural history. Fisher's remarks in the New Naturalist Journal make it clear that he wanted to get right away from the old idea of dealing with wildlife group by group, species by species, and back to what he called 'the inquiring spirit of the old naturalist'. By this phrase – and he might have been thinking primarily of Gilbert White – he was alluding to the sense of wonder which leads to a desire to find out, by observation, deduction and simple experiment. Later these words were incorporated into a sort of New Naturalist credo, which was printed opposite the title page of each book (and also on the back cover of the earliest jackets). Its words may be familiar to readers of this book, but their assertion was the key to the series and it is worth repeating them in their original form at this stage in the story:

> The aim of this series is to interest the general reader in the wildlife of Britain by recapturing the inquiring spirit of the old naturalist. The Editors believe that the natural pride of the British public in the native fauna and flora, to which must be added concern for their conservation, are best fostered by maintaining a high standard of accuracy combined with clarity of exposition in presenting the results of modern scientific research. The plants and animals are described in relation to their homes and habitats and are portrayed in the full beauty of their natural colours, by the latest methods of colour photography and reproduction.

The four seem to have had a productive lunch. It was on that day, wrote Billy Collins, that general ideas started to firm up into concrete suggestions and the rather complicated editorial arrangements for the series were worked out. Huxley and Fisher would form the nucleus of an Editorial Board (Collins usually referred to it as the 'Committee'), whose job it would be to 'plan happy marriages between subjects and authors' and nurse each book through to publication. Adprint would take charge of the production of illustrative material, including photographic plates, maps and text figures, and Collins would print the book – and, it must be hoped, provide tea and cream buns. The three sections of the New Naturalist triumvirate, editors, printer and illustrator, would meet frequently to review progress and to plan the next stage in the process. And a long and exigent process it turned out to be. Nevertheless, the concept slowly limped its way to fulfilment through all the frustrations and obstructions of wartime, and by the end of 1945 the first books were rolling off the presses. 'The New Naturalist was born.'

Expanding the Board: John Gilmour and Dudley Stamp

One matter which must have been raised at the Soho lunch, if not very soon afterwards, was the composition of the Editorial Board. It was clear that it would need to be broadened. The interests of Huxley and Fisher lay in broadly similar fields, and neither claimed expertise in botany or the earth sciences. If the library was to embrace the whole of British natural history, including the rocks, soils and landscape, it would need at least two more editors. There is nothing on record to say when Julian Huxley (for the choice seems to have been his) co-opted John Gilmour and Dudley Stamp, though it must have been at some time between June and November 1942.[2] Both men would have been on anyone's short list.

John Scott Lennox Gilmour was aged 37 in June 1942. He had contributed to the Huxley-edited *The New Systematics* (1940), and Huxley had probably known him for some time before that since both men were among the founder members of The Systematics Association, created in 1935 to promote the study of evolutionary relationships. Though of Scottish ancestry, Gilmour was born in London, and educated at Uppingham and Clare College, Cambridge, where he read Natural Sciences, specialising in botany in his final year. His conversion to the pleasures of botanising had taken place in his early teens, and in the New Naturalist book *Wild Flowers* he explained in characteristic style how it came about:

> 'When I went to a preparatory school I knew and cared nothing about wild plants. At the end of the summer term each boy had to produce fifty named

species. On the last day but one I had not collected a single plant. Desperation drove me to a high-speed tour of the lanes near the school, guided by a friend who had already made his collection, and on the following day I duly presented my fifty plants. This discreditable incident implanted in me, against every modern principle of education, a passion for the British flora which has never been extinguished.'

John Gilmour. Wild Flowers pp. 1-2

No sooner had Gilmour graduated than he was appointed Curator of the University Herbarium and Botanical Museum at Cambridge. He must have been highly regarded at Cambridge. The work plunged him into the world of naming and classification, then a stagnant and neglected field. With two fellow Cambridge botanists, William Stearn and T.G. Tutin, he did much to get that particular ball rolling again, and their work helped to establish Cambridge in the forefront of modern taxonomy. Gilmour's individual contributions tended to be based not so much on experiment as on philosophy and on his wide reading. He was naturally reflective (later he became President of the Cambridge Humanists), and it is easier to imagine him discoursing from a common room armchair, puffing on his pipe, than chasing around the world after orchids or scaling cliffs in search of alpine flora. In scientific terms, he was a pragmatist. He argued that since classification is essentially utilitarian, there was no possibility of devising a perfect system that would satisfy everybody. Nor was there any need to do so, since different classifications served different purposes. In 1939, he and A.J. Gregor came up with their own method of cutting the material to suit the cloth. Their system of 'demes' did not gain universal acceptance by any means, but it was a useful reminder that there were more ways than one of tackling the problem.

Such work is read and argued over only by specialists. Gilmour did not in any case publish much, though when he did his writing was usually clear and well thought out. His New Naturalist correspondence was generally short and to-the-point, though written in a sprawling hand. He is remembered by a larger number of botanists for his career in scientific horticulture, which began in 1931 after his appointment to the assistant directorship of Kew Gardens at the astonishingly early age of 25. After the war was over, he took up a new senior post at the Royal Horticultural Society's garden at Wisley, before returning to Cambridge as director of the University Botanic Garden. Wartime disrupted his career, as it did that of most British scientists, and at the time of the first New Naturalist meeting, Gilmour was working for the Ministry of Fuel and Power.

John Gilmour's career, then, had taken in scholarship, science and administration. There was another side to him that might have influenced Huxley. Gilmour was, by all accounts, a pleasant man, strikingly good looking, urbane (D.E. Allen mentions his 'suavity and *savoir faire*') and *cultured*, not just in a scientific sense but in his wide sympathies and love of poetry, music and philosophy. He brought to the New Naturalist Board not only a comprehension of the botanical world but also a socially cohesive quality. According to his close friend and colleague Max Walters, he was good at creating an *esprit de corps*: 'More than any other person in his generation, John was able to be *persona grata* to a very wide spectrum of colleagues ranging from the professional taxonomist to the amateur gardener.' His liberal outlook and unfailing good manners were no doubt an equal asset at the New Naturalist Board meetings.

John Gilmour (1906-1986) in 1967. (*Photo: D.A. Huggons/BSBI*)

While one could imagine alternative names to Gilmour's, there was one candidate for the role of geographical editor who was head and shoulders above any other. Laurence Dudley Stamp was aged 44 at the time of the first Board meeting. He looked older: a bluff, bulky, balding figure, generally dressed in tweeds, with a vaguely military air about him. Stamp, like Fisher, was a man of quite awesome energy. More or less continuous illness in childhood meant he had received very little formal schooling, but he achieved such brilliant results in the Cambridge local examinations in 1913 that he was accepted as a student at King's College, London at the age of 15. Having gained a first class degree in geology and endured a spell in the Royal Engineers on the Western Front, he stayed on at King's College until 1922, obtaining his doctorate in geology and dashing off another first class degree, this time in geography (by the end of his life he had amassed quite a collection of degrees). King's was Julian Huxley's London college. Although his time there did not coincide with Stamp's, the two might have met later after Stamp's return from three years of academic duties in the Far East to the Cassell Readership in Economic Geography at the London School of Economics, a post held until 1945. Huxley certainly knew of him by reputation.

Stamp is remembered for two great achievements. The first was his authorship

L. Dudley Stamp (1898-1966) on the way to a university conference in 1963. (*Photo: Nature Conservancy*)

of a remarkable succession of geography textbooks in the 1920s and 30s, written 'at a time when good texts were urgently necessary to support the great development of the subject at home and overseas'. Some of these books are enormous tomes of 600 pages or more, and ran into many editions: *The World: a general geography* (19th edition 1977); *Asia: a regional and economic geography* (10th edition 1959); *The British Isles: a geographic and economic survey* (6th edition 1971); *Chisholm's handbook of commercial geography* rewritten by Stamp (20th edition 1980); *An intermediate commercial geography* (12th edition 1965). There were many others. If Stamp was as devoted a university lecturer and family man as he was a textbook writer, he must have been a master of what today's officer workers call 'time management'. He was a one-man publications industry and, indeed, his New Naturalist contracts were made out to 'L. Dudley Stamp of Geographical Publications Ltd'.

The non-geographical world remembers Stamp for other reasons. From 1930 to 1935 he organised the great Land Utilisation Survey, the first countrywide survey to map every acre of England, Scotland and Wales at a scale of one-inch to the mile. The work was organised over three seasons from 1931 to 1933 with the help of thousands of volunteers, mainly from schools and other educational establishments, representing the equivalent of 230 man-years of effort. Stamp himself spent a great deal of time being driven about in the family car, standing on the passenger seat taking notes with his head peering out through the sun roof. In this way he covered much of the British Isles, and the varied scenery and underlying rocks covered in so short a time must have made a profound impression on him. The maps were accompanied by county reports written by Stamp and his associates, in which the use of land was analysed in unprecedented detail. It was unfortunate that the survey took place during the great depression, when many farms had gone out of production, so that the survey was, in a sense, 'untypical'. But at the least it was a valuable template for future land-use study, and it influenced Lord Justice Scott's great report on rural planning (1941-42). Stamp later wrote a book based on the Land Utilisation Survey called *The Land of Britain: its use and misuse* (1948) which has been praised as a landmark in British geography for its masterly handling of complex detail.

Stamp was one of the pioneers of postwar planning. His knowledge of land use drew him by degrees into government committee work. In 1942 he was appointed Chief Adviser to the Ministry of Agriculture on rural land use, for which work he was appointed CBE. As Lord Scott's deputy, he was responsible for the drafting of much of the Scott Report, the blueprint for the Town and Country Planning Act of 1947. The report also involved him in the planning for National Parks and 'nature reservations', all of which were to be part of the better world for which the British people had fought (see Chapter 11). Later, Stamp became much involved with the work of the Nature Conservancy, and it was he, rather than one of the biological editors, who wrote the New Naturalist book on nature conservation. In all this, a picture emerges of an energetic, practical man, a gifted textbook writer with a sure command of clear simple English, not a theorist, nor, in the narrow sense of the word, an intellectual. Stamp's interests spanned the whole of physical, social and economic geography, and he was also a naturalist and pioneer conservationist. Perhaps he attempted too much. His obituarist reminds us that, as in the case of John Gilmour, he was less important for what he did than for what he was: an open friendly man, helpful at getting 'lame dogs over stiles', and a great enthusiast. After his death in 1966, his fellow editors recalled how Stamp 'got things done by Socratic persuasion, leavened by a happy humour and the most patent love of his multifarious subjects'.

Titles and authors

The Editorial Board's complement was completed with the addition of Eric Hosking to act as photographic editor, and whose 'initiation' I describe in Chapter 3. The Board first met on 7 January 1943, and continued to meet at intervals of a couple of months throughout the war. Billy Collins and Wolfgang Foges were usually both present. It was a hard working committee. Billy Collins was at pains to point out that this was not 'a nominal board to which a fee was paid for the use of their eminent names'. According to Collins, not one meeting was ever suspended, even during the V1 and V2 raids of 1944 when the Board became very much aware that their temporary accommodation in a studio off Oxford Street contained 'an uncomfortable proportion of glass'. Recalling those days, Dudley Stamp said that 'Few of our early meetings passed without an air-raid warning; we occupied a succession of temporary offices in different parts of London and rather took the view that if a bomb were intended for us that would be that.' After the emergencies were over, the Board returned to the Collins offices in St James's Place, and it was there that Eric Hosking took the photographs reproduced here,

The New Naturalist Editorial Board, late 1945. Left to right: Eric Hosking, James Fisher, Julian Huxley, L. Dudley Stamp, John Gilmour. (*Photo: Eric Hosking*)

probably with the purpose of illustrating Billy Collins' article. For those who are familiar with the better-known portrait of the Board taken twenty years later, the relative youth of the editors in 1945 is striking. All are smartly dressed, Dudley in his tweeds, the rest in dark city suits. Fisher and Hosking have horrible 1940s haircuts and round spectacles. Stamp is examining a brand new copy of *Butterflies*. Gilmour sits there to one side with his pipe, smiling like a matinee idol. Billy Collins is recognisable only by his eyebrows.

The initial task of the Editorial Board was to draw up a detailed scheme for a set of books that would cover the whole field of British natural history. Billy Collins' original proposal was to produce a set of thirty-six titles, so that the whole library could conceivably be written and published within five years or so.[3] That list was almost immediately extended to fifty titles. According to Dudley Stamp:

> 'We soon realised that it would be better to have an elastic series giving our authors scope to develop their own particular fields of study, and that we could in this way interest the cream of scientific workers who would never agree to write within a framework laid down by others. We believed that the time had come to break away from the long-held belief that books on natural history must be written in popular style – a belief common to most publishers. We believed on the contrary that there was a large public waiting to be introduced to serious work and research in progress, provided it was presented in an attractive manner.'
>
> *L. Dudley Stamp*. Preface in *Nature Conservation in Britain* (1969)

Over the first few meetings of the Board, the five members drew up a comprehensive list of titles and, in some cases at least, prospective authors. We would like to know what all these titles were, but unfortunately no minutes of these early meetings have survived. Possibly none were made. According to Billy Collins, writing in early 1946, the range of subjects was divided into four groups. They were, in his words:

'1. Organisms; e.g. butterflies, birds, flowers, trees etc.
2. Habitats; e.g. mountains, moorlands, woodlands etc.

3. Regions; e.g. London, Highlands of Scotland, Snowdonia etc.
4. Special Subjects; e.g. Game, Art and Natural History, Conservation etc.'

In practice, of course, this rather reductive scheme was not always realised. The authors contributed their own ideas, and in most cases were wisely left to write their own book within broad and generalised guidelines. One can, by deduction, make a fair guess at the list. Table 2 is my best stab at reconstructing the original titles of 1943; and readers familiar with the titles of the New Naturalist library will see right away how far it differs from those actually printed. Some, like *The Open Sea*, *Spiders* and *The Lake District* took a decade or more to complete. A few were eventually combined, like *Mountains* and *Moorlands* or *Lakes* and *Rivers*. Others became fragmented, like *Art and Natural History* which became *The Art of Botanical Illustration*. A few, like *The Thames Valley* and *Molluscs* were never written at all.

The problem lay not so much in finding experts for particular subjects, but finding anyone with sufficient free time to write a book. In 1943 practically every potential author was engaged on war service of one kind or another. It was impossible to forecast when the war would end, although most people seem to have been confident that Britain would win it. The authors' difficulties were compounded by the removal of major reference libraries, such as those of Kew or the British Museum (Natural History). Fraser Darling, stranded on the remote Summer Isles, had only his own books and experiences to draw on. Others, like Maurice Yonge (pronounced 'Yung'), were up to their ears in work, with academic duties alternating with wartime committees and active service in the territorial army and fire service. Leo Harrison Matthews was a member of the 'secret army' of inventors, working on radio and radar installations for aircraft. At least one future New Naturalist author was locked up in a German prison camp. It was not a propitious time to ask someone in a senior position to write a book. Fortunately, many of those contacted seem to have been keen to write for the series, once things got back to normal, and most of them eventually did so. In the meantime, Collins resigned himself to a slower production rate than he had hoped for. At one time, he noted, 'only one MS was being written'.

It may seem surprising that the publishers seemed to know exactly who they wanted to write the first books of the series. This was, of course, the function of the Editorial Board, without whom the publisher would have had little idea. At that time, the worlds of professional biology and natural history were relatively close-knit, thanks to the British penchant for clubs and societies in which amateur and professional mixed freely. Huxley had tutored Alister Hardy and worked with E.B. Ford; Fisher knew Fraser Darling, Arnold Boyd, Richard Fitter and others very well. Gilmour was an active member of botanical societies, and a former colleague of W.B. Turrill and Victor Summerhayes. Stamp was a good friend of Sidney Wooldridge, and knew Gordon Manley and Alfred Steers. And one did not need to be a zoological initiate to have heard of Maurice Yonge, Leo Harrison Matthews or Brian Vesey-Fitzgerald. A few names, like Ernest Neal's, were suggested to the Board by others, but otherwise these were men who knew one another at least by repute, and were often colleagues and friends as well, through common membership of the British Ecological Society or the contemporary Nature Reserves Investigation Committee.

The first round of authors were contacted during the late summer of 1943. I have tracked down two of the initial letters in the private papers of Sir Alister

Hardy and Sir Maurice Yonge, both of them written by Julian Huxley, the first on 17th August 1943, the second the day afterwards. Apart from one or two more personal paragraphs, the wording is almost exactly the same in each, the clear inference being that this was a standard letter. Since this letter is such a clear expression of the ethos of the New Naturalist series, as well as its first significant document, I reproduce the entire letter here. The notepaper is headed THE NEW NATURALIST. A Survey of British Natural History. The printed address, that of the Adprint offices on Newman Street, W1, has been crossed out and a temporary address at 16 Queen Anne's Gate, SW1 typed in.

Dr C.M. Yonge,
Department of Zoology,
The University,
Bristol.

18th August 1943

My dear Yonge,

I have undertaken to serve on the editorial board of a new series, of about 50 books, on British Natural History, and want to persuade you to write a volume in it.

The series is being published by Messrs Collins and produced by Messrs Adprints. It is planned to have two new features: (1) it will contain a large number of colour illustrations (either 32 or 48 in each volume) executed by the most up to date techniques. Black-and-white illustrations will also be included. So far as colour illustrations go, some volumes will be illustrated wholly by reproductions of paintings, etc. but the majority will be colour photographs, especially taken for the series by a body of well known nature photographers, equipped with the latest apparatus; (2) it will not adopt the traditional method of dealing with the various plants and animals group by group, but will attempt to give what I may call a survey of the natural history resources of our islands. Some volumes will deal with particular groups, not, however, from a comprehensive taxonomic point of view, but with reference to the scientific, cultural or practical interests of the group as a whole and of various selected members of it; others will deal with habitats – e.g. Moorland, inter-tidal zone, etc.; others with regions – e.g. London, the Scottish Highlands, the Thames, etc.; and still others with the human relations of the subject – e.g. sport, conservation, art. Throughout, the geology and geography will be treated as a part of the natural history, on the one hand in relation to scenery, and on the other as affording a basis for the plant and animal ecology.

The series will be addressed to the intelligent layman, i.e. the books must not be too technical and must be interestingly written, but they must not be merely 'popular'. Each volume will contain from about 75,000 to about 100,000 words and the price is fixed at 12/6d. I enclose a preliminary note on terms; if you are interested I will send a detailed contract.

What I am hoping you will do is to write a book on the Natural History of the Sea Shore. We shall also try to secure one on the open sea as a habitat, another on fisheries – each of course being treated from quite a different angle.

There is no hurry about this book at the moment, as we have some titles fixed up for the next 12 months or so, but we should like to think that we might expect it in the Spring of 1945. Meanwhile I am most anxious to get you committed to writing the book as I am sure that nobody else could do it better, and I don't want you to get committed to other ventures. As you will see, the financial terms are really very favourable. Furthermore, I understand that you are already doing a little book on the same general subject for Britain in Pictures. While this would not in any way compete with the larger book proposed, it would doubtless help in preparing the way for it.

If, as I hope, you are interested in this, we might meet some time when you are up in London and discuss it further. It may be that you would wish to get in a botanist to help on the plant side, either as a collaborator or simply as consultant. That could easily be arranged.

Yours ever,

Julian Huxley

Although the letters to Yonge and Hardy were signed by Huxley, not all of the early books were commissioned by him. Evidently the titles were divided between the four literary editors, each taking on those within his particular field. Huxley seems to have taken the lead for the more general 'biological' books, and his name must have been a factor in persuading senior biologists like John Russell, Alister Hardy and Maurice Yonge to write for the series. At this stage, James Fisher took on the more obviously birdy titles, but after Huxley's departure to head UNESCO in 1946 he also took over much of Huxley's role as a sort of managing editor. Gilmour and Stamp respectively edited the botanical and geological books. The idea was that each editor would themselves contribute a book to the series. Stamp immediately set to work writing *Britain's Structure and Scenery* with the same speed and efficiency that characterised the production of his textbooks. Gilmour was pencilled in to write the wild flower book for the series, but failed at first to make much headway with it. Fisher was planning to write *The Fulmar* and *Sea-Birds* once he had completed his researches. Huxley never did contribute a book. A pity: he could have written a wonderful unifying book about evolution. There was also an intention for all five editors to combine in writing a book about nature conservation in Britain. This concern with nature conservation, reflected in the standard 'credo' of the series, reflected the involvement of ecologists and geographers in national planning in the immediate postwar years. Huxley, Gilmour and Stamp were members of one or other of the committees that produced the blueprint for nature conservation after the war (see Chapter 11). But a book on nature conservation written in 1943 would have been a very slim volume!

From the start, it was evident that the New Naturalist library would be a unique opportunity for someone like Hardy or Darling to write a really meaty, enthusiastic book, their popular *magnum opus*. As an additional incentive, Huxley had described the publisher's terms as 'very favourable'. Favourable is a relative term that academic scientists and writers of popular fiction might define very differently. They do not seem particularly generous. The standard contract for the early New Naturalist books guaranteed a minimum first edition of 10,000 copies. For these the author would be paid a fee rather than a royalty: £200 on acceptance of the manuscript and £200 on publication. In practice, however, the print-run of these books unexpectedly ran to 20,000 or more copies. The author was paid £40 for every 1,000 copies printed over the initial guarantee of 10,000. For the

Table 2. The New Naturalist: original title list (1943)

The list of titles below has been reconstructed from surviving documents such as publisher's contracts, commissioning letters and other correspondence. There were at least ten other titles (bringing the total to 50) which were listed at an early stage, but I can do no more than guess what they might have been [Wild Orchids, Limestone Flowers, Bogs and Fens, Woodlands, Migration, Seals, Caves, Ecological Communities, British Naturalists and British Islands are my best effort, but I could be completely wrong!].

INVERTEBRATE ORGANISMS
Butterflies (E.B. Ford)
Dragonflies (Cynthia Longfield)
Spiders (W.S. Bristowe)
Insects (A.D. Imms)
Molluscs

VERTEBRATE ORGANISMS
Fish (E. Trewavas)
Mammals (L. Harrison Matthews)
Reptiles and Amphibians (Malcolm Smith)
Sea-Birds (James Fisher)
Woodland Birds (Bruce Campbell)
Shore Birds (Eric Ennion)
Birds and Men (E.M. Nicholson)
Marsh and Freshwater Birds (R.C. Homes)
Moorland Birds

PLANTS
Biology of British Flora (W.B. Turrill)
Wild Flowers (John Gilmour)
Grasses (C.E. Hubbard)
Trees (E.W. Jones)
Ferns and Mosses (F. Ballard)
Mushrooms and Toadstools (John Ramsbottom)

HABITATS
Moorlands (W.H. Pearsall)
Mountains (W.V. Lewis)
Lakes (E.B. Worthington)
Rivers (A.A. Miller)
The Sea Shore (C.M. Yonge)
The Open Sea (Alister Hardy)

REGIONAL VOLUMES
Natural History of London (Richard Fitter)
The Thames Valley
Highlands of Scotland (F. Fraser Darling)
Snowdonia (Bruce Campbell and others)
The Broads (R. Gurney and others)
The Lake District (E. Blezard and others)
Dartmoor (L.A. Harvey and others)
A Country Parish (A.W. Boyd)

SPECIAL SUBJECTS
British Game (Brian Vesey-Fitzgerald)
Art and Natural History
Nature Conservation (all the editors together)
Geology of British Isles (L. Dudley Stamp)
Climate and Life (Gordon Manley)
Fossils (A.E. Trueman)
Natural History of Man (H.J. Fleure)

second and any subsequent edition he was paid a royalty of 10 per cent. For a book with a first edition print run of 20,000, therefore, the author would earn £800, irrespective of the rate of sales. Thereafter his income would be higher in terms of percentages, but pegged to sales rather than printings. At a time when £2,000 represented a comfortable annual income, a New Naturalist book certainly represented something more than mere pocket money; on the other hand, no one was likely to give up his day job on the strength of it. Moreover, these relatively favourable rates lasted only so long as New Naturalist titles were printed and sold in large quantity. Those authors who had not completed their books by the end of 1949 were sent an 'Addendum' to their contract which amounted to a unilateral adjustment to the terms in the light of falling sales and the rising costs of colour reproduction. In 1945, a first edition sold out in six months or so. By 1949, however, the large editions that Collins needed to print because of block-making and setting costs were unlikely to sell out in as many years. Billy Collins' solution was to pay the author the balance owed on 10,000 books, irrespective of the size of the edition. This in effect reduced an author's earnings to the £500 mark. By the mid-1950s, the sales were such that Collins could no longer guarantee to print even 10,000 books, and from then on the author was paid a 10 per cent royalty on the basis of half-yearly sales. This brought his income down to about £100 a year.

In fact some other natural history publishers offered terms that were considerably better. One reason for the delicate economics of the New Naturalist series was the high production costs. Another was the system by which the five editors themselves earned a royalty from the books instead of receiving a set fee. Essentially, one quarter of the author's earnings from each book were paid to the editors. Originally this amounted to £12 per 1,000 copies sold, but under a memorandum of agreement made in 1952 between Collins and the editors, the terms were modified to a royalty of 1½ per cent of the published price for the first 10,000 copies printed, and thereafter 2½ per cent i.e. half of a per cent for each editor. They certainly earned their half a per cent, but the system operated at the expense of the authors. The latter seem to have accepted these terms with equanimity. There was, however, considerably more friction over the sharing of costs for text drawings, and, more generally, over Collins' sometimes sloppy accounting and failure to answer letters promptly, which tended to worsen over time. This is not a matter over which we New Naturalist admirers need linger. Correspondence of the sort that fills the New Naturalist files is no doubt the lot of most publishers. It is mentioned here because to do otherwise might cast an unwarranted rosy glow over the series at odds with what readers might have learned from the authors themselves.

The editors at work

The first New Naturalist books broke into the shops in late 1945. Billy Collins had advertised the series widely, both in the trade and in countryside magazines, and there seems to have been a widely shared sense of expectancy about them. Dudley Stamp recalled that:

> 'We knew that many of the older publishers were shaking their heads and prophesying a colossal flop. But such was the obvious interest aroused, even in advance of publication of the first titles, that the initial printing order was increased from 5,000 to 10,000 and then at the last minute to 20,000 of each. The war, both in Europe and the East, came to an end while those books were passing through the press. Though publishers were rationed for paper and had to use their scant supplies to best advantage, the public were never rationed for books – if they could get them.'
>
> Author's Preface, *Nature Conservation in Britain* (1969)

Those who can remember what bookshops were like in 1945 recall the colourful impact of *Butterflies* and *London's Natural History* (soon to be joined by *British Game* and *Britain's Structure and Scenery*) on those drab shelves. They were the 'godsend' referred to by Dudley Stamp at the start of this chapter. They were topical, up-to-date, involved the amateur naturalist in new ventures and set new standards of presentation. Above all, these books were perfectly timed for the ending of the war and the hoped-for Better Britain. They had a ready market among the ex-servicemen who had spent the previous years in the desert, or on Atlantic convoys, or in the jungles of Burma and Assam. They were seized eagerly by evacuees who had had their first taste of country living during the war, and found they liked it. The interest in the series revealed by the unexpected size of advance subscriptions must have caused many publishing heads to turn in surprise. But without the stocks of paper and the team of authors and photographers built up by Collins, no other publisher was able to compete in this market. All the same, the investment in the series had been enormous. By the time the first books were on the shelves, the editors had already been busy for nearly three years. From mid-1943, their main tasks had been commissioning titles and approving the colour photographs. But as more and more manuscripts started to arrive, the bulk of their time was spent outside the boardroom, corresponding with their authors, commenting on the drafts and sometimes contributing very substantially to the completion of the book. The exact nature of their work, and the way they approached the task, is perhaps illustrated most clearly by reviewing the course of a 'typical' New Naturalist meeting. I have chosen one from October 1951, which exhibits a representative range of problems arising and a good mix of titles. First though, let us deal briefly with some of the broader aspects of the editors' work.

The size of the editor's task in nursing a title through to publication varied from book to book. One author might leave him little to do, while another might involve him in a mountain of correspondence, hours of editing and a great deal of diplomacy. Having decided on the ideal combination of author and subject, one of the editors wrote to the individual concerned and asked him to provide a synopsis of the book. This would be copied and circulated to the other editors, who might suggest modifications here, a change in emphasis there, or make suggestions of their own. The synopsis having been accepted, the author would then be sent a standard contract, tailored for that particular title. In some cases, the contract arrived long after the book was begun. The author then had one more hurdle to overcome: a specimen chapter to give the editors a better idea of the author's style and approach to his or her subject. It was at this point that an editor would be formally assigned to the book. The editors all took their duties very seriously: as far as possible, statements and references were checked carefully, and any outstanding points were discussed with the author (often directly, for at this stage in the series author and editor met face to face). Finally, the responsible editor would write a preface for the book, introducing the author and explaining the special qualities which made author and subject, in Billy Collins' felicitous phrase, 'a happy marriage'.

The Editors' Preface was nearly always left unsigned (the exception is in *Nature Conservation in Britain*, where all four surviving editors pay tribute to the departed Dudley Stamp). It was always *the* Editors' Preface, not *an* Editor's. These collective efforts help to lend the New Naturalists an intimate quality, a series of well-mannered introductions: editor introduces reader to author; author introduces reader to the book. Beneath their collective preface, the editors added words that would come back to haunt them, stating that they had taken every care 'to ensure the scientific accuracy of factual statements in these volumes', while resting 'the responsibility for the interpretation of facts' on the author alone. Confident but dangerous words. There have not been many books where an editor is willing to share in the responsibility for errors.

For the first four titles of the series, *Butterflies*, *London's Natural History*, *British Game* and *Britain's Structure and Scenery*, the editorial burden seems to have been relatively light; for the latter title, Dudley Stamp was indeed effectively his own editor. *London's Natural History* was Richard Fitter's idea. His BTO colleague, James

The Board gets down to work, Fisher checking the previous minutes, Stamp inspecting the *Butterflies* jacket. (*Photo: Eric Hosking*)

Fisher, had wanted him to write a book for the series, but on the Thames Valley. Fitter replied that he didn't know a lot about the Thames Valley but would do one on London, because that was where he lived at the time. He had also been a very active member of the London Natural History Society, and had amassed a great deal of information on London wildlife. Fisher might have been expecting a bird-oriented book, but the result was much more novel and interesting: a study of the relationship of man and wildlife in an almost wholly non-natural landscape. It was in fact the first book about urban ecology. Richard Fitter recalls writing the book methodically, two hours after supper every evening, until the first draft was ready early in 1945. He remembers Huxley enthusing about the opening of his sample chapter 'Before Londinium': 'In the year 1877, at Meux's Horseshoe Brewery at the southern end of Tottenham Court Road, a well was bored, which plunged through 1146 feet of solid rock and millions of years of London history.' That was the kind of thing they were looking for. Fitter's professionalism might have left the editors with relatively little to do, but John Gilmour did get Sir Edward Salisbury to contribute a list of flowering plants and ferns recorded from London's bomb sites to balance Fitter's list of London birds.

Frank Fraser Darling's book, *Natural History in the Highlands and Islands*, involved the team in considerably more work. As Darling pointed out much later, 'The book was very much a one-man effort, written largely during the later war years when I was living in island remoteness and when travel to libraries was difficult.' Nor did he have much opportunity to speak to specialists. The personal viewpoint is expressed in the title: it was Darling's own 'plain tale of a remarkable region' and not a textbook on highland wildlife. Darling's closest collaborator, James Fisher, had, with his usual enthusiasm, become involved in the detail as well as the outline of the book. 'His friendship has been sorely tried,' wrote Darling, rather mysteriously, in his preface. Billy Collins, for his part, relates offhandedly that he, Julian Huxley and James Fisher 'paid visits' to remote Strontian to discuss the book with the author. Strontian was very remote indeed, if you happened to live in London. In fact, I find this the most remarkable statement in the entire history of the New Naturalist series. All three of them, all that way, and more than once! These books must indeed have been something out of the ordinary.

The editors made a number of contributions to the overall style of the library. One of these was the special distribution maps, first used in *Butterflies*. The editors considered that the existing maps of the British Isles were inadequate and so a new outline map was printed by Adprint under Dudley Stamp's direction. For the purpose of pinpointing localities accurately, the map carried the National Grid for Great Britain around the margins, and it could be, and has been, mistaken for a prototype of the later distribution maps that use the standard 10km grid square (as used for the *Atlas of the British Flora*). However, the latter maps were devised at a conference held by the Botanical Society of the British Isles in 1950, apparently without reference to the New Naturalist maps. The similarity between the two is only superficial: while the New Naturalist maps represent actual sites, the much more methodical plotting on the BSBI maps is based on grid squares on a presence or absence basis. Nevertheless, the New Naturalist maps, first used to map the distribution of butterflies, were a significant advance on earlier efforts to map the flora and fauna, and enabled the editors to claim that 'nothing so effective has previously appeared'.

It was at Dudley Stamp's request that another common feature of the library appeared – the list of titles printed on the back of the dust jacket. Originally the latter had been used to advertise related or forthcoming issues, or, in the very first books, the ethos of the series itself. The first book to list all the titles (plus some forthcoming ones) was *Natural History in the Highlands and Islands*, but it became a regular feature only after the mid-1950s. At first the list contained all the main series titles, but from about 1967 those that had become out of print were omitted, so that a new generation of New Naturalist readers might never know of *A Country Parish* or *Natural History in the Highlands and Islands*. The listing of the Monographs was never as methodical. At least one title, the ill-fated *Ants*, was, so far as I am aware, never listed or advertised at all. It flitted briefly in and out of publication like a phantom in green buckram. It was only in the mid-1980s, after nearly all the older titles had gone out of print, that the full list of mainstream New Naturalist titles was reinstated – in ever smaller print as the list grows apace.

It is a measure of the dedication and team spirit of the original Editorial Board that they worked together in seeming harmony for more than twenty years, far longer than any of them had expected when they accepted Billy Collins' invitation. Together they saw some 70 titles through to publication, and their ranks were sundered only by death: Stamp's in 1966 and James Fisher's in 1970. The successful idea of a team of five, with Sir William Collins or a deputy in the chair, was adhered to. Stamp was replaced by a fellow geographer, Margaret Davies, among whose contributions was the substantial revision of H.J. Fleure's *A Natural History of Man in Britain* in 1970. Fisher was replaced by Kenneth Mellanby, another natural history polymath who had recently contributed one of the more successful latter-day titles, *Pesticides and Pollution*. John Gilmour, who retired in 1979, was replaced by his partner on *Wild Flowers*, Max Walters. The last surviving member of the original team, Eric Hosking, died in 1991, and his departure broke the last thread of continuity that had had its origin in war-torn London, nearly fifty years before.

In 1956, the New Naturalist library celebrated its 50th title and broke the sales tape at half a million, an average of 10,000 copies sold of each title. It was an appropriate time to take stock. The original aim of the series had been, as we have seen, for just that number of titles covering the whole panorama of British natural history. It had almost succeeded in doing so, but there remained significant gaps, such as bird migration, fossils and pollen history, mosses and lichens, and the Board was keen to publish more regional titles, like the long-delayed Norfolk Broads and the Lake District. One receives the distinct impression that no one really wanted to wind up the New Naturalist library. A special meeting was held in February 1956 to review the series and tighten up the publishing schedule by producing a programme of publications for the next five years. The editors now considered that 'to fulfil the original aims of the series, some 90 to 100 volumes would be required ... over the next 15 years' (i.e. up to 1971). They listed some 34 desirable main series and 'approximately' 23 monographs; in effect, this amounted to an open-ended commitment, which continues to this day.

At the same meeting, various ideas to promote the series were discussed: a series of New Naturalist lectures, sponsored (they all hoped) by the Zoology Society of London, to which editors and authors would contribute; a 'New Naturalist fortnight' with window displays in the leading bookshops; a 'New Naturalist Association' with a mailing list. None of these came to anything. The climate had changed since the 1940s, and the New Naturalist library was no longer far out in front of its rivals. The pack was catching up.

A New Naturalist Board Meeting

The Editorial Board met about five times a year, in the boardroom at 14 St James's Place, generally in the afternoon or early evening. In the early 1950s (the busiest years of the series in terms of titles) the meetings were chaired by W.A.R. Collins, or in his unavoidable absence, by his chief editor F.T. Smith. Also present were the five permanent members of the Board, the natural history editor of Collins, and a secretary to take the minutes. Frequently the Board would dine together afterwards.

The procedure followed was broadly the same each time the Board met. After the minutes of the previous meeting had been approved, Billy Collins reported on the general state of the series in terms of sales, and on the reception of recently published titles. If there was any business on the series as a whole, it was usually dealt with at this point. The business would then turn to the progress of the forthcoming titles, one by one, with the responsible editor delivering the report. The monographs and special volumes would be dealt with in the same way, and any remaining business, such as pricing, payments to photographers, foreign sales and so on would be dealt with at the end. In the early years, special meetings were held at the Adprint office to view colour slides of the latest photographs, but from about 1952 these projections took place less frequently and at St James's Place (see below). The minutes were typed on foolscap paper, and were signed and dated by Collins or Smith. In the 1950s, the minutes were recorded in considerable detail, but they became more perfunctory later on. Here I give an abridged version (for the original is too long for reproduction verbatim) of a typical New Naturalist Board meeting, that was held on 8 October 1951 with Mr F.T. Smith in the Chair. The matter in quotes is as recorded; the rest consists of my summary and explanation.

THE NEW NATURALIST

A meeting of the Editorial Board was held on Monday, 8th October 1951 at 14 St James's Place, SW1 at 5.30 pm.

PRESENT Mr F.T. Smith (in the Chair), Mr Fisher, Dr Stamp, Dr Huxley, Mr Gilmour, Mr Hosking, Miss Obee, Miss Reider.

MINUTES OF THE LAST MEETING 'Mr Fisher said he thought it was important in future that the minutes of the meetings should be signed by Mr Collins as Chairman as this was in accordance with normal practice.' Stamp had proposed, with Fisher seconding, a motion that Collins would sign the minutes of the previous two meetings at the next meeting, and that it should then be regarded as binding. [This was done; the point was that some of the previous meetings had been signed before the Board had had a chance to read them.]

EDITORS' ROYALTIES 'It was agreed that this should be left and discussed between the Editors when they dined together after the meeting.'

PROGRESS OF MAIN SERIES

BRITISH MAMMALS (L.H. Matthews). Fisher had received the author's corrected page proofs. He 'would have to spend quite a lot of time on them as the author had inserted corrections which could not be got on to the [printed] page'. Hosking reported that the black and white photographs had been proofed, but two would have to be reproofed. [*British Mammals* was published on 17 March 1952.]

CLIMATE AND THE BRITISH SCENE (G. Manley). [The line drawings had been sketched by the author for making into blocks by Miss Birch of Adprint. There were a large number, and their cost – £150 – had become an issue at the previous meeting since the Board had agreed on a limit on line drawings of a miserly £40 per book.] Fisher took the responsibility for exceeding the sum, and 'had received a very nice friendly letter from the author, asking if there was any way in which he could help over this difficulty'. Huxley suggested that he be asked to agree that a proportion of the over-estimate be set against the author's royalties. That would hold down the price. 'Printing quantity to be fixed. Mr Fisher suggested 7,500.' [*Climate and the British Scene* was published on 13 October 1952.]

AN ANGLER'S ENTOMOLOGY (J.R. Harris). Fisher had obtained the corrected galley proofs, after sending a series of telegrams to Harris in Dublin, but still awaited the final black and white selection and captions to line drawings and black and white photographs. 'Mr Smith suggested that Collins' representative in Ireland should call on Harris' and that Fisher should brief him. [The outcome seems to have been satisfactory, for *An Angler's Entomology* was published the following year.]

SEA-BIRDS (J. Fisher and R.M. Lockley). Fisher reported 'that 200,000 words had been written, 150,000 by James Fisher and 50,000 by R.M. Lockley'. Previously, Fisher had suggested to Mr Collins that the book be published in two volumes, but although Collins thought that a good idea, Huxley and Lockley did not agree. 'Mr Fisher said he had now come round to their view .... and would have to boil down his portion of the book to around 50,000 words.' Smith said he would get Raleigh Trevelyan to work out the maximum wordage for a book of 320 pages, and 'Mr Fisher would endeavour to work to this'. Smith said he would like to try to interest an American publisher before printing. Huxley mentioned that Houghton Mifflin had expressed great interest in the book, and produced correspondence to that effect. [*Sea-Birds* took another year to complete, and was published on 1 March 1954.]

DARTMOOR (L.A. Harvey). 'Mr Smith raised the point as to whether the recognition of Dartmoor as a National Park had affected the text.' Fisher agreed to look into this and arrange any necessary changes. Stamp wanted to see this title in print by summer 1952, and Smith agreed 'to expedite production as far as possible having regard to other titles'. [It took Fisher several months to read the proofs, and *Dartmoor* was published not in 1952 but on 31 August 1953. That did at least allow the authors to bring the book up to date.]

THE WEALD (S Wooldridge). The manuscript had been lost, and the author was using the carbon to incorporate Huxley's criticisms. Stamp would take delivery of the manuscript (MS) and prepare it for the press. 'The author is cutting down the number of colour from 32 to 16 and has agreed to cut the b & w.' The loss of the MS raised the matter of keeping a careful record of 'where a particular MS was at any time'. [*The Weald* was published on 16 March 1953.]

MUSHROOMS AND TOADSTOOLS (J. Ramsbottom). Gilmour produced 20 out of the 24 chapters [the published book has 23 chapters]. The author had promised 'absolutely faithfully' to deliver the remaining four by the end of the month. Once the MS was complete, Gilmour, Hosking and Ramsbottom would meet to discuss photographs. [Drastic cuts to the draft proved necessary, and *Mushrooms and Toadstools* was not published until 26 October 1953.]

THE SEA COAST (J.A. Steers). The MS was too long, and cuts had been asked for to bring the book down to 288 pages. Fisher reported that this had now been done, and the MS was ready for setting. [*The Sea Coast* was published on 18 February 1953.]

THE BROADS (A.E. Ellis). Fisher reported that the author was making progress and had nearly finished his part of the book. Stamp had been promised a contribution from 'Jenkins' (*sic.* The person referred to was J.N. Jennings). A section by the late Dr Robert Gurney was already in hand. Stamp queried whether the book might be overtaken by the National Park issue, 'but it was thought not as this area will not be taken over for some time'. [Not for another 40 years in fact, but the startling discovery by Lambert and Jennings that The Broads were artificial did mean substantial rewriting. *The Broads* was not published until 1965.]

MARSH AND FRESHWATER BIRDS (R.C. Homes). Fisher reported that the author wanted to get on with 'the London Bird Book' [*Birds of the London Area* 1957] first. Agreed. 'He would complete the MS for [*Marsh and Freshwater Birds*] in 1953. In this way we would get a really good book from him.' [The London book took longer than was envisaged and this title was never published.]

THE LAKE DISTRICT (E. Blezard and others). Fisher reported that the joint authors had after two years produced 30,000 words of script. Blezard's own work was up to standard, but the rest was no good. The best person to write the book would be W.H. Pearsall, but he could not do it for 18 months. Huxley said this was an important title and must be first class: they decided to ask Pearsall. Fisher would write to Blezard explaining the Board's decision to start again, and 'they hoped it might be possible to fit Blezard into this'. [Pearsall accepted the contract, but was too busy to write the book. W. Pennington wrote the book, using Pearsall's notes. It was published in 1973.]

DRAGONFLIES (C. Longfield). Fisher said they had narrowed down the plates to 40, but proposed to pay the photographer [Sam Beaufoy] for the full 80 he had taken. Collins felt he could not agree to this, and suggested that 'the photographer should forego part payment for the photographs on condition that he owned the copyright'. Huxley suggested that Hosking should discuss arrangements with Beaufoy, and thought 'payment of something like ½ to ⅗ of what we originally suggested might be a fair figure'. The book should be held up until the dispute with Adprint had been sorted out. [It was published on 15 February 1960.]

ANIMALS IN ART (F.D. Klingender). Huxley said the author had incorporated a lot of philosophic material and that 'Animals in Art and General Thought' described it better, and this would make it 'a more novel and interesting book'. The length must stay the same as in the original contract. [The author died in 1955, and the book was never published.]

TREES (? E.W. Jones). It was becoming clear that it would be better to concentrate on *Forestry*, for which Sir William Taylor, the recently retired Director of the Forestry Commission would be a suitable author. 'To go ahead.' [The title became *Trees, Woods and Men* by H.L. Edlin, published in 1956.]

FERNS AND MOSSES (F. Ballard). 'Mr Gilmour to write Ballard and get it settled that we are definitely not going to do this book.'

MOTHS (E.B. Ford). 'Huxley reported that the author was getting on with the book.'

MONOGRAPHS AND SPECIAL VOLUMES

FLEAS, FLUKES AND CUCKOOS (M. Rothschild and T. Clay). 'Mr Fisher said it had gone to the printer.' [*Fleas, Flukes and Cuckoos* was published on 5 May 1952.]

THE WREN (E.A. Armstrong). There was a considerable discussion about this title in relation to the amount of cutting Mr Collins was insisting on to bring the book nearer to its contracted length. The matter of the editors forgoing their royalty on this title in the interests of keeping the price down was later discussed over dinner. [After considerable argument and delay, *The Wren* was published within the prescribed limit of 320 pages on 28 March 1955. The Editors did waive their royalty.]

VIRUSES (K.M. Smith). Mr Smith circulated his reports on the 40,000-word MS, saying he thought 'it was good as far as it went' but that the book 'did not say anything about remedies'. Nor did it say anything about the nature of viruses 'and whether they were alive or not'. Smith 'queried whether there would be a definite sale for this book and was assured that there would be'. In the view of ongoing research on the subject, Gilmour was to ask the author to agree to postponing publication for 'a year or two'. Hosking found the photographs 'rather dull', but agreed that this was inevitable. [Two more chapters were added, and *Mumps, Measles and Mosaics* was published on 1st February 1954.]

THE HERON (F. Lowe). Smith circulated his report: the subject matter 'was all right' but suggested that the last two chapters 'were redundant and could at least be abridged'. Huxley thought these chapters were the most interesting. The colour photos were not very good. 'It was agreed to accept this book subject to Mr Collins' approval.' [*The Heron* was published with only one colour plate on 12 July 1954.]

THE HERRING GULL'S WORLD (N. Tinbergen). 'Mr Fisher said that Mr Collins had asked Mr Trevelyan to cost this book and that it worked out too high. It was pointed out that we were committed to this book by Minute of 4 April 1951 and the author was pressing for an answer...Mr Smith should raise this matter with Mr Collins immediately on his return. The author has supplied all photographs and line material.' [Collins honoured their commitment, and *The Herring Gull's World* was published on 28 September 1953.]

BUILDING STONES OF ENGLAND (Dr Arkell). 'It was agreed that this should be turned down.'

ANIMALS IN ULSTER (C.D. Deane). 'To be discussed...at dinner. Mr Smith thought it would be too local in appeal.' [The book was turned down.]

MISCELLANEOUS POINTS Mr Smith explained that the price of all future New Naturalist main series titles would be 25 shillings. This also applied to new editions of old titles. It was suggested that Collins should write to Adprint and suggest buying the projector which they own, and that in future the showing of transparencies should be at St James's Place. Mr Hosking though a fair offer would be £30/£40 for the projector and trolley.

MEETING TO VIEW COLOUR TRANSPARENCIES OF 'THE FLOWERS OF THE SEA AND COAST' 'To be held at Adprint office on Friday, October 26 at 2.15pm.' [This seems to have been postponed until 5 December, when slides for *The Weald*, *Dartmoor*, *Sea-birds* and *The Sea Coast* were also viewed.]

DATE OF NEXT MEETING

To be held at 14 St James's Place on Tuesday, 20 November at 5.30 pm.
Signature..........W.A.R. Collins (Chairman)
Date.........20 November 1951.

The New Naturalist Board meeting in the Collins boardroom in Grafton Street, June 1966. Standing (left to right): James Fisher, W.A.R. Collins, Sir Dudley Stamp. Sitting: John Gilmour, Sir Julian Huxley, Eric Hosking. Stamp died two months later. (*Photo: Eric Hosking*)

Table 3. New Naturalists – Proposed future programme 1956-62 (quoted verbatim from New Naturalist Board minutes 20 June 1956)

	Main series	**Special volumes**
1956	*Trees, Woods and Man (H.L. Edlin)	*The Rabbit (Thompson and Worden)
	*Mountain Flowers (Ravens and Walters)	
	*The Open Sea I (A.C. Hardy)	
1957	*The World of the Soil (Sir John Russel)	*Birds of the London Area (R.C. Homes & others)
	Insect Migration (C.B. Williams)	The Hawfinch (G. Mountfort)
	The Open Sea II (A.C. Hardy)	*The Salmon (J.W. Jones)
1958	*Spiders (W.S. Bristowe)	*Folklore of British Birds (E.A. Armstrong)
	The Broads (E.A. Ellis & others)	*Bumblebees (Butler and Free)
	The Peak District (K.C. Edwards)	Lords and Ladies (C.T. Prime)
	Shore Birds (E.A.R. Ennion)	
1959	Dragonflies (C. Longfield & others)	The Jackdaw (K. Lorenz)
	Rare Plants (D. Pigott)	The Trout (Frost and Brown)
	Fossils (H.H. Swinnerton)	The Rook (J. Fisher)
	The South West Coast (L. Dudley Stamp)	The Fox (Thompson and Worden)
1960	*The Lake District (W.H. Pearsall)	The Gannet (Barlee and Fisher)
	*Marine Molluscs (C.M. Yonge)	The Peregrine (Lees and Bond)
	Bird Migration (K. Williamson)	Nature Chronology in Britain since 100BC (D.J. Schove)
	Weeds and Aliens (Sir E. Salisbury)	
1961	*Marsh and Freshwater Birds (R.C. Homes)	The Greylag Goose (K. Lorenz)
	Wild Fowl Resources (Severn Wildfowl Trust)	The Crossbill (A. Robertson)
	The Ice Age and After (S.E. Hollingworth)	The House-Sparrow (J.D. Summers-Smith)
	Land and Freshwater Molluscs	
1962	*Woodland Birds (no author)	The Cuckoo (H.N. Southern)
	British Naturalists (C.E. Raven)	The Partridge (A.D. Middleton)
	Grasses(no author)	The Grey Seal (R.M. Lockley?)
	Pollination of Flowers (?)	The Great Crested Grebe (K. Simmons)
	The Yorkshire Dales (P.F. Holmes?)	The Pheasant (?)
	Caves (G. Grigson?)	
	Ponds, Puddles and Protozoa (A.C. Hardy)	
	Moorland Birds (J. Fisher)	
	Beetles	
	Whales and Whalers (F.C. Fraser)	
	The Art of Animal Illustration (W. Blunt?)	
	Nature Conservation (J. Fisher and J. Huxley)	
	Rivers (?)	

* contracted books

Table 4. New Naturalist Editors and senior Collins staff 1943-1993

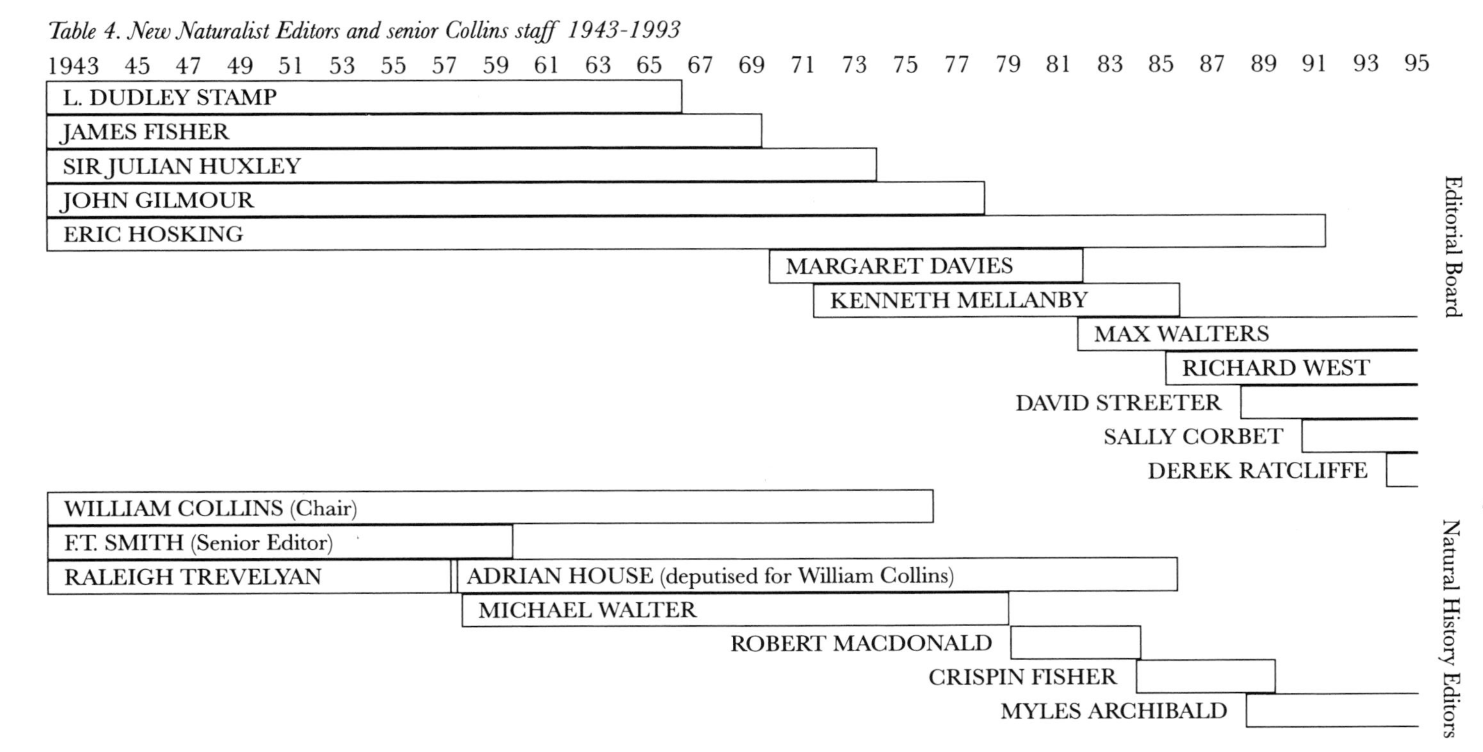

3

'The Full Beauty Of Their Natural Colours...'

If you open almost any natural history book or magazine published more than fifty years ago, one thing will strike you right away. However well produced, the book contains no pictures of living plants or animals taken in the wild and in full colour. Before 1943, all natural history photographers in Britain used monochrome. The plentiful colour plates in pre-war nature books depicted dead specimens, pinned, blown or stuffed, or artwork based on the crafts and technique of earlier generations. Nature as depicted in books bore little similarity to the colourful, vibrant world of the hedgerow or hillside. The live fish were photographed in tanks, the animals behind bars at the zoo, wild flowers in a vase, and, if they were in colour, the tints were often added afterwards by hand.

The colour photographs which made such an impact on the book world when the New Naturalist library emerged had their origin in the collaboration of the House of Collins and Adprint. Shortly before the outbreak of the Second World War, Adprint and Collins produced the first in a projected series of garden books called *The Garden in Colour*, which was remarkable for being illustrated exclusively by colour photography. Today, the pictures would go straight into any book publisher's wastepaper basket, but in the monochrome world of 1940 they were sensational. Their soft colours resembled an unusually detailed lithograph or aquatint, and barely caught the brilliance of the individual flowers. But for the gardener they were a decided advance on sepia or black-and-white. And, more importantly, they were a useful practical demonstration of what could be done with camera and letterpress.

The Collins-Adprint partnership continued with a soon-to-be famous series of illustrated books called 'Britain in Pictures'. These little books, produced in octavo size with distinctive coloured boards and wrappers, sold for only 4/6d (23p) each and covered every imaginable facet of British life. They sold in huge numbers: *The Birds of Britain* by James Fisher, for example, sold nearly 100,000 copies and remained in print for ten years. The sub-title of the series, 'The British People in Pictures', told the story. Commissioned at a time when Britain alone had stood up to the Dictators (the first books appeared in March 1941), they were essentially part of the war effort, of the great surge in national pride that had followed Dunkirk and the Battle of Britain. Seeing the gallant display they made in a bookshop window on the desolate front of some south coast town, the Collins chief editor, F.T. Smith, compared them with 'the bright banners on the battlements of our island fortress or, more modestly perhaps, the defiant cockades that a nation of shop-keepers might justifiably flaunt in the faces of their book-burning foes'.

Despite their lavish use of colour, the Britain in Pictures series made little use of colour photography. In the case of the nature books, this was because suitable material simply did not exist. Instead, Adprint ransacked collections of printed material for coloured engravings and paintings produced by earlier (sometimes much earlier) generations. Nonetheless the books were printed to a high standard, using good quality Mellotex paper and the beautiful 'Fontana' typeface that Collins had created six years earlier. The series was also notable in its parade of contributors, among them some of the best-known novelists, dramatists, poets and sports writers of the day. For the nature books the producers found James Fisher to write *The Birds of Britain* (1942), Frank Fraser Darling for *Wild Life of Britain* (1943), C.M. Yonge for *British Marine Life* (1944), John Gilmour for *British Botanists* (1944) and R.M. Lockley for *Islands Round Britain* (1945). Later, six of the nature books were bound together in an omnibus volume, *Nature in Britain* (1946), which Collins called a 'Guinea Book' after its price.

R.M. Lockley, co-author of *Sea-Birds* and whose book on rabbits helped to inspire Watership Down, with a favourite subject, in 1958. (*Photo: E.M. Nicholson*)

Altogether no fewer than 132 Britain in Pictures titles were published by Collins between 1941 and 1950. Their success was a fair indication of the appeal that colour illustration would have for the general public. Indeed, these little books still have their following, and many second-hand bookshops stock them. That they also added to the reputation of Billy Collins as a master of good timing is suggested by a contemporary review in prime Second World War rhetoric by Viola Garvin in *The Observer*:

> 'Not three months after that bad night when Hitler's bombers lit the second fire of London and destroyed the centre of the English book world, the old and famous house of Collins shakes its head, steps out of the ashes and comes forward with as timely and bold a piece of publishing as was ever planned .... These books manage to distil "the glories of our blood and state" and with neither vanity nor pomp to make clear to ourselves, as well as to the rest of the world, the full and serious nobility of our heritage.'

Something of this spirit might have infused Billy Collins' next idea for a series of illustrated nature books that would make a contribution to science as well as to popular literature: the New Naturalist library. At this stage, the Collins-Adprint partnership was concerned principally with colour printing. Billy Collins and Wolfgang Foges, Adprint's managing director, were convinced that colour photography could make a major contribution to scientific book publishing. Moreover, they were in a strong position to push ahead, with Collins' large allowance of paper, as manufacturers of bibles, diaries and stationery, and Adprint's experience in producing coloured illustrations. Because of the rarity of suitable photographic material, the photographs for the series would need to be taken specially for it. In the early 1940s, Foges made a special visit to the United States to round

Eric Hosking (1909-1991) and his complete collection of New Naturalist titles.

up supplies of quarter-plate size (4¼ × 3¼ inches per frame) Kodachrome colour film, as yet unavailable in Britain. This film, hitherto used mainly by commercial advertisers, was greatly superior to the 35mm Kodachrome film which *was* available, with better colour balance and definition. Foges also purchased specially designed Kodak cameras, plates and flash bulbs. To bring all this bulk equipment back to Britain in wartime required a special import licence from the Board of Trade. That Adprint managed to obtain one, helped by the good offices of the British Council and backing from the Ministry of Information, suggests that the project, like Britain in Pictures, was perceived as a significant contribution to wartime morale and prestige.

The idea was for Adprint to commission photographers for the New Naturalist books, and for Collins to find the authors. For the former they would need a photographic editor. In his memoirs, *An Eye for a Bird* (1970), Eric Hosking describes briefly his 'initiation' into the New Naturalist team, and he set down on tape a more detailed reminiscence for the writer Ron Freethy in 1983. I have edited this latter account slightly.

> 'As I remember it, it was on 29 November 1942, a Sunday, and I had just returned from lecturing at a school up in York. I had just got inside the front door when James Fisher was on the 'phone asking if he could come round with Julian Huxley and Wolf Suschitzky to see me about a project. Suschitzky was a very famous photographer of wildlife in those days, and he did a lot of work for London Zoo – they didn't have their own photographer then, and Julian was directing the Zoo's affairs. They came round right away, and I found out that Suschitzky wasn't prepared to take over the photographic editorship of the New Naturalists because it was going to involve him in too much work. The impression they gave me was that there were so few colour photographs in 1942 that somebody was needed to organise a gang of photographers to go out into the field and take photographs specially for the series.'
>
> *Eric Hosking*. In R. Freethy (1983) Collins New Naturalist. In the Beginning. *Country-side*, pp. 183-89

It was Eric Hosking's task to recruit the 'gang' and ration out the precious stocks of Kodachrome film. He would be the link man between the authors and their photographers, and, as it turned out, on occasion stand up for them on delicate matters of payment and copyright. And, of course, he would independently be taking photographs for the series himself.

The photographers ...

The first New Naturalist colour photographs were taken in 1943. Happily, we can date exactly the first use of American-made Kodachrome film for the New Naturalist series. Eric Hosking's diary entry for 8 April 1943 contains the following note:

> Kodachrome: Made 8 exposures on ¼ plate Kodachrome of various specimens to test for colouring. These are probably the first English coated Kodachrome of this size to be used in this country. It will be interesting a few years hence, when colour photography will be an everyday thing, to recall that I was privileged to experiment with some of the first in this country.

While the Battle of the Atlantic still raged, it was of course perfectly possible that the precious exposed film would be sent to the bottom of the sea by a U-boat while on its slow crossing back to America – and David Hosking tells me that this did happen on at least one occasion. But in time these Atlantic journeys ceased to be necessary. There was a Kodak processing laboratory at Harrow, and although its darkroom technicians were engaged on more vital tasks during much of the war (and the laboratory had also been bomb-damaged), it was eventually able to develop colour film for Adprint. Even so, the development of colour film was a chancy enough business at this time, requiring 17 different baths of chemicals, each of which had to be at just the right temperature.

Finding photographers with sufficient skill, equipment and spare time to take pictures for the series was not much easier. Collins had inserted advertisements in the photographic press asking for colour photographs of nature subjects. The response was apparently not overwhelming, though 'a number' were received and followed up. Most natural history photographers at that time were amateurs –

Eric Hosking was a rare exception – and few had had previous experience of colour work. Fortunately, in Billy Collins' words, 'England is a country of hobby men .... It is probable that such an undertaking as "The New Naturalist" series would not have been carried through had it not been for the existence of a group of amateur enthusiasts, members of the Royal Photographic Society, who eagerly threw themselves heart and soul into the project'. This was just as well, for the difficulties they faced were formidable: petrol was rationed (Hosking was allowed more than most since travel by car was a necessary part of his livelihood); 'non-essential travel' was discouraged; and access to many wildlife localities was restricted, especially along the south coast. As Billy Collins put it, 'The sight of a camera acted as a red rag to the Home Guard bulls.' The photographers had to obtain special passes or risk capture and interrogation – the fate of at least one intrepid New Naturalist photographer.

Through advertisements and personal contacts, a band of twenty or so photographers were assembled to work for the New Naturalist series. There was John Markham, then working as an air-raid warden and taking pictures in his spare time. After the war he relinquished the family business to become a full-time photographer. He covered much of the country for the New Naturalist series, contributing the lion's share of the colour photography for *Mountains and Moorlands, The Natural History of the Highlands and Islands* and *Snowdonia*, as well as some of the pioneering pictures of wild animals for *British Mammals*. He had a painter's eye for composition, and must have climbed a lot of hills in all weathers to achieve the most advantageous shot. His upland landscapes, especially in *Mountains and Moorlands*, are wonderful studies of rock, vegetation and weather, and are a fitting complement to Pearsall's text. Another important contributor was Douglas Wilson, a zoologist working at the Marine Biological Association's Plymouth laboratory. Wilson had made a name for himself before the war as a leading photographer of marine and microscopic subjects. His richly illustrated book, *Life of the Shore and Shallow Sea* (1937), was a natural history showcase, but he surpassed that monochrome work with the brilliant pictures he took for C.M. Yonge's *The Sea Shore*. Another contributor was Walter Pitt, a solicitor, 'whose abiding delight was to photograph live fish in tanks' (as Billy Collins put it). Unfortunately his fish pictures, though sensational, were never used, since the volume they were intended to illustrate was never completed. But some of Pitt's work did eventually surface in Malcolm Smith's *British Amphibians and Reptiles* (1951). The fish photographs were by no means the only casualty of the uncertainties of authorship. Who can forget the story, told in Christopher Page's recent New Naturalist book on *Ferns*, of the mysterious 'elderly Devonian' who had toured the lanes and shores of the West Country in 'about 1947', photographing ferns for a book that never appeared.

A New Naturalist commission must have been a wonderful opportunity for nature photographers like Eric Hosking, John Markham, Robert Atkinson and Brian Perkins. Their work for the first dozen or so books was evidently carried out between 1944 and 1947 and took in every part of the British Isles and almost every imaginable subject. In an entertaining celebration of the series published on 12 August 1994 in the *Times Literary Supplement*, Professor W.D. Hamilton wove a delightful fantasy about a New Naturalist outing in which the intrepid four set out to photograph aspects of a Welsh valley in an old timber-framed estate car. Dividing up their tasks with a few laconic words, the hill-men, Markham and Atkinson, set off through mist and rain in search of soil profiles, black-rumped

John Markham (left) and Stuart Smith with period equipment at Hickling Broad, c. 1945, very probably on a New Naturalist photographic foray. (*Photo: Eric Hosking*)

sheep and an endemic species of whitebeam. Meanwhile, down in the valley, Perkins is tracking down rare flowers and their insects, while Hosking has set up his hide in the churchyard to photograph pied flycatchers, not omitting the flycatcher lousefly crawling on the bird's back for Miriam Rothschild's *Fleas, Flukes and Cuckoos*. This flight of fancy may not be so very far from the truth, though the photographers would probably have been accompanied by a Bentley-full of authors, in this case Frederick North (Snowdonia), Victor Summerhayes (Orchids) and W.H. Pearsall (Mountains). In support of this thesis, I reproduce here a photograph taken at this time by Eric Hosking of John Markham and Stuart Smith in the reeds, very probably taking pictures for the long-delayed book about the Norfolk Broads.

Some subjects defied even the most dedicated photographers. One of these was *The Open Sea* by Alister Hardy (which was originally planned as a single volume). Almost the only photographer to have specialised in marine subjects was Douglas Wilson at Plymouth, but his fish tanks had been smashed to splinters in the bombing raid on the city, and it would be some time before the collection could be reassembled. In any case, the living planktonic and deep-sea life forms which Hardy wanted were beyond the technical capabilities of any camera yet devised. Wilson was of the opinion that in this field at least, old-fashioned pen and brush methods were still superior to photographic film; and Hardy was, as it happened, a talented water-colourist. That is why *The Open Sea* is exceptional in being illustrated by 'coloured drawings' by the author, while Wilson contributed the monochrome photography. Billy Collins took a lot of convincing, but for this book there was no other choice.

A related problem presented itself with W.S. Bristowe's book on *Spiders*, which was planned as early as 1945. The only way of slowing down spiders sufficiently for the slow Kodachrome film to register a reasonably sharp image was to dope them with anaesthetic. But at the first whiff of ether, a spider would tuck its eight legs under its body and remain stubbornly in this unspiderlike posture until the effect had worn off – after which it once more scuttle off across the carpet at speed. Even Sam Beaufoy's patient skill had to admit defeat on this awkward subject, and, like *The Open Sea, The World of Spiders* was illustrated mainly by drawings (and what magnificent drawings they were – some 232 of them in line and wash by the renowned artist Arthur Smith).

Despite such lack of co-operation on the part of nature, the New Naturalist library had, by April 1946, accumulated 2,500 colour photographs taken specially for the series. Billy Collins and the editors would assemble regularly at Adprint's offices on Newman Street for special viewings, using an early (and then very expensive) slide projector. In his *Bookseller* article, Billy Collins described how the proceedings would start with the lights being turned out, and the most recent colour photographs projected onto a screen, one by one, 'so that the editors and production men can weigh their respective merits both from the point of view of scientific detail and technical reproduction'. Sometimes as many as a dozen slides of the same subject would be shown, from which the best was selected. Occasionally, indeed, all were rejected and the photographer asked to try again. If so, this might 'at one stroke set the book back twelve months, since most of the subjects are seasonal'.

..... and their authors

It was of the essence of the New Naturalist library that the text and the colour illustrations should be well matched. It was important, therefore, for author and photographer (for most books had been assigned a 'lead' photographer) to work closely together. In most cases they did so, often forging lasting friendships in the process. In Chapter 5 we consider a classic instance of this on the very first book, between E.B. Ford and Sam Beaufoy. The results of that work convinced everyone that Beaufoy was the man they wanted to photograph living insects for other planned titles, which included *Insect Natural History* by A.D. Imms and *Dragonflies* by Cynthia Longfield. The insect book was in some ways an even greater challenge than butterflies, its subjects having for the most part less aesthetic appeal while being at least as prone to fly or wriggle away at speed from the photographer's lens. Some of the items on Dr Imms' shopping list were not creatures which any photographer would happily admit to his studio. Sam Beaufoy recalls being taken aback by Imms' request for a portrait of a live bed-bug! 'As chance would have it, the local Health Department was at that time demolishing some slum property, and frequently telephoned me to say that "some real beauties" were ready for me to collect. This was the sole type of insect of which the presence in our house was unwelcome. The numbers were carefully counted before and after photography, and the specimens then, with relief, carefully disposed of in the kitchen boiler-fire. The postman frequently delivered packets of various insects from Dr Imms, a brief note inside saying: "Please photograph the enclosed"!'

Bed-bugs are at least slow-moving. One of the first subjects that Beaufoy was called upon to photograph was a silverfish, whose shiny, torpedo-shape was then a familiar domestic sight as it scuttled across the hearth or along the pantry shelf.

Sam Beaufoy's apparatus for close-up work, 1944. (*Photo: S. Beaufoy*)

'Mr Beaufoy is quick on the camera trigger,' recalled Billy Collins, 'but, try as he would, these creatures beat him. The suggestion was then put to him that he might dope the silverfish in some way that would keep them still enough to be photographed; Mr Beaufoy, being a photographic purist, firmly refused to place his subject under control of this sort. After a great deal of persuasion, however, he finally agreed to do this on condition that wherever he anaesthetised the subject, a statement to that effect would appear below the reproduction.' The dose of ether (or enforced hibernation in the fridge) needed careful calculation. In praising Beaufoy's skill, Imms was at pains to point out that 'Care was taken not to kill them with an overdose, and consequently they display their natural colours and appearance.'

The colour photographs for *Dragonflies* were taken in the early 1950s, years before the book was published. A number of dragonfly portraits had been included in *Insect Natural History*, and Cynthia Longfield later recalled that 'Sam Beaufoy and I spent two years perfecting [the technique]....Billy Collins vetted every plate with me.' Some of the specimens were reared in tanks by Eric Gardner. Others were the result of joint collecting expeditions to the New Forest and elsewhere by author and photographer. When Cynthia's 'eagle-eyes spotted an interesting dragonfly, she would call out – "Sam! Sam! There is *fulva* – TAKE IT, *take it*!"' On another occasion, she telephoned him from the Norfolk Broads requesting his immediate attendance since a species of damselfly, thought to be extinct, had turned up again.

If, using the slow colour film of the day, the photography of living insects was often technically demanding, that of birds and mammals bordered on the impossible. In many cases, bird photography meant nest photography, which confined photography to a scant few months each year. And because the latitude of the film emulsion was then very narrow, the slightest error in exposure resulted in failure. Black and white photography generally yielded much better results, especially under the fickle, light-changing skies of the average English spring. By 1946, Eric Hosking had to some extent succeeded in overcoming this by rigging a remote-reading photo-electric cell on the camera lens, obtaining him an accurate reading within the darkness of the hide. But even so, there were not nearly enough colour photographs of British birds available to illustrate the half-dozen bird titles then planned. The first New Naturalist bird book, *British Game* (1946) by Brian Vesey-Fitzgerald, was illustrated mainly by prints collected by Adprint from older works, making this book the odd-man-out of the library. In a sense it was closer, so far as colour illustration went, to the Britain in Pictures series. And when colour photographs were used to illustrate E.M. Nicholson's *Birds and Men* (1951), their reproduction was far from satisfactory. Collins' insistence on using colour photographs of living subjects in wild surroundings was one reason for the delay and, in some cases the cancellation, of bird titles. It was also the reason why books like *London's Natural History* and *Highlands and Islands*, which today we would expect to find crammed full of colourful birds and animals, are illustrated mainly by landscapes.

Not that this was necessarily a bad thing. *London's Natural History* presents a possibly unique record of views of wartime London, of a cityscape that has all but vanished. In his preface to the 1984 paperback edition, Richard Fitter singled out the City gentleman in his bowler feeding pelicans (Plate 26B) and a bucolic scene of haymaking in Green Park (Plate XXVII) as examples of how times have changed. Plate 12 depicts London's last rookery, long since gone, and Plate II a panoramic view of central London with lines of barges on the Thames and not a single high-rise building in sight. Richard Fitter accompanied Eric Hosking on several of these urban natural history excursions. The entry for Hosking's diary dated 29 August 1943 has: 'With Richard and Maisie Fitter and Dorothy to Ken Wood, Welsh Harp, Harrow, Cheshunt, Broxborne and Walthamstow to obtain colour photographs for "New Naturalist" London book', noting also that many Swifts had already flown south. It was probably in the following year that Richard Fitter watched Hosking flag down a London tram in order to photograph it beneath the rookery at Lee Green – a speeding tram being beyond the capabilities of Kodachrome, even on a bright spring day.

Many of the wild flower portraits by John Markham, Robert Atkinson and others were made in the company of one or other of the authors. Robert Atkinson recalls field excursions with J.E. Lousley and V.S. Summerhayes for *Wild Flowers of Chalk and Limestone* (1950) and *The Wild Orchids of Britain* (1951). But his success in tracking down and photographing the majority of British orchids owed much to the advice of Francis Rose, who had a 'photographic memory' for botanical sites and was able to send Atkinson sketch maps of orchid localities from all over the country. *Wild Orchids* is perhaps the best illustrated plant book in the series, boasting not only Robert Atkinson's fine collection, nearly all of them taken specially for the book, but also monochrome studio pictures taken by the doyen of pre-war plant photographers, E.J. Bedford. Compared with many modern photographs taken on 35mm film, the depth of field and definition on these old Kodachrome plates is still impressive, and allows more detail of the surrounding habitat than is usual today. I had no difficulty in recognising the site of Robert Atkinson's Man Orchid photograph on a chance visit to Ipsden, Oxfordshire some years ago, even though the down had grown much scrubbier in the intervening thirty years.

In many instances, the authors themselves were keen photographers, and contributed pictures to their book. Dudley Stamp, for example, took much of the colour and black-and-white landscapes in *Britain's Structure and Scenery* himself, as did F.J. North for *Snowdonia* and Frank Fraser Darling for *Natural History in the Highlands and Islands*. Once, while studying specimens of *Ajuga* in the Lousley herbarium at Reading University, I came across the print of Pyramidal Bugle photographed by J.E. Lousley and reproduced on Plate XXIV of *Wild Flowers of Chalk and Limestone*. And what looked suspiciously like that very plant lay there beside it on the herbarium sheet! In other instances, it was the author, rather than Eric Hosking, who found the right photographers for his book. *Mushrooms and Toadstools* was one of these. The author, John Ramsbottom, had drawn on the services of Paul de Laszlo, who had produced some fine colour work of fungi while on shore leave from the Royal Navy Volunteer Reserve, and on an old friend, Somerville Hastings MP, who contributed some of the monochrome work. So good were Laszlo's colour pictures that Ramsbottom had them all duplicated to use in his public lectures at the Natural History Museum.

The Weald (1953) is unusual in that virtually all the illustration is the work of one man, Frederick Goldring, who, at the author's insistence, became a co-author on the strength of them. Goldring used to run a large guest house in the Weald, which was much frequented after the war by field courses led by Sidney Wooldridge. He had become an experienced hobby photographer with his own darkroom, and his pictures had appeared in guide books and photographic exhibitions. Almost all of the photographs for *The Weald* were taken on joint excursions by the co-authors, Goldring trundling his plate camera and tripod, while Wooldridge pointed out everything he wanted to illustrate. As a consequence, the match of plates and text in this book is a good deal better than average, and the splendid pictures have a consistency of style from being the work of a single person.

A similar instance was *The Country Parish* (1951), whose plates offer an architectural and landscape tour of Great Budworth in Cheshire, photographed by C.W. Bradley in close collaboration with the author, Arnold Boyd. Several of Bradley's pictures were later used again in *Man and the Land* and *Trees, Woods and Man*, and also for a New Naturalist calendar – to the great embarrassment of the author, who had been given permission to photograph private property on the understanding that it would not be reproduced elsewhere. This raised the matter of copyright. James Fisher had the uncomfortable task of dealing with his Uncle Arnold's complaint, pointing out to him that 'It is normal practice for photographers to hold the copyright and lease their use to publishers.' To which explanation Boyd retorted, in his down-to-earth Cheshire way, that 'As for the "normal practice" you mention, I know nowt about normal practices.'

Commissioning the pictures

A copy of the formal agreement between Adprint and the photographers dated 17 March 1944 survives among James Fisher's papers. Under its terms, Adprint would supply the photographer with sufficient colour film, together with 'all the photographic equipment, flash bulbs, plates and (other) sensitised material he may reasonably require and will process all photographs taken under this Agreement at their own request'. The copyright would belong to the producers, who would pay the photographer five guineas for each colour picture accepted by them, half on acceptance and half on publication. They would also pay a fee of one guinea each time it was reproduced elsewhere. Adprint undertook to accept 'at least one colour photograph of not less than 85% of the subjects taken', as long as they were of a suitable standard. The photographer was allowed to use 10% of the film for his own purposes. The standard agreement allowed '104 weeks' (i.e. two years) for the supply of colour photographs, but this presumably varied in individual cases. In cases where he had contributed twelve or more colour plates the photographer's name was supposed to appear both on the title page and the jacket of each book. He was to be sent one complimentary copy of the book, and allowed to purchase others at trade rates.

This memorandum establishes a procedure that may have been peculiar to the New Naturalist series. It bound the photographer to the producer as closely as the author to the publisher. It removed from him the responsibility for processing the exposed film, and though the plate would bear his name, he would take no part in its development. That the agreement undertook to include the name of the photographer on the jacket is a startling indication of the importance Adprint and Collins attached to colour photography. In practice, though, this clause was forgotten about when the jackets came to be designed – with one exception: Fred Goldring's name on the jacket of *The Weald*. So far as the title page went, however, the agreement was honoured.

The choice and responsibility for colour pictures rested squarely with the author. Once he had accepted an invitation to write a particular book, he was sent an Adprint memorandum asking him to compile a detailed list of subjects for which colour photographs were desired. Under the heading COLOUR PHOTOGRAPHS, the memo explained that 'Preliminary experience has shown the need of the most precise specification possible for colour photographs if serious waste of materials is to be avoided.' The document recommended the authors to exceed their quota of colour plates (which for the early books ranged between 24 and 48) by fifty per cent since, while some subjects would occupy a whole plate, others might be of half- or quarter-plate size. The subjects to be photographed were divided into four types of subject. Because of its documentary interest, I reproduce here the exact wording of Adprint's instructions to the author:

> **HABITATS** Precise description of the type of habitat, specially mentioning essential features, e.g. of scenery and of vegetation; the time or times of year required; whether close-up, middle distance or distant view is required. One or more precise localities would also be helpful.
>
> **SCENERY** As for habitats, but with careful notes on any geological or physiographical points to be brought out. Where possible, sketches or black-and-white photographs should be enclosed for guidance.
>
> **ANIMALS** NB: Many small animals can be photographed under controlled conditions (e.g. most insects). Specify whether live or dead specimens are: a) desirable, b) necessary; whether close-up or in a portion of natural habitat (e.g. butterfly on particular flower) or in natural habitat and if so, whether close-up or middle distance. If magnification is required, state amount; this should always be in simple multiples of natural size. If photomicrographs of prepared specimens are needed (e.g. butterfly scales, insect mouth parts) please state if a slide can be supplied.
>
> **PLANTS** Close-ups of single flowers or sprays or whole small plants can be taken under controlled conditions, indoors. Please state whether you require,

a) close-ups of one or a few flowers or fruits, b) close-ups of single plants or clumps, either under controlled conditions or in natural habitats; c) middle distance view of a group of plants demonstrating a natural habitat.
NB: If magnification is required (of flowers, grasses, mosses, etc.) please indicate amount as for animals.

That was what it was like to be on the receiving end of an early New Naturalist commission. It was an even taller order than it might at first seem, since the ordering of pictures was done in advance of writing the book, and before the author might have much idea of what he wanted to include!

The system did not always run like a well-oiled machine, and the arrangement between Collins and Adprint was riddled with ambiguities that later caused friction. Eric Hosking felt that his main photographers were being underpaid for their work, especially 'chaps like John Markham going off to Norfolk to take a few pictures for the New Naturalist series and coming back again'. According to their agreement, Collins would reimburse travel, but not accommodation, expenses, with the result that much of the work had to be done on day visits with much limited spare time being expended on travel. Although the photographer could, in theory, increase his income by the sale of duplicates (within his 10% allowance), in practice the market was non-existent since only Collins was using colour film for natural history book production.

The photographer's bill eventually caused the break-up of the Collins-Adprint partnership. Adprint took the view that it should be paid by the publisher for every colour photograph accepted, while Collins was certain that he had agreed to pay only for those that were actually used. According to Eric Hosking, the ensuing correspondence was 'voluminous and protracted', and led to the dissolution of the formal ties between publishers and producers in 1952.[4] But unhappy as this episode was, it was perhaps inevitable once high quality colour photography became available at competitive rates from other agencies. The association of Collins and Adprint was cemented by the peculiar conditions of wartime publishing, and the pioneering nature of the colour work for the New Naturalist library. The important point is that it worked well enough during the crucial formative years of the series.

The amassing of so many colour pictures of the highest quality then attainable under the conditions prevailing in 1944 and 1945 was, by any reckoning, a signal achievement. It may well have been for a time, as Billy Collins was to claim, 'the largest and most comprehensive library of nature colour photographs in the world'.

If at first you don't succeed .....

Almost every New Naturalist title contains at least a few remarkable photographs; even *British Game*, which was obliged to rely on old prints for its colour, contained the first-ever photograph of a wild British capercaillie cock (Plate Va, photographed in Abernethy Forest, by G.B. Kearney, 1943). Another celebrated picture is the frontispiece of *The Badger* (1948), which features the first photograph of a wild badger in colour. Ernest Neal told me the story behind it. By the time of *The Badger*, Neal was one of the leading hobby photographers, and had written a book called *Exploring Nature with a Camera* (1946). He had successfully photographed badgers using monochrome film many times, developing the prints himself in the laboratory of Taunton School where he taught biology. For the New Naturalist book, however, a colour photograph was desired. For that purpose, Eric Hosking had given him a few pieces of Kodachrome cut film, which Neal inserted into the plate holder of his camera and duly exposed at the sett in Conigre Wood. However, getting the right exposure was a matter of trial and error: 'It had to go to America to be processed and when it was returned it was useless ..... much too under-exposed. I asked Eric for advice, and he suggested I got nearer and used two flash bulbs instead of one.' The second attempt was made in April 1947, and it must have been with a sigh of relief that Ernest Neal opened the returned package to reveal a perfectly exposed and unique photograph. By then, the colour frontispiece was holding everything up, and production could at last go ahead.

Among the early landmarks of the series were the bird portraits by Eric Hosking and John Markham. At this stage, their best work was still being done in black-and-white. One of the species which most fascinated both photographers was the greenshank, the most beautiful of all the breeding waders and one of the hardest to find at nest. Hosking had made several visits to Speyside in the 1940s to obtain the wonderful series of intimate pictures that illustrate Desmond Nethersole-Thompson's book. Sometimes the weather conditions defeated him, as in 1947 when the light proved too poor for the colour photography he hoped to achieve. The famous colour plate showing Thompson in his army captain's uniform stroking the feathers of his favourite greenshank 'Old Glory', was taken by John Markham in slightly better weather two years earlier. Another classic series of Hosking photographs illustrated Guy Mountfort's *The Hawfinch* (1957), depicting this shyest of birds at its most bullish, arguing with a mistle thrush twice as big and defending its corner of a garden bird bath with the look of a champion boxer.

Colin G. Butler, author of *The World of the Honeybee*, photographed in 1976. (*Photo: Rothamsted Experimental Station*)

Among the most technically accomplished of all the New Naturalist photographs were the series of bee studies, taken by Colin Butler to illustrate *The World of the Honeybee* (1954) and *Bumblebees* (1959). For this extreme close-up work of fast-moving subjects, Butler was obliged to start from scratch and develop equipment expressly for the purpose. In a letter, he told me how this was done. To obtain the desired results, he needed a powerful light source used in combination with a very small aperture to obtain the necessary broad depth of field. Since the distance between his miniature subjects and the focal plane of the camera required exact measurement, he deployed a measuring telescope in an adjustable tilting cradle on top of the camera, a 3½ × 2½ inch Soho Dainty Reflex. Telescope and camera were prefocussed before use, and the light was supplied by an enormous home-made flash gun fitted with a reflector obtained from an aircraft's

Table 5. Lead photographers for New Naturalists 1-30

Sam Beaufoy: *Butterflies* (1945), *Insect Natural History* (1947), *Moths* (1955).
Eric Hosking: *London's Natural History* (1945), *Birds and Men* (1951).
John Markham: *Natural History in the Highlands and Islands* (1947), *British Plant Life* (1948), *Mountains and Moorlands* (1950), *Snowdonia*, (1950), *Life in Lakes and Rivers* (1951), *Wild Flowers of Chalk and Limestone* (1950), *Flowers of the Coast* (1952).
Frank Fraser Darling (author): *Natural History in the Highlands and Islands* (1947).
Douglas Wilson: *The Sea Shore* (1949).
Robert Atkinson: *Wild Flowers of Chalk and Limestone* (1950), *Wild Orchids of Britain* (1951).
Walter Pitt: *The British Amphibians and Reptiles* (1951).
Cyril Newberry: *Climate and the British Scene* (1952).
F.C. Pickering: mounted specimens in *Butterflies*; paintings in *The Art of Botanical Illustration* (1950).
Frederick Goldring: *The Weald* (1953).
Colin Butler (author): *World of the Honeybee* (1954).
E.H. Ware: *Dartmoor* (1953).
T.O. Ruttledge: *An Angler's Entomology* (1952).
J.A. Steers (author): *The Sea Coast* (1953).
C.W. Bradley: *A Country Parish* (1951).
Paul L. de Laszlo: *Mushrooms and Toadstools* (1953).

For the remaining titles, *British Game* (1946), *Britain's Structure and Scenery* (1946), *Wild Flowers* (1954), *A Natural History of Man in Britain* (1951), *British Mammals* (1952) and *Sea-Birds* (1954), evidently no single photographer was assigned.

landing light, and powered by a battery weighing about 30lb.! With the aid of this cumbersome but effective machine, Butler was able to use apertures as small as f64, producing the incredible depth of focus that can be admired on many of the 40 monochrome plates that illustrate *The World of the Honeybee*. To take these pioneering photographs, the camera shutter was opened and the flash gun triggered by a complex external shutter device used to cover the lens. Today, one can achieve the same results with a 35mm camera and a commercial flash-gun weighing only a few ounces. But forty years ago it was a different story.

Despite the emphasis on colour in the New Naturalist series, the monochrome photography was often far more accomplished. Among the most unforgettable pictures of the botanical books are the panoramas taken by R.M. Adam on a big plate camera of the sort now confined mostly to weddings. There are some wonderful examples in *Wild Flowers*, such as his striking composition of the dunes at Culbin Sands (Plate XX), or the massed lupins on the shingly banks of the Tay (Plate XVIII). His mountain-top scene of the 'secret' locality of *Diapensia* in *Mountain Flowers* is so sweeping and full of detail that one can, with a little care and common-sense, identify the exact site on a map!

Wild Flowers and *Mountain Flowers* boast a superb series of monochrome plant studies by M.C.F. Proctor, then a graduate botany student at Cambridge. Indeed, in both books the black-and-white pictures knock the spots off the colour ones, most of which are distinctly lacklustre by modern standards. By the mid-1950s, most serious plant photographers had switched to 35mm colour slides, and it was now good old black-and-white prints that were in short supply. Michael Proctor then owned an old German plate camera which he loaded with ex-RAF government surplus sheet film, which he cut up in the darkroom into 2½ × 3½ inch sections. These large-format plates produced prints of superb technical quality, with all the characteristic sharpness and depth of Proctor's photographs. But his most remarkable pictures were taken more than a decade later for *The Pollination of Flowers* (1973). This collection of 200 photographs, all of stunning quality, took up much of Michael Proctor's spare time over two years: 'In effect, they were my research project.' Most of them were taken using a ring-flash gun mounted on the lens of a 35mm single-lens reflex camera, which produced sharp, relatively shadow-free images of insects at work inside the flowers. Even so, the difficulty in capturing the fleeting moment of pollination was considerable, and the success rate was low. Proctor recalls that, 'Probably two-thirds of the exposures had the insect in the frame, reasonably sharp and correctly exposed, but the really worthwhile ones on each film could generally be counted on the fingers of one hand! So I got through quite a lot of film. In those days, black-and-white film was reasonably cheap if you bought it in bulk and reloaded the cassettes yourself.' On this occasion, the quality of the printing lived up to the technical perfection of the negatives. The printing was done in Japan by the Dai Nippon Printing Company, and the sheets returned to the publishers for binding. British print technology was, by then, lagging some years behind the Japanese.

In considering the virtues of *The Pollination of Flowers* we have travelled forward nearly thirty years from the first stumbling efforts at colour printing. Of comparable quality, I think, is a still more recent New Naturalist title, *Ferns* (1988) by Christopher Page. In recent times, the quality of illustrated books has been limited not so much by the camera – though the near-universal use of 35mm film seldom matches the razor-sharp definition achieved by the old plate cameras – as by the economies of modern book production. Since the decision in 1985 to integrate monochrome pictures with the text, the definition of New Naturalist photographs has depended, to a large extent, on the quality of paper on which the book was printed. The cartridge paper used for *British Warblers* proved inadequate for half-tones, since the texture was too coarse to allow good quality printing. *Ferns* was the first of the new-style books which really works, and it does so partly because all of the illustrations are the work of one man – the author. Not only does this allow a consistency of photographic quality and style, important for this particular book, but it greatly eases the task of the printer. Equally importantly, the 180 monochrome photographs were all developed and printed by the author himself from black-and-white film, producing prints strong in contrast and far superior to the indifferent results achieved by most commercial printers. Chris Page also wisely standardised their size and submitted them as a set of 8 × 10 inch glossy prints. Having had much of his job done for him by a thoughtful author-photographer, the printer did fairly well, within the limits imposed by off-set lithography and less-than-ideal paper. As a comprehensive set of portraits of living ferns, horsetails and clubmosses, this book is in a class of its own, and the editors took understandable pride in comparing it favourably with the finest fern books of the Victorian era.

In a half-century of technical advance, natural history photography has become at least physically easier. For his *Ferns* pictures, Chris Page used nothing more than a Pentax SLR camera, a tripod and a couple of extra lenses, the whole weighing no more than about 10lb (unless he used one of the excellent but very heavy steel tripods) – less than the weight of some of his predecessors' flash guns!

Nowadays one can usually be sure of getting the right exposure, and computerisation is the latest of a series of Japanese gadgets that allow the camera to do much of the photographer's thinking for him. Whether the results are consistently better than those of fifty years ago is another matter, and perhaps a matter of opinion, which I will leave to the reader. In a short account of New Naturalist photography I have omitted many favourites; indeed, I have omitted two of my own: Cyril Newberry's cloud studies, described by Gordon Manley as one of the most difficult problems ever given to a colour photographer, and Kenneth Scowen's beautifully composed English scenes. Perhaps that is as it should be: the work of the New Naturalist photographers, which usually manages to shine through indifferent printing, should be allowed to speak for itself. Their work spans the narrow window of time in which the cumbersome wooden boxes, with which the Kearton brothers waited all day for a photograph, turned into the compact automatics which a schoolchild can easily use to snap a passing butterfly.

Colour and the critics

The colour photographs for the first generation of New Naturalists were about as good as the unforgiving film emulsion, cumbersome technical equipment and travel restrictions would allow. It was not the fault of the photographers that their reproduction sometimes proved less than satisfactory. Some reviewers of New Naturalist titles went so far as to assert that, far from being the most admirable feature of the books, the colour printing rather let them down. The problem lay in a process that may have been *avant-garde* in terms of book production in 1943, but a decade later would already begin to look out-dated. In their advertisements, and on the statement opposite the title page of every book, the publishers took care to remind the reader that 'The plants and animals ..... are portrayed in the full beauty of their natural colours by the latest methods of colour photography and reproduction.' This phrase was omitted from the later titles which had few or no colour photographs, but it might have been a good idea to dispense with the phrase much earlier.

The colour printing of all the early New Naturalist titles was done in-house at the vast Collins printing factory in Glasgow. The factory contained its own experimental laboratory for testing new inks and different types of cloth and paper. In the 1940s, printing was still done by letterpress, not by the offset machines of today. To make the block, the film negative was detached from its mount, which was usually made of glass, and sealed onto a copper printing plate, on which the image was engraved by washing, hardening and etching. In the early days the block-making was done by the Sun Engraving Co., Watford, who would deliver the finished block to Collins for printing. The plate would be attached to what was called a flat-bed rotary machine. At this date, the inks used were opaque, and were confined to the three primary print colours: red, blue and yellow, printed in sequence. Other colours and shades were obtained by overlapping, but the process was too crude to capture the more subtle and elusive colours of nature. The main reason for the dodgy green on so many of these plates was that it depended on the exact mix of blue and yellow. The printer would normally print up to 16 different plates together, starting with a trial run to achieve the best colour match he could. Then, after adjustments to the pressure of the rollers, the ink supply and so on, a quantity would be run off for cutting and folding. Some of the authors, at least, were given the opportunity to inspect the colour proofs before they were printed and cut. As we will see, E.B. Ford sent his proofs straight back and told the printers to try again.

The rather soft, flat colours of these colour plates, reminiscent in some cases of pre-war picture postcards, were the product of using these primary colour inks (and omitting black) combined with conventional printing techniques. Later on, some improvements were made by using pre-mixed colours to achieve greater accuracy, while letterpress was replaced by more advanced printing machines that use the technique of offset lithography. Some of the earliest colour plates to be printed by the latter method were for *Mountain Flowers* (1956), and in some books you can spot the characteristic 'halo' where the colour registration has slipped. The more modern method uses a screen which produces a sharper, more accurate image. By the time colour printing had improved significantly, however, the original generous allotment of colour had shrunk to only 4 or 8 plates, and finally, for some, to none at all!

The standard of colour printing did vary from title to title, and the close-ups were consistently more successful than the landscapes – implying that the colour film of the day might have had trouble registering the subtleties of panoramic views. The relative excellence of *Wild Orchids*, *Mushrooms and Toadstools* and *The Sea Shore* probably owed much to having first-rate negatives to work from. Some of the pictures of smaller insects, like heather beetles and damselflies, were less successful, producing blurry prints. But it was the landscapes in which, as one reviewer put it, the colours are 'not exactly those which one knows'. And he was putting it kindly. Cyril Connolly found them 'strangely unreal, and cold as landscape wrapped in cellophane'. A still more cantankerous reviewer (of *Trees, Woods and Man*) wondered 'how much of the high cost [30*s*] is due to the inclusion of 27 atrocious colour photographs, with crude, hideously distorted colour and poor definition. They add nothing of real value and there are no page references to help the hapless reader to find them.....It is high time the publishers and general editors of this series revised their policy about plates.' Eric Hosking himself was critical of the colour printing in this particular book, which had borrowed extensively from *A Country Parish* and *The Weald*. The editor, Raleigh Trevelyan, felt bound to agree, though he pointed out that they had at least found a good spread of subject matter and avoided the visual monotony of a succession of trees.

Reviewers found the same faults in other titles. The view of Box Hill in *British Plant Life* reminded E.F. Warburg not of Surrey but of the Mediterranean coast. H.N. Southern noted that the mountain landscapes of *Natural History in the Highlands and Islands* lost detail in the shadows, and that the tones in general were 'muddy'. In reviewing *Mountains and Moorlands*, P.F. Holmes complained that the 'lakes are too blue, distant hills too purple, yellows and browns too bright'. For *British Plant Life* the errant pigment was green, casting a yellow blight upon the leaves of laurel and cowberry as if they had grown up inside a darkened cupboard. Holmes was not alone in suggesting that 'Many people would probably prefer a higher proportion of black and white plates in these volumes, until a better technique of colour reproduction can be evolved.' This was true, too, of the hand-coloured plates in *British Mammals* which, as Derek Steven quietly suggested, 'would have been better in their original form'.

It is hard to disagree with these assessments, and there were many others like them. One has only to glance through the plates of a book like *Wild Flowers* (1954) to see how far colour reproduction still lagged behind the black-and-white. Another criticism, less often expressed, was that for some titles there were *too many* colour plates. One sometimes has the impression that Billy Collins or his editors were determined to cram in their quota of 32 or 48 coloured plates, whether the subject matter justified them or not. Such lavishness was eventually curtailed by the rising costs of colour printing. Collins himself might have come round to the view that the use of colour had sometimes been excessive. At any rate, when *The British Amphibians and Reptiles* (1951) appeared with fewer plates than most of its predecessors, he commented that this book was one of the best balanced with regard to illustrations: since it was not over-illustrated, each picture stood out the better.

It was one of the frustrations of book publishing that just as colour printing began to improve, it became too expensive for the series. It had been the large sales of the earlier titles that had allowed each one to be filled with colour. When sales fell, economies had to be made and the series entered its monochrome 'twilight zone'. Looking through the early titles half a century on, we can perhaps enjoy the colour pictures more than did the contemporary reviewers. We appreciate them for what they are: examples of early colour printing from photographs, for which we would no more expect to find contemporary standards of printing than in a Victorian nature book. And, jaundiced as we have become today by colour photographs, we might find in these early New Naturalist books a period charm, like old calendars or cigarette cards, and all the more so when they depict beautiful scenes that have since been uglified or have disappeared altogether. To some extent, it is because of their technical shortcomings that we are not blinded by detail, and can appreciate the composition and the empathy with the subject that lay behind the best of these pictures. They are early, nicely engraved milestones on the freeway of modern colour photography.

4

The Graphic Image: The New Naturalist Dust Jackets

On 1 June 1994, the first dust jackets of *Ladybirds* rolled off the press and I was there to watch. We – that is, the artist, Robert Gillmor, the printer and myself – were standing in the printing house of Radavian Press on the outskirts of Reading. I was there ostensibly to see how the process worked, but also to enjoy the occasion of the printing of the one hundredth illustrated jacket in the New Naturalist library. The press itself was smaller than I had expected – about the size of a large van. A metal plaque on the side told us that it was a Heidelberg Offset machine, constructed in the late 1970s. It uses a technique called offset lithography which has produced book jackets and coloured plates for this series since the late 1950s, when it replaced the older method of letterpress. Offset lithography is particularly good at creating fine quality prints on smooth paper. Since I am not at all mechanically minded, I hope the reader will forgive me if I do not describe in any detail the process whereby the image is transferred to paper via a series of rollers rotating in opposite directions. The method is widely used in modern printing and is in no sense unusual. There *are* some unusual aspects of the New Naturalist jackets, but I will come to those later.

First, though, the occasion. The printing machine makes a whumph-whumph noise as the engine drives the aluminium rollers, and the jacket proofs collect in the box-like 'delivery unit' below. Tin drums of printing inks line the walls. A tin of yellow ink is open (it will colour the smaller ladybird); it looks like thick, oily custard, and the slightly acrid smell permeates the warehouse. A list of points to be checked hangs from the controls. Some are of a technical nature – 'Are you sure it's not catching up?' 'Have you checked the star wheels?' 'What about the fount?' – but the most prominent note is of more universal application: 'Don't be proud, check again!' This machine has printed all the New Naturalist jackets since 1986, starting with *British Warblers*. Because Robert Gillmor lives no more than a mile away, he is able to visit the warehouse on printing day and supervise the colours as they are printed. This means that no proof stage is needed and any necessary adjustments can be made there and then. It is all very convenient and saves a lot of messing about. Whumph, a pilot jacket flips into the box. Robert examines it, with the printer looking over his shoulder. He spots right away that the yellow is not strong enough, so that the nettle leaf on which the larger ladybird sits does not stand out from its background as it should. Today's printing inks come in a large range of pre-mixed colours and getting the colour right is the work of minutes. By referring to a Pantone chart, the artist can decide there and then on the optimum combination of inks which will deepen the background without distorting the colours of the beetles to an unacceptable degree. We try out a range of progressively deeper yellows, and finally adopt a fairly deep one. It is not the true colour of that particular ladybird, but that scarcely matters, since it is not the function of a jacket to be scientifically exact. The difference is surprising:

the nettle leaf seems to snap into focus and the whole design gains in depth, becoming much more eye-catching. 'Go for that one,' decides Robert, and within ten minutes the machine has whumph-whumphed its way through a couple of hundred more jackets. When the 1,500 or so jackets needed are ready, they will be baled and sent to Somerset to the printer of the text pages for cutting and fitting around the hardback books. The latter operation is still done by hand.

The printing of the New Naturalist dust jackets has changed a good deal over the years, as any inspection of a complete set will suggest. The earlier jackets are in softer, matt colours (especially if you have the first editions); the later tend to be brighter on average, and one might assume, rightly, that the printing technology has moved on since 1945. The jackets were first printed by contract presses, and later by the Collins printing factory in Glasgow. The range of coloured inks available in the early days was much smaller than today, and the sequence in which they were applied was very important for the desired results. The older inks were also more opaque and tended to change tone on drying. Since the artists were restricted to three or four colours, their designs turned on exploiting the overlaps of the colours to gain a greater range of colours and tones. They were so much masters of this method that it is often difficult to tell what was in fact the colour of the original printing inks.

In this chapter we will look at the design of the jackets in some detail, as far as possible through the eyes of the artists themselves. I will take the reader through the origin of the jackets and their various manifestations over the past fifty years, emphasising the commercial and technical restraints within which the artists had to work. I have also sketched in some biographical background, particularly of Clifford and Rosemary Ellis, in the hope that it sheds some light on their approach to the jackets, and why the New Naturalist library appealed to them so much.

Clifford and Rosemary Ellis

There can be few more satisfying sights for the bibliophile than a complete collection of New Naturalists in their dust jackets. 'Seen *en bloc*,' wrote the bookseller Dr Tim Oldham in 1989, 'they quite transform a collection and grace a room as well as any Ming vase.' Designed to be eye-catching, these jackets also succeed as works of art; they have passed the most critical test of all: the test of time. There is no doubt at all that the jackets are a very important part of the reason why these books are so widely collected. Booksellers who specialise in the series will tell you that most collectors absolutely insist on books in their jackets, preferably in bright, shop-fresh condition. That is rather unusual in the field of natural history publishing (though it is very much the rule for modern first editions). Of so much importance are the jackets that the books are deemed imperfect without them.

The designers of these distinctive jackets, Clifford and Rosemary Ellis, were art teachers. For a quarter of a century, Clifford Ellis was Principal of the Bath Academy of Art based at Corsham Court in Wiltshire, with his wife, Rosemary, an active member of the teaching staff. The Academy's distinctive way of teaching reflected Clifford and Rosemary's wide interests and conviction that art education should stimulate an enquiring attitude of mind. They believed that a rounded education could be achieved through art, and that, to quote from an early Academy Prospectus, its study should be 'associated with a constant and first-hand experience of the greater richness of form and colour in Nature'.

Clifford and Rosemary Ellis at work in their studio at Lansdown Road, Bath, 1937. (*Photo: Kate Collinson/Norman Parkinson Studio*)

Their posters and dust jackets reflect that belief, and generally use lithography to achieve the maximum effect.

Clifford Ellis was born in Bognor (plain Bognor then) on 1 March 1907, the eldest son of a commercial artist John Wilson Ellis and his wife Annie Harriet, whom he had married two years earlier. Artistic talent ran in the Ellis genes. His grandfather William Blackman Ellis was an artist and, as it happened, a very keen naturalist. Clifford also saw a lot of his uncle Ralph, a third Ellis artist who made inn-sign painting his speciality. A stay of a few months duration with 'Grandfather Ellis' at Arundel during the Great War made a great impression on the nine-year-old Clifford, and kindled his interest in nature study. Clifford 'brought a touch of the country to his bedroom' when he returned to his parents in Highbury. His younger sister remembers stick insects, lizards and three toads or frogs called Freeman, Hardy and Willis. Grandfather Ellis's knowledge of taxidermy had inspired in Clifford an interest in animal anatomy as the basis of animal drawing and painting. At Highbury he 'often boiled up small dead creatures to produce little skeletons, so that he could study their bone structure ... he once caused alarm by fainting while dissecting a rabbit.'

Given all this, it was fortunate that when Clifford went to a school in Finsbury, he found himself within reach of London Zoo, which he began to use as a kind of living reference library. He was given a free pass, and thereafter was often to be seen there, sketching the animals on the spot or memorising details of their form and behaviour. At least one New Naturalist jacket, that of *British Seals*, was based on first hand observation at London Zoo. It was obvious that young Clifford was going to be an artist. He was, needless to say, a clever boy.

Clifford attended full-time courses at two London art schools before taking postgraduate diploma courses in art teaching and art history, in 1928 and 1929 respectively. In the former year, he joined the staff of the Regent Street Polytechnic, where he taught perspective and was in charge of the First Year in the Art School. One of the students there was his future wife, Rosemary Collinson. Like Clifford, Rosemary comes from a family background of talented artists and craftsmen. Her grandfather was a partner in the well-known firm of furniture designers Collinson and Locke, and her father was a skilled cabinet maker. Through her mother, Rosemary was related to Edmund Clerihew Bentley, a prolific writer and inventor of the clerihew, the best known form of poetic doodle since the limerick. Their marriage in 1931 was the start of a creative partnership in which Rosemary, with her instinctive eye for colour, tone and composition, her abilities as artist, designer and teacher, and her talent for perceptive criticism, played a very active part. Nearly all their freelance work was signed jointly. Sometimes the initials R & CE were used for work initiated by Rosemary, C & RE for work initiated by Clifford, though the latter, alphabetical, version was increasingly used to indicate joint authorship. Either way, the joint signature indicated the nature of their collaboration. As a close friend, Colin Thompson, wrote in an unpublished memoir in 1986, it symbolised 'the kind of partnership that is nowadays almost unknown in art and occurs only very rarely in any sphere. It was as much part of the essence [of their teaching] as it was of the rest of Clifford's work.'

The Ellises excelled in that epitome of popular art, poster design. Between the wars a number of organisations, both in the public and private sector, made a practice of commissioning artists to design their advertising posters. Many of the leading artists of the day contributed to such work, and superb original designs

C & RE poster advertising Whipsnade Zoo and BP Petrol, 1932. (*Private collection*)

would trail past on the sides of trolley-buses or on the tailboards of lorries. One of the earliest Ellis commissions advertised Whipsnade Zoo by Car, and was produced for BP Limited. It was designed to catch the eye at a distance: a pack of wolves stare intently at the onlooker from the darkness of the trees with great round eyes. Like so much of their work, the design was based on first hand observation. Rosemary told me the story behind the picture. In those days the Ellises were great walkers and they thought little of walking to Whipsnade Zoo, out on the Bedfordshire downs, from the centre of London and by night. Arriving exhausted just before dawn, they tumbled into a dry ditch full of leaves close to Wolf Wood and fell asleep. When Rosemary and Clifford awoke out of a surprisingly comfortable cocoon of leaves, it was to find the wolves staring back at them. The poster was designed with that arresting image fresh in their minds.

The wolf poster was one of many advertising posters C & RE designed for different clients during the 1930s, among others for the Empire Marketing Board, the Post Office and Shell Mex. Particularly striking, and in some respects reminiscent of their later New Naturalist jackets, is a series of four posters they did for London Passenger Transport Board, entitled 'Wood', 'Heath', 'Downland' and 'River', this last depicting a wise old heron hidden in the reeds, while above is the watery reflection of a tea party in a punt. Several of C & RE's posters were shown at the New Burlington Galleries, London, in 'Pictures in Advertising by Shell Mex and BP Ltd' opened by Kenneth Clark in 1934. Various of their posters and other designs have since been included in other exhibitions, among them 'Historical and British Wallpaper' at the Suffolk Galleries, London in 1945, 'Art for All' at the Victoria and Albert Museum in 1949, and, more recently, in the Arts Council

Top C & RE poster for London Passenger Transport Board, 1939. (*Private collection*) *Above* C & RE design labelled 'Bus streamer "Summer is Flying" for London Passenger Transport Board' 1938. (*Private collection*)

exhibitions 'Thirties' and 'Landscape in Britain 1850-1950' in 1979/80 and 1983 respectively. What characterises so much of their work is their mastery of lithography, a difficult medium which demands a good eye for colour and a boldness of design. They were to use similar techniques for the New Naturalist book jackets. Though smaller than the posters, their main purpose was the same: to catch the eye.

The Second World War brought a temporary halt to the Ellises' freelance design work (which had by then broadened into other fields like wallpaper design, mosaics and some very striking dust jackets for Jonathan Cape). In 1936, the couple moved from London to Bath where Clifford took up a new post as assistant headmaster of the Bath Art School (then part of the city's Technical College) and was appointed headmaster two years later. The Art School remained operational throughout the war, despite several changes of premises. As Clifford put it at the time, 'the arts provide a deeper and richer life that we can all share, and the war with its upheavals has provided opportunities which have made this sharing more possible than it has been for several generations'. Besides running the Art School, Clifford's wartime activities included work as a camouflage officer, service in the Home Guard and, in different official capacities, making pictorial records in Bath of bomb-damaged buildings and of architectural ironwork threatened with removal for salvage. Rosemary continued teaching at the Royal School for the Daughters of Officers of the British Army even after it was evacuated from Bath to Longleat. This meant having to get up at 5.30, take a bus to Frome where her bike was hidden behind a telephone kiosk, and cycling all the way to Longleat – all war long. It says something about the England of those years that the bicycle was never pilfered or vandalised.

Among the acquaintances of the Ellises at Bath was the venerable Walter Sickert, the grand old man of British painting, who, at his own suggestion, gave a weekly lecture at the Bath School of Art, mostly on the work of artists he particularly admired, Degas and Daumier among them. In the 1960s Clifford Ellis gave a lecture about Sickert, broadcast on BBC's Third Programme, which was frequently moving as it brings that great artist vividly back to life. Another of the friendships they made at this time was with Sickert's pupil, Lord Methuen, and this was to have important consequences. After the war, Methuen's beautiful country house, Corsham Court, ten miles east of Bath, was returned by the war department. In the belief that such houses had an important contribution to make to the postwar future, Methuen offered the main part of the house to the Bath Academy of Art. Clifford was thereupon faced with a watershed choice, for he had been offered the Chair of Fine Art at Newcastle. He opted for the challenge of creating a new residential art school at Corsham, with courses designed to train students to meet the needs of postwar Britain. The Bath Academy of Art opened at Corsham with Clifford Ellis as Principal in October 1946. The Ellises established themselves in the top floor of one wing. In the meantime, C & RE had started to design jackets for 'an important new series of natural history books' to be published by Collins.

The Bath Academy of Art became the practical manifestation of C & RE's views on art education. In the early years, the school offered a training course for art teachers and a four-year course leading to the National Diploma in Design. The range of subjects taught was unusually wide. It encompassed dance, drama and music; dyeing, weaving and fabric printing; lettering and typography; textile and stage design; as well as sculpture, pottery, drawing and painting. Many of the staff were practitioners who taught part time, and a number became leaders in their particular field. Natural history was also on the syllabus. In keeping with C & RE's belief in the value of 'constant and first hand experience' of the richness of colour and form in nature, Clifford created a bog garden, an alpine garden and several aviaries, where a range of exotic bird and plant species could be studied. Various other birds and beasts were kept there at different times, notably chickens, bantams, geese, ducks, goats, a pig and even a couple of crocodiles. All did duty as teaching aids as well as contributing to the general ambience.

Opinions seem sharply divided on the place of Bath Academy of Art in the development of art education. One future teacher there had been warned that 'Clifford had a lot of funny ideas. Corsham was trying to teach too many things all at once ... and was not a serious art school at all – the students were only dabbling in drawing and painting and sculpture'. Others place him high on the list of the more influential postwar art educationalists. The point was that Clifford Ellis taught as a practising artist. Like his mentor, Marion Richardson, he believed in teaching by suggestion, by opening windows and encouraging the student to develop his or her own talent. He wanted to draw from them the same sense of vocation that he had always felt himself, 'like a gardener tending his plants'.

This is not the place (and I am not the person) to try to analyse how far he succeeded. It is worth observing though that Clifford's educational aims were analogous to those of the New Naturalist library, and both were part of the mood of the time. I remember reading an article in *Country Life* where Clifford and Rosemary's colourful wallpaper designs were said to 'brighten the postwar gloom'. Gloom in terms of austerity perhaps, but the prevailing mood was far from gloomy. The immediate postwar era was a time of optimism, when it was hoped that the cooperative spirit generated by the war effort would be mirrored in peacetime by a collective will to build a better Britain, with health care available to all and secondary education guaranteed to every child in the land. There was a major investment in education, and the Academy's teacher training course was created in response to the urgent national demand for teachers. Clifford Ellis and James Fisher had something more in common than a love of birdwatching. They were both ardent popularists: just as Clifford's mission was to bring art into the lives of ordinary people, so Fisher wanted to introduce the latest fruits of science and natural history to a wide audience, and to encourage mass participation. Neither saw any reason to debase their subject by doing so. This happy convergence of aims – for the Bath Academy of Art and the New Naturalist library came into being at about the same time (after a period of planning in wartime) – might help to explain why Clifford and Rosemary Ellis readily accepted the invitation from the Collins Board to design jackets for the new books. It might go also some way to explaining why these designs were so successful. It was a question of rising to the occasion.

C & RE: the standard bearers

Not everyone admires the New Naturalist jackets. People who have grown used to associating nature with photographic realism can be puzzled by them, and ask openly what the fuss is about. I have seen them described as coloured daubs or (not inaccurately) 'smudgy hieroglyphs'. Like most good things, the Ellis jackets are an acquired taste. There were evidently periodic crises of confidence on the New Naturalist Board itself when someone wondered whether the use of photo-

Showcard advertising the New Naturalist series designed by C & RE in 1947 for use at book exhibitions. (*Private collection*)

graphic jackets might increase sales (though when it was tried in 1965, they found that it did not). Perhaps the least impressed critic on record was The Spectator's reviewer of *The Greenshank* (1951), who regarded the rather stylised jacket design as 'vulgar' and 'a brutal shame' and advised purchasers to remove and burn it.

Those who admire the Ellis dust jackets probably understand the difference between a visual statement and an illustration (which Clifford Ellis himself will explain better than I can). Working within tight technical constraints, which varied from title to title and are discussed below, C & RE designed each jacket to suggest at a glance the book's content. It was never their intention to represent any subject in precise anatomical detail. The jackets are best regarded not as illustrations but as mini-posters which require a fairly simple, strong image – as Clifford put it, 'not too banal, nor too original' – to make an impact in the shop, and entice the onlooker to open the book. Lithography lends itself to bold designs conceived to make imaginative use of a very few ink colours. Invariably working from first hand observation, C & RE would select for their designs those aspects of a bird, beast, plant or landscape that would best communicate the spirit of the book in a single clear visual image. The reader might be interested to learn which jackets they themselves considered had achieved that aim most successfully. Those which Rosemary Ellis chose for two exhibitions in 1989 and 1990 were *Trees, Woods and Man, The Rabbit, The Herring Gull's World, The Folklore of Birds* and *Insect Natural History*. It is notable that C & RE put as much effort into the design of the rather unsung New Naturalist monographs as they did the main series. For an exhibition in York, they picked out the preliminary artwork for *Ants* (i.e. the second *Ants*, published in 1977), *British Thrushes* and the recently completed *Natural History of Shetland*, as well as the jackets of *Moths, Pedigree, Herring Gull, The Wood Pigeon, The World of Soil, Fish and Fisheries, The Trout* and *The Rabbit*. Clearly they were aiming to present a balanced series of contrasting designs, but their choice must also reflect a personal preference. Among Rosemary's favourite jackets are those of *The Wren, Finches, The Heron, Inheritance* and *An Angler's Entomology*. Perhaps the reader has his or her own favourites (as I have). But the level of consistency achieved by the Ellises throughout the 86 jackets they designed for the series is remarkable.

Their earliest fan was Billy Collins himself, who always made sure that the latest Ellis jackets were among those displayed at the annual National Book League exhibition. He was delighted when, during a visit to the Collins factory by the Queen, the latter picked out the New Naturalist jackets as designs which she particularly liked. Immediately after the Queen's departure, Collins was on the phone to the Ellises to report Her Majesty's words. Billy Collins made a habit of writing to them whenever they produced a design he considered outstanding. Typical of many others is a letter written in 1960 about the *Dragonflies* jacket, which had been much admired by the New Naturalist Board: 'I think it is amazing how you go on year after year thinking out such lovely designs. The series makes such an effective display in the shops, largely owing to the designs'.

The Ellises' association with the New Naturalist library began in July 1944. The Collins publicity person, Ruth Atkinson, had in the previous decade worked for the publishers Jonathan Cape and remembered the attractive dust jackets that the couple had designed for a number of Cape novels. (Two of these are illustrated on Plate 2.) At that stage, more thought had gone into the purpose and breadth of the New Naturalist series, and into commissioning titles for it, than in the appearance of the books on the shelf. The Editorial Board may have taken it for granted that the jackets would incorporate a colour photograph, perhaps one taken from the body of the book. That would certainly have fitted in with the 'philosophy' of the series, printed opposite the title page of each book, with its emphasis on 'the latest methods of colour photography and reproduction'. The surviving correspondence seems to confirm this supposition. In passing, Billy Collins refers to the notion of placing 'small photographs at the centre of [the] botanical or biological sections [of the library?]', while Ruth Atkinson advised Clifford Ellis to try to win over James Fisher 'to the idea of non-photographic jackets'. We might suppose that the photographic editor Eric Hosking was particularly keen to get a photograph onto the jacket, while Billy Collins, with his interest in contemporary art and still greater interest in sales, might have preferred something more adventurous to make a dramatic impact and proclaim an important new series of books. In 1944, colour photographs of British wildlife in natural surroundings were still hard to come by and of indifferent quality; it was not until the 1960s that laminated photographic jackets became the norm for natural history books. Dust jacket design, on the other hand, was then in its heyday, and book collectors often judge the postwar years as the high point of that particular art. At any rate, no clear decision seems to have been taken. Ruth Atkinson decided to take matters into her own hands.

> Dear Mr and Mrs Ellis,
> It is such a long time since I have seen you and I wonder if you are now feeling at all inclined to do book jackets.
> I left Cape's about six months after the beginning of the war, was in the Ministry of Labour three years .... and am now working at Collins as you see.
> We are going to do an important series of books dealing with various Natural History subjects, they will have a great many colour illustrations and I thought that you would do lovely jackets for them. There are however a great many people whom the jacket must please: besides Mr Collins, the editorial committee of this series and the producers of it, Messrs Adprint. If from this brief note you are at all interested in the idea, could you send me some of your work so that I may show it to Mr Collins; and if you would like to work out a rough, say for the first title, I will send you a lot more information about it all. I don't imagine you will be coming up to London but of course if you were it would be better still to discuss it.
>
> *Ruth Atkinson.* unpublished letter, dated 20 July 1944

The Ellises' response was to invite Ruth Atkinson down to Bath for a weekend to discuss the project. She warned them that they were not the only artists that had been invited to submit designs for the New Naturalist jackets, but in the event the others dropped out and the Ellises were alone in sending finished artwork. Billy Collins had had the great good luck of finding the right artists at first pitch. On her return to London after 'a deliciously comfortable and peaceful time' in Bath, Ruth sent them the authors' synopses of the first titles and some 'pulls' of printed colour photographs taken by Sam Beaufoy, Eric Hosking and others. On the basis of these guidelines, the Ellises set to work on jacket designs for the first two books in the series, *Butterflies* and *The Naturalist in London* (eventually published as *London's Natural History*).

If the Ellises were given any further artistic guidance it does not survive in the correspondence. It seems likely that the basic design of the now familiar New Naturalist jackets, with their broad title band and oval, and the circle containing the number of the book in the series, was the Ellises' own. What appears to have been the original colour 'rough' for *Butterflies* still survives in perfect fresh condition, and is reproduced on Plate 1. It is exactly twice the size of the printed jacket, executed in gouache paint on thick Whatman paper. Although there were minor modifications to the finished design, this is already the familiar *Butterflies* wrapper of swallowtails floating in a Broadland landscape. It is a characteristically bold design, in bright colours designed for lithographic reproduction and reminiscent of some of the Ellises poster work for London Passenger Transport Board. Already we see some of the hallmarks of the future New Naturalist jackets. The design is boldly conceived, divided not horizontally, as might be expected, but into two vertical planes of focus: the magnified caterpillar on its foamy foodplant on the left, the under life-sized butterflies flying away from the viewer to the right, and, in the distance, a characteristic Ellis motif: a vignette of a Norfolk landscape with windmill, willows and the open Broad. It was probably the contrasting colours of the adult butterfly and its caterpillar that attracted the artists. The weakest point of the design is the repeated motif of the caterpillar on the spine. For most of the later designs, they would find often ingenious ways of integrating the spine with the main design. The depiction of different life-stages and the surrounding habitat is suited to E.B. Ford's text, and the graphic style suggests that this book is not intended to be a traditional identification guide. That 20,000 readers bought a copy of this book in less than a year suggests that the jacket did its job; the excellence of the text and the attractiveness of the subject did the rest. By chance, Eric Hosking's camera caught Dudley Stamp studying the jacket at a New Naturalist editorial meeting some time in 1945. He, at any rate, looks well pleased with it.

The colophon for the series was also devised at this time. Clifford Ellis tinkered with the initial letters of New Naturalist until he found a satisfactory solution: the familiar, beautifully rendered capital letters, transfixed by an emblem of a natural object appropriate to the particular book. In the first instance, the emblem was a stickleback, which may have been intended for use on each title. By the time the jacket of *Butterflies* was printed however, the stickleback had been replaced by a caterpillar (though it does eventually swim back into view on the jacket of *Life in Lakes and Rivers*, published in 1951).

The original stickleback colophon designed by C & RE for the series in 1944. (*Private collection*)

The *Butterflies* rough was accompanied by a colour sketch of ducks on a pond, intended for the London book. Though not adopted in that form, the Ellises developed the idea of reflections in water for the gull design eventually used, with the dome of St Paul's reflected in muddy waters. Possibly the Tufted Duck on the spine was inherited from the earlier design.

Billy Collins liked the designs immediately, and on the strength of them commissioned the Ellises to design jackets for the first six books of the series. It seems that he did so before inviting the views of the editors, although he might have had a word with Fisher or Huxley first. To aid the printing process, the Ellises were asked to produce each design at the exact scale of reproduction. They also decided to hand-letter the title, with singularly beautiful results (though some may not have noticed that the early titles were hand-lettered at all!).

Printing costs became the first problem and a recurrent one. Rather than printing the jackets themselves, Collins and Adprint decided to commission Thomas E. Griffits of Baynard Press, well known for his ability to transcribe artists' designs. Lithography was a relatively expensive process, requiring good quality paper to absorb the inks. Moreover, since each colour had to be laid on in sequence, the costs rose for each new colour printed. The economic limit was a mere four colours, and at a later stage, Collins was pleading for three or even two. For an artist striving to capture the colourful world of nature, this presented a severe limitation. Clifford Ellis had initially suggested that 7 or 8 colours would be ideal. With only half that amount available, an exactness of colour tone would become a critical matter for the printer, and the artists would have to come up with imaginative designs that make the best use of overlapping colours to produce additional hues and tones. In short, the more niggardly the colour allowance, the greater the demand on the artist and printer. There were alternative ways of printing that offered a wider range of colours, but only at the expense of their brightness. The vivid colours and general effect of the Ellis designs depended on lithography. 'Please let it be litho,' wrote Clifford Ellis to Collins. 'We will make things as easy as possible for the printer.'

As an illustration of how the artists were able to obtain an arresting design using

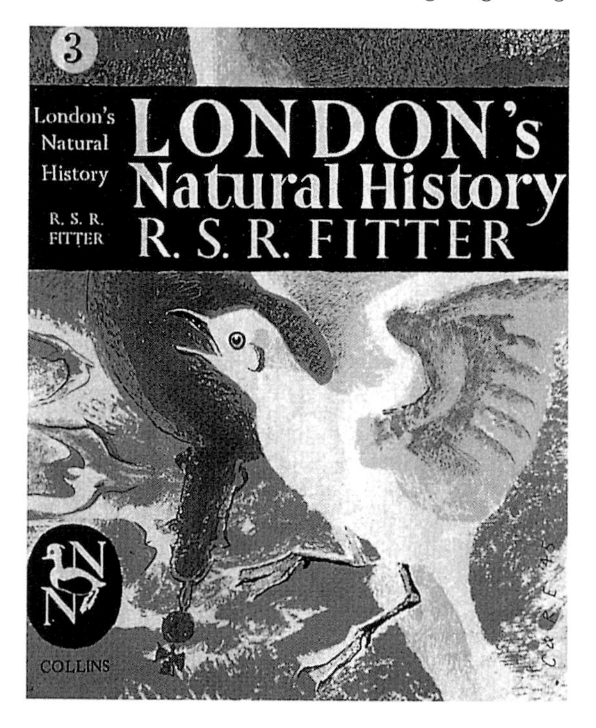

Jacket of *London's Natural History* (1945).

Jacket of *Trees, Woods and Man* (1956). [see also Plate 11]

such limited palettes of colours, we could do worse than examine a dust jacket that they, and others, regarded as one of their best: that of *Trees, Woods and Man.* Lesser artists might have conjured up a scene of logs and lumberjacks, but the Ellises approach is more tangential. We are in the midst of a plantation, that much is obvious. The trees are rooted in plough lines and are set close together, the dappled green of the canopy contrasting with the sepulchral gloom of the woodland interior. There is no need for human figures: each tree bears a mysterious number and the hand of Man is clear. The scene has the sturdy geometry of many of the Ellises poster designs, the vertical lines of the trunks and the diagonals of the boughs harmonise nicely and are pleasing to the eye. Even the broad title band does not detract unduly. What is not so obvious is that this intriguing scene is made up of only three colours: black, brown and 'fresh green', printed in that order. All the rest is professional trickery, exploiting the overlaps to gain extra colours and creating further tonality by stippling. As to what the picture means, the Ellises are not saying. It might contain a comment on modern forestry methods or it might not. It is not even clear whether the trees are broadleaves or conifers. The reader will have to open the book. In a letter congratulating them on the excellence of this cover, the author, Herbert Edlin, explained what the jacket actually represents. It depicts a Forestry Commission sample plot, in which each tree is numbered and measured every few years to determine the rate of growth. Personally, I had always assumed that the numbers meant that the trees were 'doomed' for felling; on the contrary, they are 'saved'!

The second hurdle facing the Ellises in 1944 was to win over the editors, who, as we have seen, were predisposed to use photographs. James Fisher was delegated to visit the Ellises to gain a first-hand impression of their ideas and techniques. 'I hope you get on well with Fisher and win him over to the idea of non-photographic jackets,' wrote Ruth Atkinson, perhaps a little nervously. It seems the meeting was a success, on both counts. Fisher was a connoisseur of illustrated books and bird art and became a devoted admirer of the Ellises' work. Moreover, by this stage they had new designs to show him, including the Gould-inspired cover of *British Game*, depicting a grey partridge, its head turned dramatically above the title band. Billy Collins was bowled over by it, 'I am absolutely delighted ...... I think it is quite lovely in every way. I wonder if some day you might do some big illustrated book of individual birds on the lines of Gould?' Fisher returned a convert. Whatever lingering doubts might have remained were soon dispelled by the series of original and exciting designs produced by the Ellises through 1945: the closely observed anatomy of *Insect Natural History*, the lovely colours of *Natural History in the Highlands and Islands* (with the only bright colour reserved for the diver's eye, which lies at the optical centre of the picture), the melting cliffscape of *Britain's Structure and Scenery* and the semi-abstract collage of fungal patterns for *Mushrooms and Toadstools*. They would light up the drab shelves of postwar bookshops like a burst of sunshine, and proclaim the New Naturalist books to be something altogether new, and of challenging quality. The original commission for six designs was extended indefinitely. The Ellises were henceforth as much a part of the New Naturalist 'family' as the editors and authors. At the end of 1945, Ruth Atkinson wrote to tell them that the jackets had been a great success. The editors liked them and 'so does everyone else. I am delighted with them and feel very god-motherly about them.'

The design that caused the most argument was the very first: that of *Butterflies*. The editors felt that the choice of the swallowtail was unfortunate. It was rare and, they said, unfamiliar, and one of the purposes of the series had been to guide

Jacket of *British Game* (1946).

the naturalist away from his traditional obsession with rarity. The Ellises patiently began work on a new design of a more 'familiar' butterfly, the dark green fritillary, though Billy Collins made it plain to them that he was quite happy with the original: 'Personally I have always liked the Swallow Tail (*sic*) and I hope you will do this. I do not think the criticism of its being rare matters and I do not think one could get anything more lovely.' And, because time was getting short, the swallow-tail it was to be. Clifford Ellis was critical of the printed jacket. The blue was not deep enough, and hence the title failed to stand out as it should; and the yellow was too orangey. The problem of pale colours grew worse on the reprints of this very popular title, and when the printers attempted to redeem the jacket by using a screen they merely succeeded in making it look rather grubby and a far cry from the brilliant colours intended. Even so, it was a good start and the *Butterflies* jacket set the style for the whole series.

The *modus operandi* developed very quickly. The Ellises would be sent 'pulls' from the plates for each title, and, usually, a portion of the text or the author's synopsis in order to gauge the book's content and style. They would produce a coloured rough, based on various preliminary sketches, which would be shown to the editors and returned with their comments. Sometimes the author's opinion was also sought. Occasionally a design would be turned down, but more often, modifications were suggested, usually with regard to scientific accuracy. The original generalised gull on the front of *London's Natural History*, for example, was transformed at Fisher's suggestion into a specific black-headed gull in winter plumage by simply adding a dark crescent behind the left eye. The feet on the *Yellow Wagtail* jacket were redrawn to conform with Fisher's reminder that 'the whole of the claw should be flat on the ground ... the claw is lifted off in one movement, not as in [human] walking, and the back leg is a little higher'. Occasionally pedantic insistence on strict realism stood in the way of a promising design, as with *The British Amphibians and Reptiles*. To the author, who grumbled that the colours of the adder chosen for the design were all wrong, and in particular that no adder ever showed pure white, Clifford explained that:

> 'We are aware that adders are not white, but, new-sloughed, they gleam in the sun. Our problem was to find something that would gleam in a bookshop and could be printed in four colours......Our idea of "accuracy" is to base our design on first hand experience of the subject, selecting such aspects of it as are appropriate and possible for the job – but it is not our intention to make a coloured diagram, nor is this a practical possibility in four colours.'
>
> *Clifford Ellis.* Unpublished letter to Raleigh Trevelyan, 1 January 1950

The business of strict accuracy versus graphic impact is best expounded by Clifford Ellis himself, in a reply to a letter from Professor H.J. Fleure proposing a particular design for *The Natural History of Man in Britain*. Fleure had envisaged a hill fort 'rising like an island above wooded slopes' with armed sentries standing on top of it (to conform with a request from the Sales department to incorporate a man on the cover). Clifford's reply was as follows:

> 'I will begin, I hope not too impertinently, by saying something about book jackets. A book jacket is by way of being a small poster; it is part of the machinery of book selling. Though, obviously enough, the jacket should be in keeping with the book it contains, it is unwise to consider it as an opportunity for an additional illustration. An illustration, as against the jacket, can be seen

at leisure and free from the competition of not always very mannerly neighbours. The jacket should be immediately interesting; its forms and colours should make a very clear and distinctive image. If it does its job, the book will be taken down and opened, and the proper illustrations will be seen. We suggest, therefore, that the hill fort should be seen from the air. Though one would lose the sentries, who would be too small to show, one would gain a more striking view of the earthworks and (speaking now as a sometime camouflage officer) there would be a remarkable colour contrast between the smooth grass of the hill and the dark tree foliage of its surroundings.'

Clifford Ellis. Unpublished letter to H.J. Fleure, 4 August 1948

In the event, the jacket dispensed with most of the trees as well as the sentries by adopting Fleure's further suggestion of the Uffington White Horse on the bare Berkshire downs and the autumn colours of All Souls, as one of 'the two outstanding dates of the prehistoric calendar'. They agreed on a bill-hook for the colophon, though Clifford later changed his mind and substituted an arrow-head as a more appropriate symbol of prehistoric technology.

The Ellises often went to considerable lengths to obtain a first-hand acquaintance with their subjects. Special trips were made to Shetland and Orkney in the 1980s in search of material on which to base jacket designs. The jacket of *Mushrooms and Toadstools* is based on dozens of exquisite pencil sketches in which Clifford experimented with fungal form and texture before deciding on a design (see next page). For the animal titles, his experience of observing and sketching inmates at the zoo proved invaluable. Clifford once confided that the beautiful

Jacket of *A Natural History of Man in Britain* (1951).

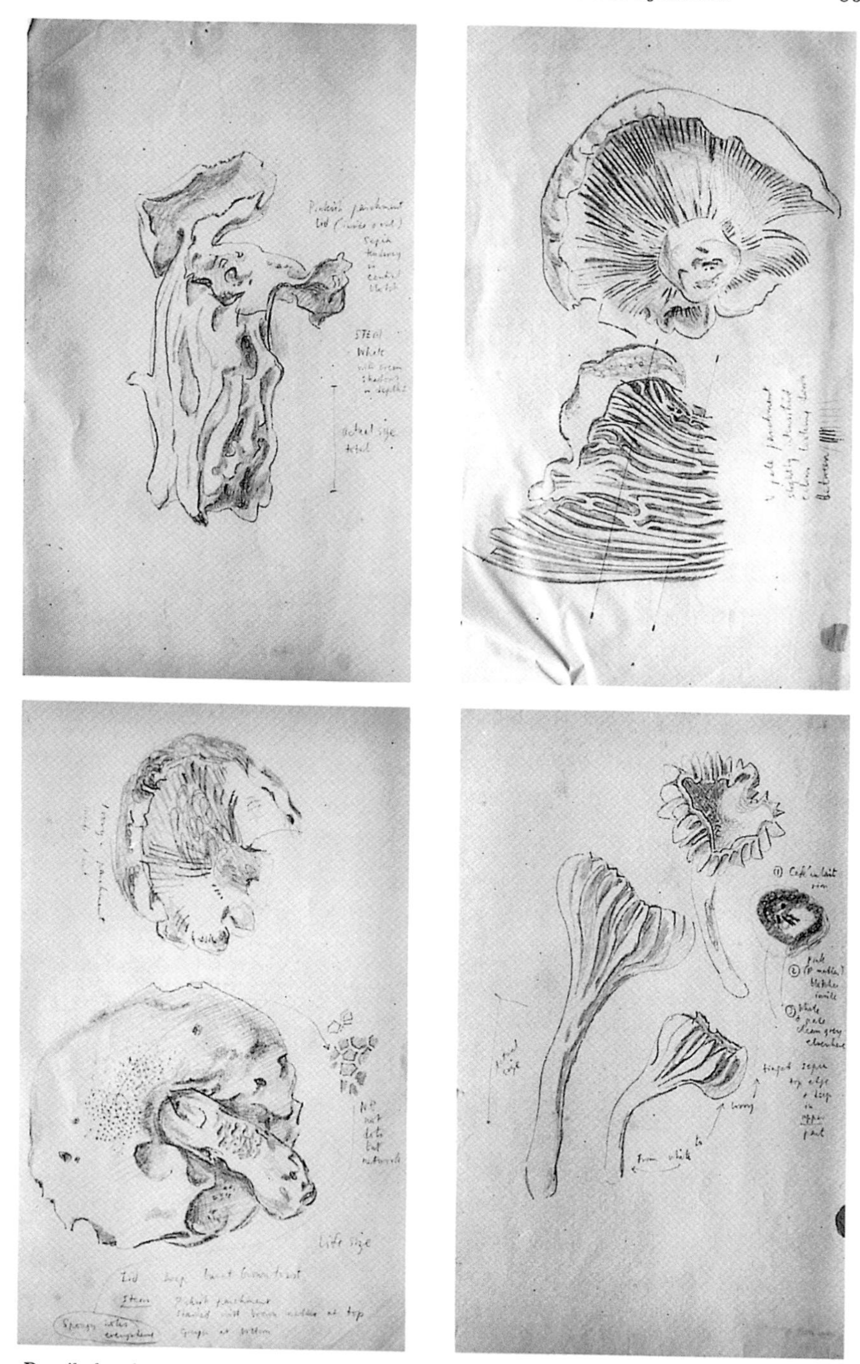

Pencil sketches of fungi from life by C & RE, elements of which formed the basis of the design of the *Mushrooms and Toadstools* jacket. (*Private collection*)

pellucid jacket of *British Seals* was

> 'a consequence of a visit to the London Zoo where I could be sure of seeing a pike in the Aquarium, to stimulate further thoughts about *Patterns in Nature*. From time to time I went out into the bitter wind to watch the seals and how they swim – like monstrous trout, an extraordinary adaptation .... The author might like a present of the French *accouder* which says for the fore limbs what *to kneel* says for the hind ones. He will know better than I that the seal doesn't use only the flipper but, with great versatility, the whole limb, and when ashore, to see what is going on, may support itself on its elbows like an old lady at her window-sill. Habitual but, in English, nameless.'
>
> *Clifford Ellis*, unpublished letter to Michael Walter, 17 January 1973

While in many cases the Ellises were given a free hand with the design within the technical limits imposed, in others they worked within a framework of suggestions by the authors or the editors or by both. W.B. Turrill wanted to have sea campions on the wrapper of *British Plant Life*, which rather circumscribed the choice of colours and setting. For his *Wild Flowers of Chalk and Limestone*, J.E. Lousley had suggested 'rolling downs and a chalk face in the background', which they broadly adopted but went one better by substituting the Westbury White Horse for a chalk pit. With the text of *Climate and the British Scene* came the idea of 'sky and cumulus cloud, giving the suggestion of changeable weather, with typical scenery included'. For *An Angler's Entomology*, James Fisher suggested 'a mayfly and a lure mayfly'. With *The Sea Coast* came ideas for lighthouses and windblown trees. And so on. Perhaps the tallest order was for the jacket of *Life in Lakes and Rivers* for which C & RE were asked to produce 'an impression of a lake in section, with a fish in it, and an insect or two, and a fisherman above, and

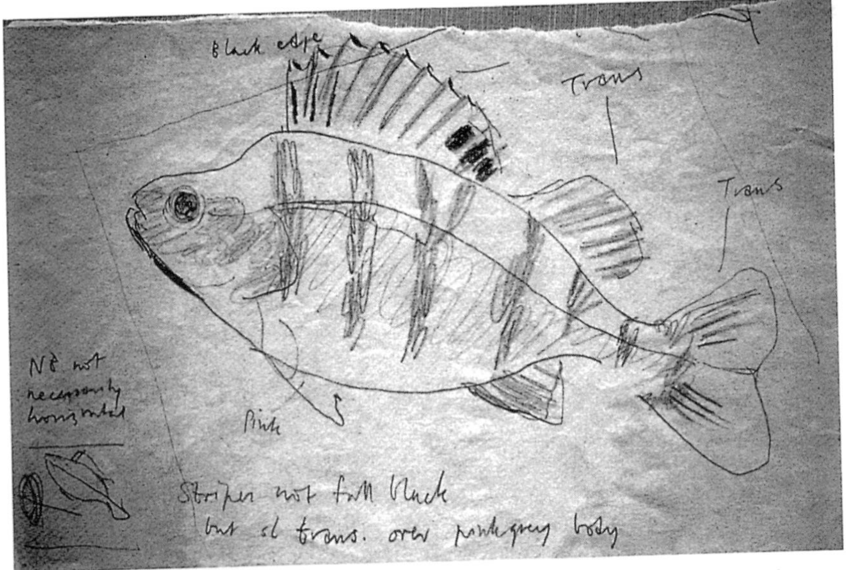

Pencil sketch of perch by C & RE used in the design of the *Life in Lakes and Rivers* jacket. (*Private collection*)

perhaps a factory in the background, and some plants and some reeds'. With a characteristic touch of wit, they got around the clutter of fishermen and insects by substituting an angler's float, which became the motif of the spine. The factory was transformed into a bridge and weir. But they remained faithful to the overall concept envisaged by the editors.

Very often, as in *Lakes and Rivers*, a key part of the design was fulfilled by the spine, which is, of course, the first part of the book most customers will see. If it failed in its purpose of persuading him to remove the book from the shelf, then all the rest of the design would have been in vain. The spines of postwar dust jackets were often left plain in order to display the title, the publishers' imprint and other information. In one of the most attractive features of the New Naturalist jackets, however, the spine is not only an integral part of the jacket design, but also often incorporates a key part of the subject – the angler's float, the woodpecker's red skullcap on *Woodland Birds*, the red blotch on the beak of Tinbergen's herring gull, the trawler's sail on *Fish and Fisheries*, the delightful begging chicks of *The Wren* and *The Heron*. If anyone could resist taking *The Heron* down to feast their eyes upon that riot of beaks and eyeballs he must have been a resisting sort of person. In other cases, the spines echo the theme of the front jacket: the wing on *Sea-Birds*, or the dragonfly's body on *Insect Natural History*. Sometimes the spine symbolises the book's contents, like the fork on *Man and the Land*, or the battlement and flag on *Man and Birds*. Often these 'echoes' contain a touch of humour: the brilliant red 'No Entry' sign on *Nature Conservation in Britain*, or the array of birds and beasts that peep round from the front on many of the later titles. There is a gaiety in designs like *Finches, British Tits* and *Hedges* which surely reflects the artists' enjoy-

Artwork of jacket for *A Country Parish*. (*Private collection*)

ment in producing them.

Not all the Ellis designs were as immediately convincing as *The Heron* or *Finches*. The first dust jacket to make the publishers scratch their heads rather than leap in the air with joy was for *A Country Parish*, Arnold Boyd's hymn of joy to his native Great Budworth. The Ellises had based the design on the weathercock of a church steeple. The rest is all sky, clouds and whirling swallows. Ruth Atkinson's reaction was 'Yes, but ..... The idea is fresh, so it needs imagination.' You can almost hear her pause, say 'um', then pluck up courage and ask, rather plaintively, '*Why are there rounded feathers on the weathercock?*'

The Collins sales staff made a more serious objection to the original jacket design for *The Open Sea: The World of Plankton* by Alister Hardy. Both Hardy and his editor James Fisher had wanted planktonic animals on the jacket, and the Ellises obliged with a beautiful design based on real microscopic organisms. Unfortunately, for all their accuracy, the animals might have been from outer space so far as the book department was concerned. 'I must ask you one thing,' wrote Raleigh Trevelyan. 'Did you invent these animals, or are they ones that exist?' They would not do in either case. The sales people were convinced that the public were not ready for plankton designs, and the artists were asked instead to 'convey a sense of mystery and movement in the open sea'. Curiously enough, at the request of Collins they provided a third cover for a projected reprint of *The World of Plankton* some twenty years later, for which they returned to the original idea of microscopic life. Unfortunately, although the author congratulated the artists on capturing so well the spirit of living plankton, and a colour proof was soon afterwards made, this too was put aside as rising costs had rendered the book uneconomic to reprint. '*The World of Plankton* jackets have had a sad history,' remarked Clifford Ellis in 1983, as another classic title went out of print. Here, at least, the reader can compare all three on Plate 8 and decide for himself which is the most effective at expressing the spirit of Hardy's book.

In a few instances, the printed wrapper failed for various reasons to live up to the original concept of the artists. *Squirrels* caused a shock, and a post mortem was held with the printers and blockmakers over the unexpected heaviness of tone on the squirrel's face, which ruined the artists' conception. The problem in this case lay in the technical complications of a design which relied on stippling rather than the solid tones of earlier jackets. For *Grass and Grasslands*, the artists had intended to create a contrasting dark tone for the trees and dairy cows by overprinting two different tones of green. Unfortunately these were printed in the wrong order, producing a more wishy-washy colour than was intended, thus losing the expected impact. It was to prevent casualties of this sort that Clifford Ellis insisted on inspecting the printer's colour proof before allowing the jackets to be printed. As a memo to the printers remarked, 'He is very fussy about matching tones and knows a great deal about it.' When sending the proof of *Inheritance and Natural History* to Ellis, the editor had commented, a bit lamely, that 'we think it looks very nice'. Clifford returned it with a characteristic annotation: 'It is *nearly* right. The grey, though the right tone, would have a more lively effect in the colour scheme if it matched the specified Pantone, i.e. if it had slightly more "Reflex Blue" and slightly less "yellow". I am glad you like it. PS. Perhaps 'COLLINS' should be a little larger and bolder.' For *Hedges*, in which the design turns on the contrasting colours of an orange-tip butterfly, his instructions were to render the orange 'brilliant'. He repeated the word at the end of a telephone conversation with someone from Collins. 'Remember, the colour must be bril-

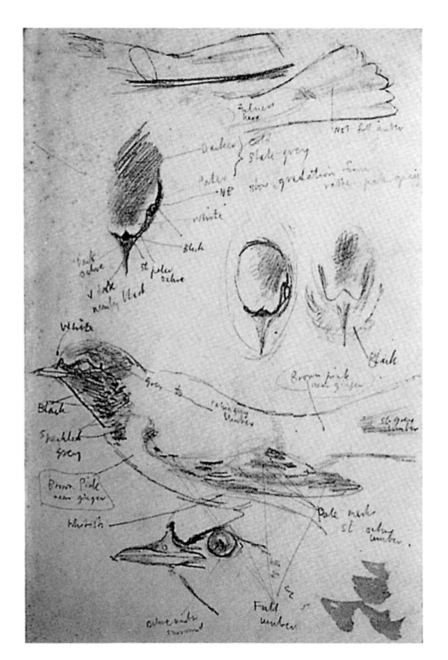

Pencil sketches of a redstart, part of the preliminary artwork for *The Redstart* jacket. (*Private collection*)

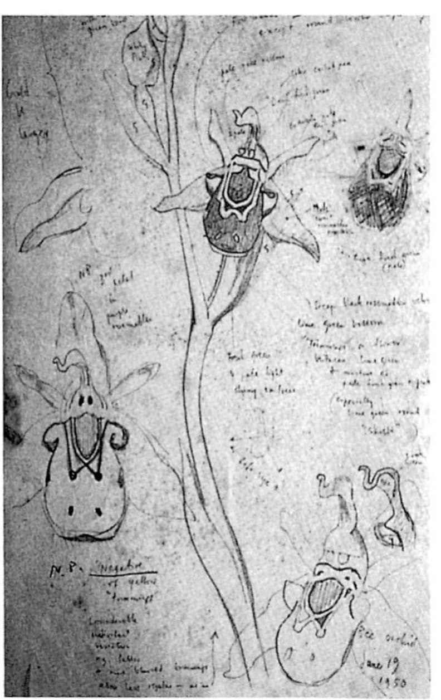

Pencil sketches of bee orchid flowers by C & RE, for the original design of *Wild Orchids of Britain*. (*Private collection*)

liant,' he reminded them, '*brilliant.*'

When so much turned on the ability of the printer to interpret the artist's intentions accurately, it was vital to have a blockmaker with the ability 'to see', as Clifford put it, 'with the artist's eye'. So long as the contract remained with Thomas Griffits, the process seems to have run smoothly enough. It was when the costs of colour printing rose steeply in 1950 and Billy Collins decided to try to make the blocks more cheaply elsewhere that problems started to mount. Hitherto, C & RE had produced a finished gouache painting on rough-surfaced Whatman watercolour paper (which produced their characteristic 'speckling' effect on the colour boundaries) for the blockmaker to copy. Now they were asked to produce separated colours for each new design on Bristol Board or transparent plastic sheets to allow the printer to photograph each in turn and then superimpose them. They tried out both methods for the next design, *Wild Flowers of Chalk and Limestone.* The printer's proofs were wholly disappointing, the intensity of the colours being reduced to a shadow of the original. When Ellis tried out the remaining alternative, drawing directly onto the block, the returned proof was even worse, losing much of the fine detail and chalk work.

These flops coincided with a period when New Naturalist production was in danger of running out of control. There were some dozen titles in the queue, and the Ellises were becoming overwhelmed with work. Several of their recent designs had been returned. The original jacket of *Wild Orchids of Britain* had depicted a bee orchid but the editors were of the opinion that 'the public will mistake [it] for an animal in the shop. I think it would be better if [they] chose another orchid which has a sensational shape and beautiful colour but which bears no resemblance to an animal.' Neither was the original of *Birds and Men* liked much, since it repeated the *London's Natural History* motif of gulls. C & RE managed to replace some of the gulls with lapwings, but they were clearly getting fed up. 'The whole business is on the point of becoming a bore,' Clifford wrote on 15 August 1950. 'Each design calls for first-hand research and, as often as not, long journeys, although the final selection of material suitable for a jacket may give little indication of what has been discarded. There is therefore little financial advantage in doing the jobs at all and the unsatisfactory handling of the reproduction at the end of it all is discouraging.' Billy Collins saw the warning light and recognised that the recent experiments in cost cutting had been a false economy. 'Your designs are so much admired by everyone, and have become so much a part of the series, that we must get the best possible results when they are reproduced.'

For the moment, then, the designs continued to be printed in the established way, with the Ellises producing gouache paintings on Whatman paper. A year or two later, however, the contract was given to Odhams Ltd, who presumably offered a cheaper rate using a new technique. Unfortunately the method failed to reproduce accurately some of the subtle colour gradations produced by the Ellises, and created harsh black edges and inaccurate tones. The first such jacket, that of *Ants*, was passable, although the blockmaker had decided for some reason to smother the original design in grey paint. But for *Flowers of the Coast*, the effect was well below par:

> 'The blocks themselves are good' wrote Clifford. 'We give bad marks only for the ineffective filtering of the sky in the blue block. But the honest-to-goodness ability to mix inks and match a colour, and especially the *tone* of a colour, is lacking. When, as in this design, we have tried to do with 3 colours what used to be done with 4, it is imperative that the *exact* colours of the original should be matched. Both blue and yellow are far too weak.... If it really worries you, have white let in round the leaf – but only on the blue block, and as we have indicated, *there*, not as the mechanical line the blockmaker has so very improperly added to the original.'
>
> *Clifford Ellis*. Unpublished letter to Raleigh Trevelyan, 9 April 1952

The blockmaker did his best, but the result was still a mess, the leaves of the sea-bindweed plant being all but invisible and the sky seemingly laden with smog. This was also the last jacket (with the exception of *Sea-Birds* which had been printed earlier) to bear the pretty individual symbols which the Ellises had been designing since 1944. 'We have decided that it would probably be best to have a general New Naturalist colophon instead,' wrote Raleigh Trevelyan by way of explanation in May 1951, 'instead of a special one for each book.' He gave no reason for this regrettable decision, but it was probably another cost-cutting exercise. The customary date was also now omitted from the designs since the jacket of *Mushrooms and Toadstools*, printed in 1953, had embarrassingly included the date of its design: 1945!

There was a fuss over the jackets of both the next titles, *The Weald* and *Dartmoor*, since neither conveyed the spring-like freshness which Collins wanted, though Raleigh Trevelyan admitted that the former 'certainly grows on one .... after all the hard things I have said about it'. But the sales department objected to the *Dartmoor* jacket on the grounds that it presented a distant view and was too gloomy, failing to meet the 'hit-you-in-the-eye' requirements of the trade. Falling sales of the New Naturalist titles meant that the jackets were becoming 'a real advertising factor'. They therefore requested some foreground animals to bring good cheer. 'But what would sell Dartmoor?' asked Clifford. 'Most visitors come by car and view it from the car or from a motor coach. Hence we didn't do a close-up.' But in deference to the sales people, the clapper bridge was moved up slightly more to the front of the design.

The problem lay partly in the printing method, partly in the reduced colour range. There was in fact a world of difference between working with four colours and working with three. Of the first 20 titles, all but *Britain's Structure and Scenery* and *Life in Lakes and Rivers* had been printed in four colours. But of the main series titles between No 21 *British Mammals* and No 50 *Pesticides and Pollution*, the majority are in three colours only. For *The Greenshank* jacket the Ellises had even experimented with two, relying on a scribble of line to provide additional texture and tone (that technique was also used, rather more successfully, for *An Angler's Entomology*). Colour limitations may be one reason why many of these later jackets are in darker tones with a black title band. For *Pesticides and Pollution*, the Ellises had initially designed the colourful jacket depicted on Plate 9, depending on a remarkable range of two and threefold overprinting to squeeze a rainbow of colours from only three inks. However, this was rejected by the printers presumably because it was too complicated. The adopted design in only two colours presents an apocalyptic vision of smoke and spray in a dark sinister landscape awash with pollutants. It works on its own terms, but it is more appropriate to Rachel Carson's *Silent Spring* than Kenneth Mellanby's sober and objective text, and, in any case, is far from what the Ellises originally intended.

In the four-year delay between *Pesticides* and the next book, *Man and Birds*, Clifford Ellis and the natural history editor Michael Walter worked out a new way of printing the jackets using colour separations rather than coloured artwork. The new approach was necessitated by 'economies in blockmaking with subsequent difficulties', but it did at least allow the artists to revert to a norm of four colours and inaugurate the series of bright and beautiful jackets produced by C & RE in the 1970s. The process was helped by the availability of improved semi-transparent printing inks in a much wider range of tones, and the artists could now choose between transparent and opaque inks. Moreover, they were now able to devote more time to their designs, having retired from full-time teaching. The snag with using colour separations of black on white was that no one could see the full glory of the picture until the block had been made and the printer's proof run off. The visible artwork is no more than patches of black paint on white paper with the Pantone number pencilled below. Clifford compared the business of 'seeing the picture' with that of reading the music score of a quartet. Even the accompanying colour sketch can give no more than an impression of the finished jacket since it was painted with opaque pigments, while the jacket is printed in more lucid, transparent inks. From *Man and Birds* onwards, then, there is no 'finished artwork' to exhibit and frame. The picture is the jacket itself.

The jackets designed by C & RE between *Man and Birds* (1971) and *Inheritance and Natural History* (1977) are surely among the peaks of their craft. Only a master of colour could have created the aqueous jacket of *British Seals*, or the gay colours of *Finches* against a flat print-like background, or the superb looming *Ant* in its grassblade jungle. More jackets were designed during this period than were actually used. Many expected titles were, for one reason or another, never completed. Moreover the brand new jackets prepared for planned new editions of *Insect Natural History*, *Mushrooms and Toadstools* and *The World of Plankton* were also

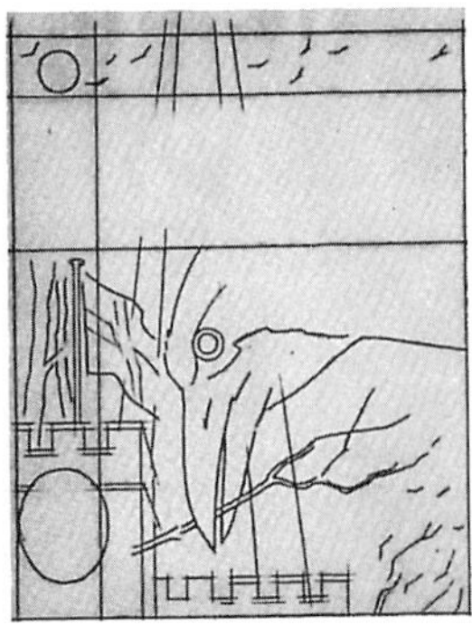

Artists' pencil outline and printed jacket of *Man and Birds* (1971), the first to be prepared from the artists' colour separations.

wasted since Collins had decided after all not to reprint them. Colour separations exist for some of these titles, but because printers proofs were never made, and are expensive to produce, we may never see them in their full glory. The colour sketches reproduced for the first time on Plates 12 and 14 can give only an indication of what we have missed by the non-appearance of *Waysides, Lichens, Seaweeds* and *Ponds, Puddles and Protozoa.*

If the last Ellis covers from *British Tits* (1979) to *The Natural History of Orkney* (1985) fail to match the dizzy heights of their immediate predecessors it is less the fault of the artists than the editors: too many birds and beasts, for which a repetition of motifs was perhaps inevitable. Neither does the thick plastic wrapper in which these books are indelibly sealed work to their advantage. And as for the laminated jacket of *Reptiles and Amphibians*, wholly unsuited to the surface texture of the lithographic print, we must avert our eyes and be thankful for glories past. The design for *Orkney* was the result of an expedition to Orkney by Clifford and Rosemary in 1983, during which they drew from life the native North Ronaldsay sheep feeding on seaweed at low tide. The jacket contains a characteristic Ellis device, the title band cutting two planes of view, so that beyond the scene of sheep munching weed lies a distant view of the Orkney coast: terns in flight over an intricate pattern of rock and water. As usual, a vital part of the design – the curling horn of the ram – spills over onto the spine of the jacket. Unfortunately this cannot be seen on the paperback copies of the title which were given plain spines. This was the last of the Ellis jackets. After a short illness, Clifford died in

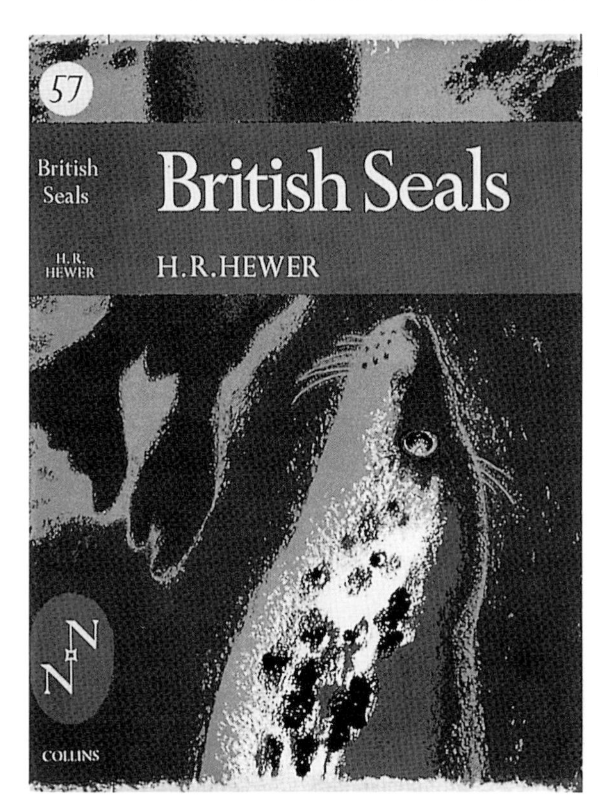

Jacket of *British Seals* (1974).

1985 before he and Rosemary could begin work on the next title, *British Warblers*. Rosemary was invited to continue alone, but she declined to do so in a graceful and charming letter to the then natural history editor, Crispin Fisher. The Ellis jackets were the work of a husband-and-wife partnership, and with Clifford's death the partnership must cease. To her, Crispin had written that 'you and Clifford have been part of my upbringing .... His death marks the end of an era. Your work has a freshness and quality as modern today as it was 43 years ago – and that can't be said for any other graphic designer I know.'

Postscript: the Gillmor jackets

The Ellises were a hard act to follow. It was the unenviable task of the natural history editor, Crispin Fisher, to find either an artist who could produce work in a similar spirit, or to change the design of the New Naturalist jackets altogether, perhaps using colour photographs. He would have had some justification for taking the latter course. Because of rising costs and falling sales, he had been forced to switch the main production from hardback to paperback, leaving a small edition of casebound New Naturalist books for the hard core of collectors who would buy them at almost any price. Most of the new paperbacks were to be given photographic wrappers. To the gratitude of New Naturalist admirers everywhere, Crispin decided to maintain the hardbacks in the Ellis tradition. He does not seem to have hesitated long in his choice of an artist. As a fellow bird artist, Crispin had known Robert Gillmor for many years, and admired his combination of accuracy and liveliness, as well as his habit of sketching from life. Robert told me that he was nervous of accepting, 'but Crispin Fisher was persuasive ..... I am most interested in the Ellises work and enormously admired their NN covers from the first – never thinking I could ever follow in their footsteps.'

Robert Gillmor is, of course, one of this country's most prolific and successful bird artists. He has illustrated shelves of books and bird magazines during the past 35 years, and his meticulously observed watercolours always find a ready market. He is a remarkably versatile artist, having used a wide range of materials to capture the essence of a bird, most characteristically with ink and brush, but also with silk screens, lino cuts and other more experimental techniques. Some of his watercolours are painted in the hide, by watching the bird with one eye and the sketch pad with the other. At home in Reading, he works in equally cramped surroundings in a shed in the garden, piled high with magazines, box files and artistic impedimenta of all kinds. In recent years, he and his wife, the landscape artist Susan Norman, have held successful joint exhibitions in East Anglia, whose shores are the base for many of Robert's watercolours of gulls, terns and avocets.

It must have been a daunting task, designing a jacket which would sit harmoniously on the same shelf as the Ellis jackets, and knowing that a small army of book collectors would be inspecting your work very critically indeed. Wisely, I think, Robert Gillmor did not attempt to emulate the Ellises' style but by using the same technique of colour separations and overlapping colours he succeeded in producing jacket designs that sit on the shelf with their forebears without serious disharmony. The texture of Gillmor's work is quite different, however, perhaps reflecting his background in book illustration rather than that of the Ellises in posters and art education. Combining line and colour, and the occasional use of textured paper, he has produced some quite brilliant results, like the rocky background of *Caves and Cave Life* or the tree trunks on *The New Forest* jacket. Like the Ellises, he has become expert at exploiting colour overlaps, and perhaps few would tell at a glance that he is limited to only three colours (plus black). For sheer virtuosity, the *Heathlands* jacket is remarkable in managing to create three birds, three flowers, a butterfly and a sand lizard, all in reasonably lifelike colours, with such a restricted palette.

The nature of Gillmor's designs is to a large extent suggested by the title of the book: some demand landscapes, others require recognisable close-ups of birds, flowers and fish. With his first jacket, *British Warblers*, his intention was to produce a simple bold pattern suggestive of life in the tree-tops, using areas of solid colour (on the paperback the effect is spoiled by the omission of the female blackcap peeping round the spine). *The New Forest* called for a different approach. Robert told me he sketched the rugged old trees from life during a day in the Forest, and had a lot of fun mixing the colours, obtaining two tones of green for the leaves, and suitably coloured tree trunks by mixing blue with orange! On the whole, I think Robert Gillmor's style lends itself best to the use of contrasting textures and strong designs. The jacket of *Ferns* is surely his masterpiece so far. The bold and varied patterns of the ferns themselves are ideal for lithography, and the combination of their browns and greens against a uniformly textured background is irresistible. When the artist aims for closer realism, as with the horseshoe bat on *Caves and Cave Life*, or the various birds on *British Larks, Pipits and Wagtails*, the effect seems to me less successful, though in the case of the bird titles it is hard to imagine an alternative (though the jacket of *The Hebrides* is very successful at combining birds and a seascape).

Once the final design has been worked out, Robert Gillmor draws the line artwork and the blocks of colour on separate sheets of transparent plastic, overlaying the design. These elements can then be combined to check that they register accurately. Before this stage is reached he will have produced a colour sketch of the envisaged design, which is sent to the author for comment. This often results in minor amendments: the omission of a second mole on *The Soil*, or of a trellis above the garden gate in *Wild and Garden Plants* in order to reveal more meadow beyond; or a discussion with Colin Tubbs about the exact colouring of fallow deer in late summer. The original design for *Freshwater Fishes*, featuring trout with water crowfoot flowers, was replaced because it gave the false impression that this was an angling book. Niall Campbell, one of the authors, suggested the theme of the replacement design: 'a selection of fish, coarse and game, either as a mixed or varied group ..... a sinister pike hovering behind a group of small colourful fishes. Something like that.' The sticklebacks of *Freshwater Fishes* were certainly colourful – more like guppies than sticklebacks – but, if one accepts, as did Clifford and Rosemary Ellis, that the job of a book jacket is to produce a distinctive, interesting image, not a coloured diagram, then their departure from strict accuracy hardly matters.

His series of eleven New Naturalist jackets so far (not counting the one surrounding this book) has surely established Robert Gillmor as a worthy successor of the Ellises. The New Naturalist hardbacks still gleam in the bookshops (when they are allowed to do so) as well as ever they did; or perhaps even more so, now that the art of dust jacket design has elsewhere been all but eclipsed by glossy, unimaginative pictures created by the camera.

5

E.B. Ford and *Butterflies*[5]

As an introduction to Professor E.B. Ford FRS, Fellow of All Souls, author of what many consider to have been the best butterfly book ever written, it is impossible to better Miriam Rothschild's account of her first meeting with the great man. The two shared various academic interests and it was natural that, on visiting Oxford some time in 1956, Dr Rothschild (then Mrs George Lane) should wish to call on Ford to pay her respects. Ford suggested that she came to his office in the zoology building at about 11 o'clock the next day:

> 'I duly did so, knocking on his door punctually as the clock struck. After a moment's silence there was rather a plaintive long drawn out cry: "Come in!" I opened the door and found an empty room. I looked round nervously – not a soul to be seen, but an almost frightening neatness pervaded everything. Each single object, from paper knife to *Medical Genetics*, was in its right place. Each curtain hung in a predestined fold, and you felt that if a slight breeze or an unexpected earth tremor had disturbed one of them, it would have automatically resumed its rightful position. An unkind fate seems to have decreed that I share all my rooms with Typhoon Agnes: the sight of all this distilled essence of neatness and order took my breath away. I stood there, probably with my mouth open, trying to reconcile this vacant room with that ghostly cry – had I dreamed it? – when suddenly Professor Ford appeared from underneath this desk like a graceful fakir emerging from a grave. Apparently he had been sitting cross-legged on the floor in the well of his writing table, lost in thought, but he held out his hand to me in a most affable manner. His explanation for this rather startling welcome was: "My *dear* Mrs Lane – I didn't know it was you." I'm sure Henry Ford won't mind me saying that the really distinguished butterfly people are usually a trifle eccentric, and you never have to ask a great man for an explanation. But only the great men find time to sit and *think*.'
>
> *Miriam Rothschild.* Dedication: Henry Ford and Butterflies in *The Biology of Butterflies* (1984)

This meticulous sense of order, raised at times to a level approaching preciosity, is a vein that runs through all Ford's written work and, still more, his recorded sayings. Many scientists today might envy the opportunities granted to E.B. Ford to sit and think. His whole working life, spanning seven decades, was based at Oxford. His administrative burdens were, for the most part, light; he had no wife or children to occupy his time; he never watched television and rarely read the newspapers. Rarely did his scientific work involve lengthy technical preparation or the use of complex laboratory equipment; on the few occasions when it did, he left the operation of 'the engine' to assistants. To an extent seldom found today, but not unusual in an earlier generation of academics, his was a science based on the observation of nature and on pure intellect. The stability of his life enabled him to plan ahead with precision, and this gave his work a shape, and a logical

Edmund Brisco Ford (1901-1988) with four (jacketless) New Naturalist titles on the shelf behind him. (*Photo: John S. Haywood*)

progression. In the preface to his magnum opus, *Ecological Genetics*, published in 1964, Ford revealed that 'This book was planned in 1928, and in considerable detail. At that period I believed it would be necessary for myself and others to work for a quarter of a century before it could be written. I was over-optimistic; more than thirty years were in fact needed.' It was largely as a spin-off from the great undertaking of his life – the demonstration of genetic theory in the field – that he wrote the two books for which he is most widely remembered: the New Naturalist volumes on *Butterflies* and *Moths*.

An account of E.B. Ford might usefully begin with his name. His initials stand for Edmund Brisco, but he usually signed his letters E.B. Ford, and evidently disliked the name Edmund. To most friends and colleagues he was 'Henry Ford'.

A natural assumption would be that his nickname was borrowed from the American motorcar mogul, but Ford's close friend John Haywood denies this. Evidently Henry was a family name which gradually became his adopted cognomen. To students and other inferiors he was 'the Professor', and a few have referred to him as 'E.B. ', though I suspect that the latter is no more than a book-name, like some of the rarely used fanciful names given to species of Lepidoptera. Because most New Naturalist readers know him as E.B. Ford, he will so remain in these pages. Calling him Henry would be an assumed familiarity of which he would certainly have disapproved.

The earliest and perhaps the most important of the influences on E.B. Ford was his father, the Rev. Harold Dodsworth, rector of Thursby, a village near Carlisle. The Fords were a Cumberland family descended from an eighteenth century baronet. E.B. Ford had a familial connection with Charles Darwin, who had married his first cousin, Emma Wedgwood (of the pottery family). Another connection was the Briscos of Crofton, from whom Ford owed his unusual second name. Ford once claimed that his line had made a habit of marrying their cousins, explaining that one reason why he had not married was that there were no cousins left for him. The Fords were evidently one of those close-knit country houses, where sisters, cousins and eccentric bachelor uncles all lived under the same roof. Near the end of his life, Ford recalled that 'We lived in one of the family houses [the Manor House at Papcastle, near Cockermouth] until I was ten. The long front was of the William and Mary period, though parts of the house were much older. It was built on the views of a Roman fort on the hill above, and Roman coins and pottery were constantly turning up in the garden' – to which Ford attributed the start of his life-long passion for archaeology. Ford was born there on St George's Day, 23 April 1901. His father was an educated man, a graduate of Wadham College, Oxford. Ford describes him as 'an outstandingly good speaker, and very much a figure in society'. There were strong ties of affection between the male Fords. Late in life, E.B. Ford enjoyed nothing more than to talk about his father, and also his uncle, a church organist, to whom he owed his taste for classical music. Of particular interest to us, of course, is that Harry Dodsworth encouraged his son to collect butterflies. E.B. claims though that this was a matter of the father following the son, not the other way round: '[My father] was not a naturalist and had never collected Lepidoptera. When I started to do so, on 27 July 1912 [aged 11], he almost at once was delighted to join me, and we gradually developed our entomological studies together: each starting without previous knowledge.' (Ford 1980; Note the characteristic Fordian precision about the date.)

The butterfly-collecting expeditions of the Fords *père et fils* had results of importance for the history of science. By then, Harry Dodsworth had moved from Papcastle to the rectory at Thursby, and among the local collecting grounds were a few damp fields at the edge of woods near Great Orton, an area long known to collectors for its isolated population of the marsh fritillary. One collector had left detailed notes over 36 years on the mysterious changes in abundance of this butterfly, from its appearance in 'clouds' to seasons when hardly a specimen could be found. The Fords decided to continue this work, beginning in 1917 when 'We had to work for several hours each day to obtain a few specimens.' Seven years later, however, the butterflies were back in enormous numbers, 'a dancing haze in the few fields they occupied, and one could catch several specimens with a single stroke of the net'. We now know, of course, that these

E.B. Ford, equipped for the field in the early 1970s. (*Photo: John S. Haywood*)

fluctuations are due to parasitism, to which the marsh fritillary is prone. What especially interested the Fords was that the sudden increases in numbers were accompanied by an equally dramatic outburst in variation. In 'normal' seasons, the wings of this pretty insect are fairly uniform in pattern, and of much the same size. After the explosion in numbers in the mid-1920s, however, it was hard to find two specimens alike, and the more extreme departures from the norm were in fact deformed, in some cases so much so that they could hardly fly. An ordinary collector might have blessed his luck, bagged a large number of them and exhibited them proudly in his cabinet as 'aberrations'. But Ford's natural curiosity had been stimulated and refined by his growing interest in heredity. He realised that the butterflies had taken advantage of an opportunity for evolution, and that evolution could therefore take place much more rapidly than Darwin had supposed. By 1930, the Fords had collected enough information to write a joint paper for the London Entomological Society: the landmark, 'Ford and Ford, *Fluctuation in Numbers and its Influence on Variation in* Melitaea aurinia'. In his New Naturalist book, *Butterflies*, E.B. Ford depicts some of these specimens on Plate 39. They have as good a claim to being 'historic butterflies' as the Camberwell beauty and Bath white on Plate 1. The point was that such work of world importance was well within the province of any amateur collector. What the Fords had done was to open people's eyes to the scientific possibilities lurking behind the fun of a collecting expedition. They continued their marsh fritillary work until 1935. Harry Dodsworth died in 1943, and E.B. Ford dedicated *Butterflies* to the memory of the father 'with whom I collected butterflies for thirty years'.

The marsh fritillary story has taken us ahead of the tale. E.B. Ford's schooling at St Bees in Cumberland leaned towards the classics, perhaps following in the steps of his father. However, his extra-curricular interests were already pulling him in the opposite direction, towards experimental science: 'Working on archaeology introduced me to cultivate the habit of deduction from my own observations, a habit I began to apply in a scientific direction through collecting Lepidoptera. As a result I had begun evolutionary studies before going up to Oxford. I was a convinced Darwinian. I had read *On the Origin of Species* in bits, as a boy, but was indirectly rather than directly influenced by it.' In 1920, Ford went south to his father's Oxford college to study for a degree in classics. At some early stage he decided to switch to the zoology course, but herein lay a problem. Wadham was not a science-oriented college, and Ford evidently had to arrange his own tuition. He managed to persuade men of the calibre of Gavin de Beer and Julian Huxley to give him private lessons. He did not think much of the zoology school, which then specialised in comparative anatomy and embryology, and had turned its back on field studies. Late in life, he recalled that 'As an undergraduate I could not find any naturalists who were geneticists (or the reverse). I wanted to study the genetics of wild populations..... I was troubled because though zoology seemed orientated towards evolution, evolution did not seem to be studied genetically.' Ford was therefore obliged to study the subject from books, working his way through all the standard textbooks and published papers in journals – not too demanding a task at this date, for genetics was still a young science. Already Ford's sharp and lucid mind was hurling the chaff from the wheat:

> 'I thought Darbishire's book the best. It was obvious even then that Punnett writing on mimicry had not properly looked at the butterflies he was writing about.... Punnett's book on poultry was trivial. I thought Morgan's book, *The Physical Basis of Heredity*, one of the worst written books I had ever encountered..... How dreadful are the blots on the pages that are called figures. If anything could have stopped me taking an interest in genetics, it would have been that book of monumental dullness and incompetent presentation.' [And so, gleefully, on.]
>
> *E.B. Ford.* In: *Some recollections pertaining to the evolutionary synthesis* (see references)

At this point, another facet of Ford's many-sided personality comes into play. He had the confidence and ambition to seek out the brightest minds in his chosen field, and those who were likely to be of greatest help to him. They, in turn, found Ford remarkable. In some of his later books, Ford tells stories about his friends, either because of something interesting they had told him, or to illustrate a particular point. There is no doubt that Ford took a great pride in his acquaintances, who included men of letters, politicians, clergymen and dukes, as well as fellow scientists. Among the great friends of his undergraduate days were Leonard Darwin, son of Charles ('I frequently asked him what Charles Darwin [had] said on various topics'), the aged Edwin Lankester and the great Cambridge geneticist R.A. Fisher. Later friends included A.L. Rowse, M.R. James (with whom he shared a taste for ghost stories), F.A. Lindemann, Churchill's wartime scientific adviser, the archaeologist Sir Mortimer Wheeler and John Sparrow, the Warden of All Souls. He knew Thomas Hardy, who told him the Dorset legend that a white moth escapes from a man's mouth at the moment of death. He was on at least nodding terms with Winston Churchill. And, having once arranged an audience with Pope Pius XI, he even referred to 'my friend, The Pope'!

After his father, Julian Huxley was the next great influence on Ford's scientific outlook. It was part of Ford's usual good luck that his early years at Oxford coincided with the brief period (from 1919 to 1927) when Huxley was lecturing and researching in genetics there. The two first met in 1921. Although Julian Huxley is best remembered as an ornithologist and a student of evolution, he was at this stage in his career concerned more with genetic physiology and the development of body processes in the embryo. Huxley suggested that he and Ford collaborate on a study of the brackish-water amphipod 'shrimp' *Gammarus chevreuxi*, whose eyes are sometimes shades of red and sometimes black. By breeding the shrimp and studying their mysterious eyes, Ford and Huxley (for the generous Huxley insisted that it be that way round) were able to show that the black pigment and its rate of development was controlled by a gene. They called this a 'rate-gene', and its existence has a crucial role in evolution for it means that, since rates of development vary genetically, they are therefore open to natural selection; or, to put it another way, genes are able to control the onset of processes in the body. This work shed new light on human evolution. Ford was later to demonstrate that human blood groups were affected by natural selection, and predicted (accurately) that particular blood groups were prone to certain diseases. It may have been at Ford's suggestion that the *Gammarus* work was not confined to breeding programmes in the laboratory but was also demonstrated in the field. Several times, Huxley and Ford visited Plymouth to sample the frequency of different genotypes of shrimp in the wild salt-marshland of Plymouth Sound. Much of Ford's work involved lengthy field observation, and sometimes entailed camping out in remote places and uninhabited islands.

In the 1920s, Ford also worked for some years with John Baker and Charles Elton on a study of the health and parasites of the wood mouse. The work involved dissecting hundreds of corpses, and it was important for the mice to be captured alive and in good condition. This meant inspecting the traps in Bagley Wood in the small hours of the morning. Baker and Elton, fearing that the task might not be congenial to the fastidious Ford, decided one night to hide out in the woods to check on whether he was fulfilling his side of their arrangement. They had, of course, misjudged their man. Ford was on his way. The first they heard of him was a plaintive voice in the darkness, then an elegant figure emerged from the shadows, shouldering a lantern on a pole and exclaiming in his high-pitched voice: 'I am the way, the truth and the light......' – probably a show put on for their benefit.

Huxley departed from Oxford in 1927, the year in which Ford gained a Masters degree to add to his B.A. in zoology. Ford began to work with R.A. (later Sir Ronald) Fisher, the most brilliant geneticist of his day, whom Ford had known since his days as an undergraduate. He described in characteristic fashion how they met.

> 'Julian told him about me, and he decided to travel to Oxford to see me. It did not occur to him to let me know he was coming, so when he arrived at Wadham College, I was out. Accordingly he settled down in my rooms to wait for me to return. Fisher often visited me at Oxford, and I was constantly going to see him at Harpenden [the Rothamsted research station], or at Cambridge. He never visited anyone at Oxford but me.'
>
> *E.B. Ford. Some recollections*, ibid.

Fisher had used mathematics to predict how genes operated in the wild. In 1927-28, he demonstrated how natural selection influenced genes through his famous theory of dominance. It was this work, wrote Ford, fifty years later, that 'led me to plan my book *Ecological Genetics*. [Fisher's] 1927 paper opened up the possibility that I had had in mind for some years of taking genetics into the field.' It chimed in well with Ford's observations of the powerful selection forces at work on the marsh fritillary. From 1928 onwards, Ford decided to devote his life to the study of evolution in the wild, seeking conditions in which change could occur rapidly, and analysing the variation of wild populations in conjunction with genetic experiments in the laboratory. He was to call this technique 'ecological genetics'. In so doing, he founded a new science, based on a new form of natural history.

We have reached the period which Ford describes in great detail in his New Naturalist books. Ford was the first person to notice that butterflies and moths were ideal subjects for the study of evolution in the wild, with their annual generations and variable and easily observed wing patterns. He perfected a technique of estimating the size of a colony of butterflies and moths by marking, releasing and recapturing specimens. Fisher's statistics did the rest. An early choice of subject was the scarlet tiger-moth *Panaxia dominula*, which occurs in two or three fairly constant forms, and of which there was a suitably isolated population nearby at Cothill Marsh. It also helpfully appeared on the wing just after the start of the long summer vacation, when Ford was freed from his teaching duties. His work on the scarlet tiger, which was continued by P.M. Sheppard after 1947, was described by Ford, with his habitual modesty, as the most thorough quantitative study ever made of any wild population of animals upon earth. It is best read in Ford's own words in the important chapter on evolution in *Moths*, where he describes the work as 'the first practical test of a problem which has proved a storm-centre in modern biology, the relative importance of natural selection and of random survival in the evolution of small isolated communities'. His other work of this period included the first scientific study of industrial melanism in moths, that is, the production of dark pigment in certain species to match the darkening of tree trunks and other vegetation from soot and other airborne pollutants. Another important study was an analysis of the variable wing patterns of the meadow brown butterfly, based on Fisher's mathematical principles. In both, cases Ford was assisted by a promising medical student, H.B.D. Kettlewell, who subsequently took over the work.

To find isolated populations of butterflies in which evolution could be studied, Ford made regular camping visits in late summer to the Isles of Scilly, accompanied by his schoolmaster colleague and friend, W.H. Dowdeswell. The two became adepts at camping out on uninhabited islands, so much so that in his book *Understanding Genetics* Ford suddenly launches into a long digression on camping techniques. For those who have read his accounts of butterflying on Tean in his New Naturalist books, the following extract may be an interesting glimpse of the (I suspect, enjoyable) scene behind the science.

> 'We belonged to the Primus-stove age; perhaps this type is still best in remote conditions. There should be a small oven, with a shelf inside, to fit on it. Enjoyable food and good cooking are most important. Conditions nearly intolerable can be gaily supported if one can look forward to one's meals. A benison goes up to him who returning after a hard day, perhaps on a neighbouring island, prepares with quick efficiency hot scones for tea. The camp must not be polluted by the presence of a drunkard or a total abstainer.
>
> 'Think not to live off the land: occasional augmentation, yes, but nothing more. After breakfast; clearing up the camp; preparing a picnic lunch; scien-

Camp on Tean, 1950. (*Photo: E.B. Ford*)

> tific work, perhaps in various habitats, for hours; cooking; a quiet drink before dinner; washing up; probably analysis of results in the evening; and planning for the future, there will be no time for food gathering or fishing. As to fresh meat and vegetables, *solvitur ambulando*. Remember disinfectant tablets for the drinking water; simple remedies; sunglasses; rat traps; and this great rule: semi-permanent camping begins with an unstinting visit to a really good grocer.'
>
> *E.B. Ford*. In: *Understanding Genetics* (1979)

While he was always dressed punctiliously, in dark suit and tie, in Oxford, Ford dressed appropriately for the field in hard-wearing tweeds or even jeans. He usually wore a trilby hat, having lost most of his hair in his youth. In the well-known painting, reproduced as the frontispiece, Ford appears as one of a quartet of leading ecologists, though all that can be seen of his head is this brown trilby hat. His posture, scrabbling down on all fours to examine some insect, is said to be well observed and absolutely characteristic. It was this that gave the painting its unofficial title of 'the New Religion'. When it was not obscured by a hat, he had a characteristic way of poking his face forward and tilting it back, as if to express simultaneous interest and scepticism. He had a big, beaky nose, a pointed chin, and permanently surprised eyebrows hovering above round, tortoiseshell spectacles. He was, or could be, an affable man and a cordial host, introducing at least one admiring American scientist to what the latter called 'the finest traditions of England'. The same visitor remembered some of his expressions: 'My dear Lincoln, carefully controlled observations *are* experiments,' and 'You *do* know, Lincoln, that good science is an art; unfortunately the reverse is not true.' Ford's rather mannered discourse was not, of course, as unusual then as it would be today. Miriam Rothschild found him exclaiming 'My *dear* Mrs Lane, you really *must* write a book about this,' as he poured out another glass of white Cinzano. He had a large fund of stories and anecdotes, 'each with a long-polished patina acquired in the telling'. When Miriam mentioned that her sons and daughters all had carrot-coloured locks, Ford recalled the red-haired daughter of a distinguished entomologist who had been chased across a sand dune by a swarm of sex-starved male burnet moths. When their conversation moved on to the taste of various cryptic and noxious forms of life (Ford made a practice of 'sampling

Butterflying near Thorverton, Devon, 1972. E.B. Ford on the left. (*Photo: John S. Haywood*)

the body fluids' of moths to see whether they were noxious or whether they were bluffing), he recollected that 'When I was in Texas, I used to eat rattlesnakes and I recall they tasted of cold scrambled eggs.' At the end of their conversation, he repeated, 'Indeed, my dear Mrs Lane, I think you really *should* write a book about all this.'

Ford stood on his dignity, and was easily offended, but in certain odd ways he could be more tolerant than most. He seemed not at all surprised to see moths crawling around in Miriam Rothschild's son's hair, merely remarking that one of them was a frosted orange, and that it was remarkably well-adapted to its chosen environment. He was not so amused, however, by the duster that had been treated with ammonium nitrate and exploded when he banged the blackboard with it. He refused to deliver his lecture until the culprit had owned up and apologised – the malefactor was, in fact, not a student but a colleague. 'He liked to be a terror to his students,' noted one colleague, and exploited his eccentric persona to the full as a means of controlling them. His lectures were always prepared to the last detail and delivered in his most formal voice. Thomas Huxley (Julian Huxley's nephew) told me that he began his lectures on the hour and stopped just as precisely. 'The feeling was that he stopped in mid-sentence and then began (several days later) by completing the unfinished sentence of the previous lecture. No introductory pleasantries. And if someone was late ... he would stop and "eye" the person until they had found a seat, and the silence while this went on was terrifying.' John Pusey, another of Ford's students, recalls his curious way of referring to animals in a genetics context as 'the *snail* animal' or 'the *thrush* bird'. Ford seems to have approached his lectures in the same spirit as his father's sermons.

Off-stage he was usually more approachable. Some of his students used to take him out to dinner, where he could be friendly and entertaining. Ford either liked people or he didn't, and those in the latter category he simply ignored. As he used to remark when someone for whom he had no regard was mentioned, 'You can't like everyone.' To the student who was on the same wavelength as Ford, clever, dedicated, well-mannered and preferably male, he could be an inspiration and be generous with his time and encouragement. He had the trick of making people feel more intelligent and amusing than they really were. He also had a gift for selecting good research workers and giving them free rein, often in promising areas that he himself had originated. From the 1950s onwards, his genetics laboratory was inhabited by some of the brightest scientific talents in a generation, among them Bernard Kettlewell, Philip Sheppard, Kennedy McWhirter and Robert Creed.

It is easy to accuse Ford of snobbery, but his likes and dislikes were not necessarily based on wealth and position. We could cite, for example, the case of the distinguished geneticist J.B.S. Haldane, whom for some reason Ford could not abide, though the thick-skinned Haldane did not seem to notice. The story has it that Ford heard that Haldane was looking for him while on a visit to Oxford:

> 'Very agitated, Ford sought out a colleague, Robert Creed: "Take me home, I do not want to talk to that man." At that time, Creed had a Brooklands Riley competition car which was a very small open two-seater in which one sat four inches from the ground. When Ford had inserted himself, with hat and briefcase, and Creed was at the front cranking the engine, Haldane turned up, bent down and said, "Ah Henry, I wanted to talk to you." To this Ford replied from his position near the road, "I am so glad to see you, Jack. I would be delighted to talk to you – but we are very busy. In fact, as you see, we are so busy that we have to use a motor racing car in order to get about."'
>
> *Footnote* to Appendix C in Berry, R.J. (1990) Industrial Melanism and the Peppered Moth, *Biological Journal of the Linnaean Society* ***39***, p. 320

His misogyny was well known and notorious – he seems to have regarded women as though they were not so much a different gender as a different species. There is a famous Oxford story, probably dating from the early years of the war, when one by one, Ford's male students were called up for war service. Soon there was only one left, and Ford's customary preludial 'Gentlemen!' was replaced by 'Sir!' Eventually, he too was taken away, and Ford turned up to find himself facing a room full of girls. Always master of the occasion, Ford faced the lecture theatre as though it were empty. 'Since there is no one here,' he announced, 'there will be no lecture.' He was, of course, opposed to mixed colleges, and when in 1978 his own college proposed to allow women guests to lunch, Ford complained bitterly, believing that 'it would tend to break down the unique style and ethos of All Souls', and fearing that this would be but the first step towards admitting women students to the college (as, indeed, it proved). On the other hand, there are women who recall Ford with great affection. One student remembers him as 'an excellent tutor' and her 'only good friend at Oxford', and another recalls him as the kindest and friendliest of men. He was devoted to Evelyn Clark, his partner on archaeological expeditions to Cornwall, and with Miriam Rothschild he had an obvious empathy, a matching of equally brilliant and original personalities.

Perhaps he was not so much prejudiced as old-fashioned. Ford disliked gadgets of any sort, and was ham-fisted when it came to mechanical matters. John Haywood remembers a terrible noise in an adjoining room as Ford attempted to nail something together, later emerging to inform him, with evident pride, that 'I have been doing some carpentry.' Ford genuinely admired those skilled in technical matters, especially John Haywood's and Sam Beaufoy's mastery of what he called 'the photographic camera'. He did not learn to drive a car until late in life (though he usually found someone to take him to wherever he wanted to go). His scorn of television and radio extended also to daily newspapers. Neither did he approve of fast food. What he might have thought of pizza shacks and hamburger huts can only be imagined, but he deeply disapproved of fish-and-chips. When the first fish-and-chip van appeared in Oxford, some wag, to tease him, cried out, "Look Henry, fish-and-chips on wheels!" To which Ford remarked that he would never have believed it possible that there could be so much wickedness in this world.

Writing and illustrating *Butterflies*

E.B. Ford wrote *Butterflies* at the mid-point of his career, while still plain Dr Ford, Reader in Genetics (and his doctorate was very recent), at a pause in his work imposed by wartime activity. He had written four books already: *Mimicry, Mendelism and Evolution* (with G.D.H. Carpenter, 1931), *Mimicry* (1933), *The Study of Heredity* (1938) and *Genetics for Medical Students* (1942), all of them weighty and characterised by his crisp, lucid style. Two of them remained in print as standard textbooks for more than thirty years. His published output in the scientific journals had been relatively small, but each paper was, in its way, a landmark. In the whole of his life, Ford published nothing trivial. In 1940 he had written perhaps his most influential paper about the idea of genetic polymorphism, based mainly on field studies of butterflies and moths. Briefly, he had recognised that individuals in a population occur in a number of distinct and recognisable forms in fairly constant proportions, and that these forms must be genetic since they could not be the result of geographical or environmental factors. Through polymorphism, natural selection maintains a delicate balance in the natural world between conformity and diversity. This means that minority forms can redound to the advantage of the insect by allowing it to adapt – for example, when the peppered moth found that its favourite resting places were covered in soot.

All this was fresh in Ford's mind when he wrote *Butterflies*. That the book would be something quite outside the ordinary run of natural history books would have been obvious to anyone who knew Ford and his work. What we would like to know is how he managed to write such a big, important book with such evident speed during wartime, when nearly everyone was working harder than they had ever worked in their lives. To broaden the question, we would very much like to know what E.B. Ford got up to during the war years. Ford being Ford, this is a subject of boundless and often fantastic rumour, but, unfortunately, very little fact. In the former category is the story of Ford on a vital intelligence mission to the United States; Ford attending meetings at 10 Downing Street; and Ford in the secret service in Istanbul. These may be examples of what his biographer Bryan Clarke called 'romantic misinformation of the kind that he did not tend to discourage'. When a colleague asked him why he had discontinued his studies on the meadow brown butterfly after 1939, Ford had replied mysteriously that 'Some of us were needed at once.' Rather better documented (because they are told by eye-witnesses) are stories of Ford knitting woollen balaclavas and socks for the Navy during his wartime tutorials! What are the facts? That he continued his pre-war investigations on insect pigments (publishing papers in 1941 and 1942), described as one of the first successful attempts to relate chemistry to classification; that he extended his work on polymorphism from butterflies to human blood groups, and wrote a book about it; that for part of the time at least he was lecturing and tutoring in genetics, much as usual; and that towards the end of the war he wrote *Butterflies*. On the basis that there can be no smoke without fire, we

E.B. Ford (left) and Sir Alister Hardy, posing beneath their joint portrait in the Zoology Department, Oxford. (*Photo: John S. Haywood*)

might suppose that he was engaged in some kind of clandestine war work, perhaps at the behest of his friend Lindemann, now the most influential scientist in the land. Whatever its nature, whether it was too secret or because it was another opportunity to add to the Ford legend, he never spoke about it. But it seems clear that, for some of the time at least, university life went on much as normal. Perhaps Ford took advantage of the absence of male students to write the book.

As to the genesis of *Butterflies*, Ford was not an assiduous hoarder and filer of correspondence, and the relevant years are missing both from his private papers in the Bodleian Library and from the contemporary internal minute books and files of Collins. Julian Huxley had sent letters to a number of distinguished scientists in August 1943, asking if they would contribute a book to the series, and Ford's might well have been among them. The one document still extant is his contract with the publishers to write *Butterflies*, which is dated 20 January 1944 and calls for the delivery of the manuscript by the end of March. As explained elsewhere, this does not mean that Ford was allowed only two months to write the book. It was normal practice for Collins to issue a formal contract, known as a 'memorandum of agreement', only when the book was in an advanced state of preparation; and in any case, the specially-produced New Naturalist contracts were not ready for sending until November 1943. Ford's contract arrived later than his colleague Alister Hardy's, suggesting that his book was not the very first to be commissioned. It seems that *Butterflies* was number one in the series not because Ford was ahead of the game but because he managed to write his book more quickly than anyone else could. By mid-1944, it was clear that *Butterflies* would be completed first because it was for that title and no other that the Ellises were asked to design their first New Naturalist dust jacket. From what we know or can infer, *Butterflies* was written after August 1943 and possibly completed by March 1944. It was as if the book was already formed in E.B. Ford's mind, rolled up and awaiting its time, like one of the Dead Sea scrolls.

Illustrating *Butterflies* took up much of Ford's spare time in 1944. The plates of mounted specimens were relatively straightforward to produce. Ford selected and arranged specimens from the large collection that he and his father had built up over the previous thirty years, augmented with others from the collections of the Hope Department of Entomology in the University Museum, including the historic early collection of J.C. and W.C. Dale. He was also lent specimens from the British Museum (Natural History) and from Dr H.B. Williams and W.H. Dowdeswell. The arrangement of the plates was probably facilitated by the nature of the Ford collection, in which the layout was designed to 'demonstrate geographical variation and a number of other general principles'. He arranged these 31 plates of several hundred specimens with great ingenuity, contriving to include all the British species and their most important variations, but rather than doing this in the traditional taxonomic sequence, he laid them out under thematic headings: Butterflies of Woods and Rough Ground; Seasonal Forms and Geographical Races; The Re-introduction of the Large Copper; Sexual Abnormalities, and so on. Of particular interest is the plate of Historic Butterflies from the Dale collection, illustrating the first-caught specimens of a number of rare butterflies, including a 250-year-old Bath white. Each plate was photographed at approximately life-size by F.C. Pickering, engaged for the purpose by Adprint, using a large-format plate camera. 'We encountered numerous technical difficulties', said Ford in his preface, 'which he satisfactorily overcame.' What he did not mention was that the high quality of the colour printing was entirely at his own insistence. The proofs were appalling, and there was barely a single plate that had reproduced the colours and tones of butterflies exactly. When he complained, he was told that this was the best the printers could manage. Unimpressed, Ford purchased, at considerable expense, a set of the Royal Horticultural Society colour charts, which he cut up into squares and decorated the margins of the proofs with, indicating the exact colours he wanted. His comment to Adprint was that if the RHS could print the correct colours, so could they, that the technology to do so was obviously available, and that they should feel ashamed of themselves. He got his way. The plates of set specimens are easily the best examples of colour printing in the first half-dozen New Naturalists, and, fifty years on, they remain, in my opinion, the most comprehensive and interesting collection of British butterfly pictures ever made.

But plates of dead museum specimens were nothing unusual. It was part of the ethos of the New Naturalist library to illustrate *living* animals and insects in colour, and here the difficulties were much greater. The practical problems of obtaining and photographing specimens were onerous enough in the England of 1944, with limited supplies of petrol, even more limited colour film, and many of the best collecting grounds sealed off by the military. To attempt the task, Eric Hosking approached Samuel Beaufoy, then head of the Electrical Engineering Department at Ipswich School of Technology, who had, over a number of years, succeeded in photographing the life-histories of many of the British butterflies. All of his work, however, had been in monochrome. No one had yet succeeded in photographing living butterflies in colour. With the technology available, the difficulties were immense: the emulsion of the only available colour film, Kodachrome, was painfully slow for such lively objects as butterflies and there were additional problems of lighting and colour balance. No wonder Sam Beaufoy expressed due caution, especially since the colour producers, Adprint, were assuming that everything could be done in a single season, regardless of the biological facts of insect life-cycles. A year afterwards, W.A.R. Collins wrote that the work 'nearly broke Beaufoy's heart .... He started work and test after test was returned to him by Kodak as a complete failure.' The delay between photographing and processing the film proved yet another difficulty, since by the time the film had been processed and returned, a particular stage in a butterfly's life might be over for another year.

Sam Beaufoy told me in a letter that:

> 'The New Naturalist work "took over" all my leisure time, and involved help from my wife and young daughter in rearing the butterflies. Additionally, the only type of camera then available which fitted my technical requirements was the Kine-Exacta, using 35mm film. Mr Stemmer, of Adprint, was able to obtain the loan, for weekends only, of such a camera. During the week it was in constant use for medical photography at one of the London Teaching Hospitals. It was collected thence on Fridays and put on the train at Liverpool Street for Ipswich station, whence it was collected by me or my wife. Its return rail journey was made on Monday mornings.'

He eventually worked out a method which produced almost complete success by standardising the distance between light and subject, and determining the correct combination of filters and exposures. It had proved impossible to photograph butterflies out of doors to a satisfactory standard. Most of the living insects and their early stages were therefore reared from eggs or caterpillars collected from the field by Sam Beaufoy, or sent to him by Ford from the latter's own breeding cages. Each adult butterfly was placed carefully on its natural foodplant and photographed within a few hours of its emergence from the chrysalis, after its wings had dried but before it took flight. This is one thing to state, but quite another matter in practice. Some species were more co-operative than others as Beaufoy recalled:

> 'I have had a Painted Lady behave so well that the whole operation was over in five minutes. On another occasion, a sweltering summer's evening, in the closely-shuttered studio, I spent nearly three hours on one of the Blues. The

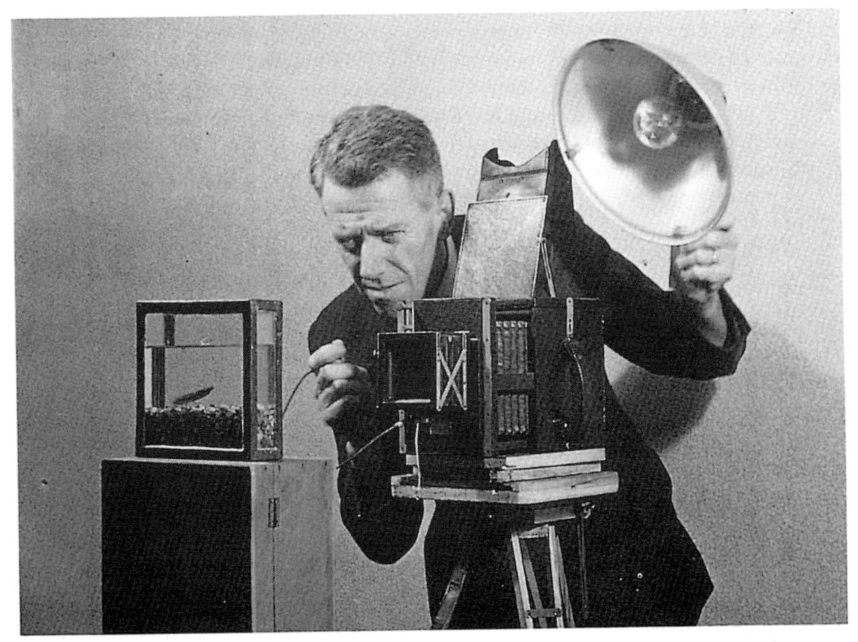

Sam Beaufoy photographing a dragonfly nymph using a plate camera and flash gun in 1947. (*Photo: S. Beaufoy*)

> specimen was the only one available; the photograph was urgent. Each time that it flew from its setting, it hid itself effectively in the oddest and most awkward corners, and finding it again meant an exhaustive search of the room each time. In the end, it did consent to pose long enough for me to take the photograph.'

Sam Beaufoy recalls that the luggage for his summer holiday in Devon that year was encumbered not only by heavy photographic equipment, but also by numerous breeding cages in which the earlier stages of some species of butterfly travelled a return distance of some 550 miles. Fortunately, the house had ample space in which he could set up his 'studio'. As a result of working together on the illustrations, Beaufoy and Ford became great friends, the former admiring Ford's unique approach to studying butterflies, the latter in thrall to Beaufoy's mastery of 'the photographic camera'. They sometimes met in Cornwall and Devon in the 1950s while Ford was studying clinal variation in the meadow brown, and Beaufoy joined in with enthusiasm in the research, contributing field studies of Lepidoptera and primroses, and reading several of Ford's books in manuscript for his opinion as an interested 'layman'.

Probably the last part of *Butterflies* to be written was the Preface, which, like so much of Ford's writing, is lofty, lucid and phrased with care. The key to the book lies in the fifth paragraph:

> 'I have written this book in the hope that it may be useful to scientific entomologists and biologists in general but, in addition, I have especially kept before my mind the needs of butterfly collectors and of all those who love the country. Perhaps it may increase their pleasure by widening the scope of their

interests. Many would, no doubt, wish to go no further than this, for there must be a large number of collectors and naturalists who have no intention of becoming amateur scientists. Indeed I should not wish all of them to do so; but I hope that some of them may, for they would add to their enjoyment. Accordingly, I have pointed out numerous interesting lines of experiment and observation which could be undertaken by anyone using the simplest means.'

Ford was not deliberately writing in a 'New Naturalist' style, for no such style yet existed, and he had evidently been given only the broadest and most generalised of instructions. The publishers were, in fact, exceedingly fortunate that his was the first book of the series to be printed. Not only were butterflies next to birds in popularity and appeal, but Ford was, in many ways, the archetype of the New Naturalist. He was an outdoors man and loved the countryside and its wildlife; he had reaped the benefit of the post-Darwin generation's work in evolution, ecology and behaviour; he was a leading exponent of combining laboratory science and field study; aided by his classical training, he wrote well (you would have to search hard to find an error of grammar in any of Ford's work); and he had the learning and imagination to see things in the round. In *Butterflies*, one can find Ford the historian ('deeply impressed as I am with the importance of the past in interpreting the present'); Ford the collector; Ford the leading geneticist; Ford the anatomist and physiologist; and Ford the scientific thinker: Ford among the grass stalks and bracken, and Ford beneath his study desk thinking about evolution. The book succeeds not only because of its wide learning and originality, but because, like all the best New Naturalist titles, it is a personal book and could not have been written by anyone else. If we treasure *Butterflies*, we do so not only because of its wisdom and insight, but because when Henry Ford tells us that he once caught one of Britain's rarest butterflies, he adds that he was 'tempted to bite into it to determine if it were unpalatable'.

The bestseller

What makes a bestseller? Presentation, promotion, preferably a television series, and serialisation in the popular Sundays, all go into the magic pot; one hopes that the quality of the writing may sometimes play a part, too; but the most important factor of all is good timing. *Butterflies* was launched at exactly the right moment, as demobbing had begun, and a nation weary of war was dreaming about butterflies. The greatest surprise was not the actual sales but the size of the pre-publication subscription list, which caused Billy Collins to double the in-

Table 6. The numbers of copies of Butterflies printed for the various editions are as follows:

First edition 1945	Fontana paperback 1975
Second edition 1946	
Third edition 1957	20,000
reprint 1962	20,000 [bound in 1946 and 1947]
reprint 1967	3,000
reprint 1971	2,000
New edition 1977	3,500
	2,500
Readers Union edition, 1977	2,500 [based on the Fontana text and
reprint 1977	black and white plates]

tended edition to 20,000 copies. At 16*s*, the book was quite expensive for those days, yet it sold out within a few months of its appearance. It was to remain in print for 35 years, and in that time sold about 53,000 copies in hardback, sales that would do credit to a successful novelist and far exceed most natural history titles today.

The success of *Butterflies* 'showed the world that there was a new wind blowing across the pastures of the British naturalists', as the Editors were to put it in their preface to *Moths*. Whatever the origin of the wind, it was certainly not evident in the reviews. In later years, the New Naturalist library attracted a number of loyal reviewers, among them Cyril Connolly, Geoffrey Grigson and Brian Vesey-Fitzgerald, and the books were reviewed in the quality newspapers and magazines as well as in scientific and naturalist journals. But this following took some time to build up and the few, rather bland, 'puffs' quoted on the flyleaf of later editions of *Butterflies* suggest a scraping of the barrel for this first book of the series.[6] I myself have trawled through dozens of microfilm copies of contemporary newspapers without finding more. I was also unable to find the review quoted from *Illustrated London News*, which in 1945 was little more than a picture gallery of bombed-out towns and war heroes. The *Yorkshire Post*, one of the best regional papers with a traditional interest in natural history, praised the way in which the book encouraged the naturalist 'to engage in investigational work'. Cyril Diver's more critical review in the *Journal of Ecology* might have disappointed the editors. Though properly appreciative of the ambitious scope of *Butterflies* and the lucidity of its writing, he rightly drew attention to the problem of writing about general issues in biology and evolution while being cramped for lack of good examples among our small number of native butterflies.

The textbook stuff on genetics and the butterfly natural history do, indeed, sit together rather uneasily: for half of the book butterflies are themselves the subject; for the other half they serve as examples. The key chapter of the book is the one on Evolution, but this can be understood fully only when the previous three on theoretical and practical genetics have been read and absorbed. Surely no more than a minority of readers stayed the course. Most naturalists and collectors would have been far more interested in the chapters on butterfly behaviour, distribution, geographical races and relations with other insects, all of which had something new to say. One skipped the 'dull bits', as one did the Agnes Wickfield chapters in *David Copperfield*. In so suggesting though, I may be doing less than justice to the thousands of people who read *Butterflies* in 1945. Attention spans seem to have been longer then, and the war had created a cultural revolution among intelligent people of all classes which bore fruit in the immediate postwar years. Perhaps Ford's was one of those rare books that satisfied nearly everybody. As Miriam Rothschild once exclaimed, 'E.B. Ford is the author of the best book ever written about butterflies – a redoubtable achievement, for there is general agreement about this – and to weld entomologists into a coherent body must in itself be something of a *tour de force*.' Ford himself was astonished at its success. He was proud of the book, though he might not have agreed with his biographer, Bryan Clarke, that it was the best thing he ever wrote.

It might interest readers to learn how much Ford earned from *Butterflies*; I, for one, had assumed that it was a great deal. His contract, in the standard format specially designed for the New Naturalist series, stipulated £400 for the first 10,000 copies printed, half on acceptance and half on publication. Payment on the second 10,000 copies was at £40 per thousand copies sold, and thereafter at a ten per cent royalty. I estimate that he would have earned about £1,000 in the first year, and thereafter perhaps £75 to £150 per year. The former was quite a respectable sum in 1945, and was considered by the editors to be a generous rate. But it was not exactly a fortune, and the financial inducement to write a New Naturalist title, never very strong, grew weaker as the sales declined. Most of the books were necessarily written in the authors' spare time, or in retirement. That is the main reason why the series proceeded at the sedate rate of only two new titles per year from 1945 to 1949, and why there was a sudden glut of completed manuscripts in the early 1950s, when everyone finished together.

The second (slightly revised) edition of *Butterflies* in 1946 was, like the first, of 20,000 copies, and this time Collins did not run out of stock until 1956. Thereafter, with greatly reduced sales affecting every title in the series, the title was reprinted in much smaller numbers, while the price rose from a guinea to 35*s* in 1962 and to 45*s* in 1967.

While there were minor revisions made to the second and third editions, Ford was not allowed to revise the text thoroughly until the early 1970s, when the new setting for the Fontana paperback enabled him to do so. Indeed, that may be the reason why he allowed a paperback edition at all, for up until then he had resisted it. The paperback was an ill-starred enterprise. Ford explained that, 'I have long since ceased to be an entomologist in the ordinary sense, having years ago passed over to invent and develop the science of Ecological Genetics. Thus British butterflies have long passed outside the scope of anything I am concerned with or know about.' What made the job a peculiarly depressing one was the exclusion of the colour plates, which meant that the whole text had to be gone through to eliminate all references to the original plates and substitute for the monochrome plates new ones taken by Ford's technical assistant, John S. Haywood. It was a labour of Sisyphus. 'It is utterly miserable', wrote Ford, 'going through this book which has become famous and has been an outstanding success, and cut it about in this wretched way. I hate doing it, and I feel, *and know*, that all this work I am doing is merely to turn something good into something bad.'

The paperback was eventually published in 1975, after numerous complications and misunderstandings, but it sold poorly. The Fontana paperback series had had a brief period of success in the late 1960s, but the sales had declined, and *Butterflies* was one of the last titles to be reprinted in that form. The correspondence between publisher and author grew acrimonious after the former had failed to answer two queries about royalties. One can imagine Ford's eyebrows climbing above his spectacles with astonishment and indignation. He wrote: 'That I should be so treated by a Publisher has of course given rise to surprise in Oxford.'

The last edition of *Butterflies* in the traditional New Naturalist format was published in 1977, without any of the colour plates that had once been its crowning glory, and at an inflated price of £8. By then, the blocks from which the plates were made were badly worn – through an oversight, no duplicate plates had ever been made – and the original colour slides were by now dirty and scratched. New colour blocks could not be made without raising the price beyond the pockets of the average book buyer. In retrospect, it would have been better to declare the book out of print. But Collins wanted to keep this book in print, not out of hope of further profit (for there was none in reprinting the older titles) but to keep a famous and still useful text before the public. Unfortunately, they neglected to warn the reader of the change, and kept the original preface which laid such emphasis on the now non-existent colour plates. Ford's reaction reflected that of many readers: 'What a poor thing it looks .... The statement in the original Editors' Preface about the use of colour plates has rightly caused amusement and contempt when read in conjunction with the last edition, containing no colour plates at all.' Numerous purchasers returned their copies to the bookshop with expressions of anger at having been, as they thought, taken in by Collins. It was a sad finale to one of the natural history classics of the twentieth century. Collins' then natural history editor, Robert MacDonald, admitted that the monochrome edition had been a mistake: 'I think the lesson is, either to leave the books out of print or to reprint them with the colour at whatever price is necessary.' *Butterflies* was declared out of print in 1983, having exceeded the sales of all the case-bound New Naturalist books, with the sole exception of *Britain's Structure and Scenery*.

Moths

The second New Naturalist book by E.B. Ford was intended to expand, as he put it, 'certain of the concepts laid down in *Butterflies* and applying them, by way of illustration, to moths'. He had hoped to make a start soon after completing *Butterflies* and his contract for *Moths*, dated 23 February 1945, was for the delivery of a book by the end of the following year. But the post-war years were a particularly busy time for Ford, and it was not until 1951 that he was able to get

Cothill Fen, near Oxford, where for many years E.B. Ford studied the genetics of the scarlet tiger moth. (*Photo: Peter Wakely/English Nature*)

The printer's proof (*left*), pencil and colour sketches (*below*) of the *Moths* jacket. (*Private collection*)

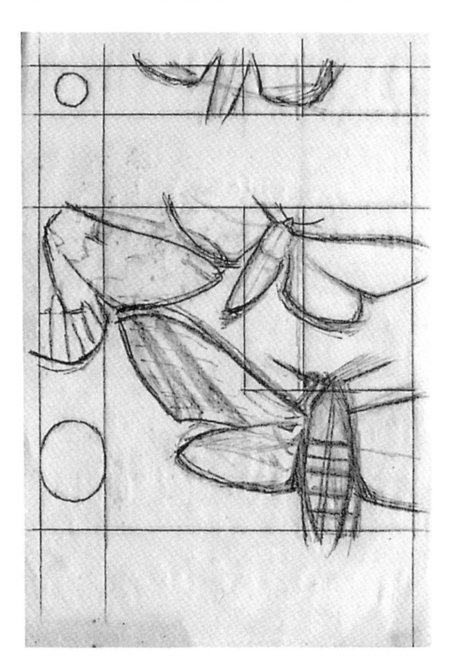

on with the sequel. The manuscript (which Ford evidently typed himself) was completed in late 1953, and was read by both Huxley and James Fisher who pronounced it 'very good, similar in treatment to *Butterflies*'. Because of printing delays, the book was not in the shops until February 1955. *Moths* had been eagerly awaited by those who had read and enjoyed the earlier volume, and this time the book was widely and enthusiastically reviewed. It was seen, rightly, as a companion volume that showed, as Miriam Rothschild expressed it in the *Sunday Times*, 'how we can graduate from being a mere collector to being a scientist – without losing any of the thrill and the pleasure of collecting in the field', a view echoed in a detailed review in the *Times Literary Supplement* and by John Moore in *The Observer*.

The illustrations were in similar format, all of them taken, with great skill, by Sam Beaufoy, including the plates of mounted specimens, arranged by Ford from his own collection and from those of his department and of colleagues. They had photographed 48 colour plates, as with *Butterflies*, but increases in the cost of colour printing had by now reduced the original allocation to 32. Although improvements in film emulsion and close-up equipment had made insect photography accessible to a growing band of naturalists, Beaufoy achieved at least one more break-through in producing the first pictures ever taken of the life-stages of the waved black moth, an aberrant little insect whose caterpillars live on fungi and whose pupae are slung like hammocks. It was unfortunate therefore that the latter, unique, picture was the victim of the only transposed caption in the second edition, which rather took the skin off Ford's delight in finding a virtually error-free text.

Moths was never the commercial success that *Butterflies* was, nor was it ever likely to be, given the lesser appeal of its subject. But there were other reasons. After an introductory chapter on the anatomy and physiology of moths, which was already fairly tough going, Ford launched straight into four 'heavy' chapters on theoretical and practical genetics which, since he took a knowledge of the material in *Butterflies* for granted, tended to be more 'difficult' than the earlier work. The later chapters reverted to traditional subjects, but overall something of the fresh air and sunlight of *Butterflies* was missing. One problem Ford could do nothing about was that there were too many moths, with the consequence that their use as examples, rather than subjects, was intensified. So keen was Ford to write about his own genetics research that he even brought in a lengthy section on the meadow brown butterfly. Even so, *Moths* is, overall, a worthy companion to *Butterflies*, written in the same lucid, readable style and full of rainbow flashes of insight into the evolution and behaviour of these secret insects of the night. There is a brief account, of which he might have made more, of Ford's ascents in an RAF barrage balloon with Alister Hardy to determine whether or not moths fly towards the moon. In terms of new information, such as the sections on extinct moths and local races, this book perhaps outweighs its predecessor, and, like it, *Moths* remains a useful book, both for serious study and for enlightening one's forays with a lantern and net. The first edition of 8,500 sold rather slowly – *Moths* was one of the slow-but-steady titles of the series – and by the time it went out of print in 1981, it had been through three editions and sold about 14,000 copies. Whether it might have sold better had more of the book been concerned with the life and habitats of moths is an open question.

E.B. Ford: the later years

Since we have followed Ford's scientific career in some detail up to the publication of his best-known book, let us now round off the story by detailing, necessarily more briefly, the latter part of his life. Shortly after *Butterflies* was published, the originality and importance of Ford's work was recognised by the academic world with a series of honours. He was elected President of the Genetical Society of Great Britain and a Fellow of the Royal Society in 1946, and in 1954 the latter awarded him its Darwin Medal. In 1958, his name was proposed by Miriam Rothschild for the first scientific fellowship of All Souls College. Ford was duly elected, and thereafter he was jokingly referred to as 'Mrs Lane's fellow'. All Souls, with its magnificent library and dining facilities and, at that time, masculine ethos, was at the centre of Ford's life from then on: he served his college as Dean, where his responsibilities included reading the sermon in the college chapel. In 1977, he received the singular honour of being elected Distinguished Fellow and Senior Dean.

In 1964, his life work *Ecological Genetics* was published, the magnum opus he had been preparing for more than thirty years. Though the book created a great surge of interest in the subject, many of its themes had already been rehearsed in *Butterflies* and *Moths*: the collecting days with his father; the August camps on Tean with 'Bunny' Dowdeswell; 'tiger-hunting' in Cothill Marsh with Ronald Fisher and Philip Sheppard. By this time, Ford had his own small department of genetics at Oxford, for which a specially designed laboratory had been built in the 1950s – ironically, on a patch of waste ground, known as 'Henry's weed garden', on which he had monitored caterpillars. He was appointed Professor of Ecological Genetics in 1963, not so much an administrative position as a recognition of his achievement in founding a new scientific discipline with a band of gifted disciples, several of whom were to be appointed to university chairs of their own. As a scientist, Ford had risen to the top of the tree, and been honoured in the way he would have wished, by the Royal Society and by Oxford. His ungenerous country failed to follow their example with a knighthood, as it should have done.

During the 1940s and 1950s, E.B. Ford was much involved with the development of nature conservation in Britain. He served on the Wildlife Conservation Special Committee under Julian Huxley and Arthur Tansley, whose famous White Paper 'Cmd. 7122' led to the founding by Royal Charter of the Nature Conservancy in 1949. He joined the board of the Conservancy and served there for ten years, longer than any of his scientific peers, with the exception of J.A. Steers and W.H. Pearsall. He was a valued member not only for his contacts and influence but for his unrivalled knowledge of off-shore islands and insect localities. Ford's interest in nature conservation had grown as he watched so many of the best butterflying and mothing grounds disappear under the plough, both during and after the war. He approved of, and helped to establish, nature reserves, including his scarlet tiger research area at Cothill Marsh. He also took part in the Nature Conservancy's administrative work, which included interviewing candidates for research Fellowships and scientific posts. Derek Ratcliffe recalls his own ordeal in front of three Fellows of the Royal Society, including Ford, whom he remembers staring at him, beady-eyed, and occasionally firing off staccato questions. When Derek admitted that he had switched from zoology to botany out of boredom, one of the other Fellows turned to Ford and jokingly remarked: 'Yet another deserter, Henry!', to which Ford answered with a snort of disgust.

E.B. Ford, Julian Huxley and John Baker in Bernard Kettlewell's garden, 1971. (*Photo: John S. Haywood*)

Ford never really retired. Though he ceased his formal duties as professor in 1969, he retained an office in the zoology department which he visited almost daily. His latter days were filled with travel – he was invited to supervise the design of genetics laboratories in many countries, including Finland, Jordan, Canada and France. He continued to study Lepidoptera, and among his last projects was an investigation with Sir Cyril Clarke of the unusual genetics of the gipsy moth. He wrote several more books and pursued his antiquarian hobbies of field archaeology, heraldry and the exploration of medieval churches. The latter formed the subject matter of his last book, *Church Treasures in the Oxford District* (1984), where each 'treasure' is described with the same meticulous attention to detail that he devoted to his scientific observations. Once, while taking notes on the interior of Cumnor Church, he and John Haywood heard the crunch of footsteps on the gravel path outside which came to a halt by the porch. They assumed that it was the old man who locked up the church each night – 'We'd better leave, Henry, before we are locked in.' But on stepping outside, they found no one. Ford's scientific training rose to the occasion: 'I think we ought to write this down, right away.' Evidently they failed to do so, for Ford makes no mention of the episode in *Church Treasures*. Haywood tells me that instead they headed straight for the nearest public house bar to steady their nerves.

In his last years, Ford returned to butterfly collecting, probably to give him something to do on his walks around Oxford as much as anything else. He was one of those people who never like to waste time: strolling along country lanes just for the pleasure of it was not his style. I have had the privilege of being shown a new collection of butterflies by E.B. Ford, made in the final decade of his long life, each one set with mathematical precision, of a piece with the rest of his work. Sad to say, the earlier collection from which Ford selected specimens to illustrate *Butterflies* and *Moths* has been destroyed. John Haywood told me that the cabinets containing the collection were kept in Ford's attic, and were hardly looked at from one year to the next. When, too late and after Ford's death, the cabinets were inspected, almost the entire collection had been devoured by mites and beetles. In view of the part they played in Ford's work, they are a loss to science; Ford, on the other hand, was always willing to sacrifice specimens for scientific purposes, and also, it is said, made a habit of destroying such material after it had served its immediate purpose. He was not, it seems, a sentimentalist.

It was only in the last year or two of his life that Ford was much bothered by serious infirmity, although he suffered from asthma and had recovered from a heart attack in 1971. As late as the mid-1980s, he and John Haywood were planning another Church Treasures book, this time on the Cotswolds where Ford had lived and which had long been his favourite corner of England. He hoped to see the century out and to write a book about evolution and extinction. In about 1985, his housekeeper, to whom he had been devoted, died, and he also lost some treasured possessions in a burglary. He was never quite the same again. E.B. Ford died on 22 January 1988 at the age of 86. As he had requested, his body was cremated, and the ashes scattered on a stretch of downland hillside near Birdlip, where chalkhill blues and brown arguses dance among the summer flowers.

The originality and influence of E.B. Ford as one of the century's most eminent geneticists are beyond dispute, notwithstanding the fact that not all of his work has withstood the test of time. As a scientist, he must have seemed like someone from an earlier age, perhaps a clergyman-naturalist like Ray or Gilbert White. What is one to make of him? Bryan Clarke, who knew him well, confesses to feeling a mixture of admiration and infuriating annoyance. While his work was always original, brilliant and well-written, it was also often parochial to an extreme degree. Ford was a paid-up member of Oxford's mutual admiration society; the only Cambridge man he ever paid much attention to was R.A. Fisher, and as for the work of our cousins across the Atlantic, he barely acknowledged its existence in the pages of *Ecological Genetics*. So far as the New Naturalist books are concerned, that is at once his strength and weakness. They are not so much natural history books in any generalised sense, as a record of the work of E.B. Ford and his closest colleagues. At the same time, it was because they were written from first-hand knowledge, and because that knowledge was so wide, that these books are brimming with fresh insights for anyone prepared to accept the challenge of the new scientific natural history.

Of Ford the man, opinions are almost as many as witnesses. That many of his students and colleagues did not enjoy being ignored by him, or given a hostile glare as they arrived late for a lecture, or enduring his sarcasms, is evident; others remember a kindly man, whose supercilious look and veneer of biting wit could not disguise his fundamental good nature. He was certainly unusual. From several of those who knew him well, I have heard the phrase: 'You never forgot Henry Ford.' Nor shall we. I end by quoting part of two tributes, the first by Michael Majerus, a geneticist and the author of the preceding volume in this series, and the second by the scientist who perhaps understood him best, Miriam Rothschild.

> 'In 1964, for my tenth birthday, I was given a copy of Ford's *Butterflies*. I used my pocket money to buy the companion volume *Moths* and these two books undoubtedly influenced the rest of my life. From that summer I began to do more than make a child's haphazard collection of butterflies and moths. I began to run a moth trap, and to record religiously, all species taken, forms, when

I could identify them, notes on behaviour, and finally I began trying to rear broods from pairings between different forms, following the advice given in Ford's books. My own experience is testament to the truth of Ford's contention that the basic elements of Mendelian genetics can be understood by a child of eleven in an afternoon and thereafter applied (although I was only ten).

'I think the likely course of my career was laid down with the gift of that book. My interest in polymorphic Lepidoptera was certainly conceived in 1964, and has persisted now for 25 years. I read Ford's *Ecological Genetics* in 1972, and the idea of working on green versus brown lepidoptera larvae for a PhD came from that reading and subsequent delvings into the bibliography. Although I never met Ford or Kettlewell (I did hear both of them speak on several occasions, but in those days I was a shy retiring youth), I think it is true to say that Ford influenced my interest very considerably.'

Michael Majerus. In: Berry, R. J. *Industrial Melanism and Peppered Moths*, (1990) ibid.

'It is sad that the English language lacks the French adjective *genial*, for this word could have been coined to describe Henry's versatile mind and original ideas, coupled with his incisive attention to detail and persistent pursuit of a fruitful line of research. Furthermore he has attracted innumerable students – professionals as well as amateurs – into the butterfly field, recognising how the aesthetic appeal of these matchless insects, which caught our imagination as children, and are associated with the nostalgia of the golden age, can provide us, now we are older and possibly wiser, with one of the best biological tools ever invented. It is not only a rewarding field in itself, but one which carries with it a love of natural history and ageless delight. I will always be grateful to Henry Ford for suddenly endowing my fleas with wings.'

Miriam Rothschild. Dedication: Henry Ford and Butterflies in *The Biology of Butterflies* (1984)

REFERENCES

Much of this chapter draws on the memories of surviving friends and colleagues of E.B. Ford mentioned in the acknowledgements and from unpublished memorials by Bryan Wilson and Peter Placito. I have also inspected that section of the correspondence of E.B. Ford held in the Bodleian Library which refers to Collins publishers and to *Butterflies* and *Moths*. The following are published references mentioned or alluded to in the text.

Berry, R.J. (1990) Industrial melanism and peppered moths (*Biston betularia* L.. *Biol. J. Linn. Soc.* ***39***: 301-322. Contains anecdotes about Ford and Kettlewell, and the appreciation of *Butterflies* by M. Majerus.

Creed, R. (ed.) (1971) *Ecological Genetics and Evolution. Essays in honour of E.B. Ford.* Blackwell Scientific Publications, Oxford. Includes a full Ford bibliography compiled by Miriam Rothschild and H.B.D. Kettlewell and a foreword by Professor J.W.S. Pringle.

Diver, Cyril. (1945) 'A Faunal Monograph' (review of *Butterflies* by E.B. Ford). *J. Ecol.* ***33***: 204-205.

Ford, E.B. & Huxley, J.S. (1927) Mendelian genes and rates of development in *Gammarus chevreuxi. Br. Journal of Experimental Biology,* ***5***: 112-134.

Ford, H.D. & Ford, E.B. (1930) Fluctuation in numbers and its influence on variation

in *Melitaea aurinia. Trans. Royal Ent. Soc. of London*, ***78***: 345-351.
Ford, E.B. (1940) Polymorphism and taxonomy. In: J. Huxley (ed.) *The New Systematics*. OUP, 493-513.
Ford, E.B. (1979) *Understanding Genetics*. Faber and Faber: London.
Ford, E.B. (1980) Some recollections pertaining to the evolutionary synthesis. *In*: Mayre, E. & Provine, W.B. (eds.) *The Evolutionary Synthesis: Perspectives on the Unification of Biology*. Harvard University Press. A fascinating account of his life and work by Ford himself, based on a 1974 lecture.
Ford, E.B. (1981) *Taking Genetics into the Countryside*. Weidenfeld and Nicholson: London.
Ford, E.B. & Haywood, J.S. (1984) *Church Treasures in the Oxford District*. Alan Sutton: Gloucester.
Ford, E.B. (1989) Scientific Work by Sir Julian Huxley FRS. *In*: Keynes, M. & Harrison, G.A. (eds.) *Evolution Studies. A centenary celebration of the life of Julian Huxley*. Macmillan: Basingstoke (for the Eugenics Society).
Rothschild, Miriam. 1984. *Henry Ford and Butterflies*. In: Vane-Wright, R. I. and Ackery, P. R. (eds.) *The Biology of Butterflies*. Royal Entomological Society symposium No. 11, Academic Press: London, xxii-xxiii.

Obituaries

Clarke, Bryan C. (in prep.) Edmund Brisco Ford. 1901-1988. *Biographical Memoirs of the Royal Society.*
Clarke, Cyril. (1988) In *Nature* **332**: 20.
Jones, D.A. (1988) In *Tree* **3**:115-116.
Oxford Today, Vol.**1**:54
The Times, January 23 1988.

Full Bibliography of books by E.B. Ford

Mendelism and Evolution. (1931, 8th ed. 1965) Methuen: London. Available in paperback; translated into Spanish.
Mimicry. (1933) (with G.D.H. Carpenter). Methuen: London. Translated into Spanish 1949.
The Study of Heredity. (1938) Thornton, Butterworth: London. 2nd ed. 1950. Oxford University Press.
Genetics for Medical Students. (1942; 7th ed. 1973). Chapman and Hall: London. Translated into Italian 1948.
Butterflies. (1945; 3rd ed. 1977). Collins New Naturalist: London. Fontana paperback ed. 1975.
British Butterflies. (1951) King Penguin Books: Harmondsworth.
Moths. (1955; 3rd ed. 1972) Collins New Naturalist: London.
Ecological Genetics. (1964; 4th enlarged ed. 1975). Chapman and Hall: London. Translated into Polish, French and Italian.
Genetic Polymorphism. (1965) Faber and Faber: London.
Evolution Studied by Observation and Experiment. (1973; 2nd enlarged ed. 197?). Oxford Biology Readers, Oxford University Press.
Genetics and Adaptation. (1976) Institute of Biology Studies. Edward Arnold: London. Also in paperback.
Understanding Genetics. (1979) Faber and Faber: London.
Taking Genetics into the Countryside. (1981) Weidenfeld and Nicolson: London.
Church Treasures in the Oxford District. (1984) (with J.S. Haywood). Alan Sutton: Gloucester.

6

A Question of Style

Reviewing the New Naturalist library in *The Sunday Times* on 8 March 1953, Cyril Connolly wrote: 'The New Naturalist .... is a most interesting enterprise. Unlike so many popularising ventures, it consists almost entirely of the work of specialists, and of specialists who seem to be writing for advanced students rather than for the general public. Much consequently depends on the felicity of style if such books are to break through and become general reading.' As successful examples of the latter, Connolly cited Miriam Rothschild's *Fleas, Flukes and Cuckoos* and James Fisher's *The Fulmar*. Though many might agree with him about the former title, I doubt that I am alone in agreeing with Maurice Burton that *The Fulmar* is 'so emphatic and precise that nothing is left to the reader's imagination'; and that Fisher should have employed his worst enemy to prune it.

The first dozen or so New Naturalists evidently did 'break through and become general reading', judging from their sales, whereas we are told that some of the later titles were bought mostly by university students. It is hard to think of any comparable books today that succeeded so well in their broad appeal. Part of the reason must have been that these early classics were so readable. As we have seen, the original editors of the series were notably successful in persuading people of the highest calibre to write for it, and the marriage of subject and author was also singularly happy. Many New Naturalist authors of that time had a natural bent for writing: Ford, for example, with his classical training, or Yonge with his journalistic leanings, or Harrison Matthews with his natural gifts as a storyteller. The subjects were also well chosen. None were over-specialised: each book covered a broad or at least a popular subject, and the eclectic mixture of titles avoided undue repetition or a formulaic approach. Even *A Country Parish* represented an approach familiar to English naturalists since the eighteenth century.

Equally importantly, the timing was just right. Few of these early books could have been written before the Second World War, or at least not in quite the same way. Nor could they have been written much later, for in each case the subject soon grew too large to treat with the same totality and broad scale. One no longer (unless one is exceptionally gifted) studies the beach, but magnifies the pebbles. Compare E.B. Ford's *Butterflies*, which is really a book about environmental genetics using butterflies as examples, with R.J. Berry's book, *Inheritance and Natural History*, written a generation later. Both are very well written, and information is presented to the reader as simply as possible without being *over*-simplified. But while *Butterflies* is based very largely on the work of one man, the author, and on simple field-based experiments allied to natural history, *Inheritance* knits together much more impenetrable subjects like biochemistry, cell biology and applied mathematics. It is no criticism of *Inheritance* to say that it is not as readable as *Butterflies*. It could not possibly be so if the author was to remain true to his aim of explaining the importance of inherited properties in understanding ecological problems. It is an important book and a unifying book; it would be expecting too much for it to be an easy book as well. A carefully selected reference list in

Inheritance runs to 28 micro-printed pages. When E.B. Ford started to study genetics in wild populations, the entire world literature on the subject could be accommodated on a couple of book shelves.

In 1945, the frontiers of science were still close enough for people like Ford, Yonge and Imms to see from horizon to horizon. The latest findings of biology and the earth sciences could be made intelligible, in skilled hands, to the average butterfly collector or wildfowler. 'Skilled hands' is important. So many of the early New Naturalist authors were excellent teachers, and could write about their subject with a simple, unforced enthusiasm without resort to jargon. They also, whether deliberately or not, allowed their personalities to shine through the writing. The dour formality of scientific journals was, for the moment, cast aside. It is not difficult, when reading about these men and women, to see why they wrote as they did. These postwar writers saw it as their duty to popularise their subject. It accorded with the spirit of the age, and with their own inclinations too. Almost every writer for this series was a field naturalist at heart. There are vivid memories of Pearsall striding along the high tops of the Pennines and the Lake District, of Ramsbottom emerging for excursions still dressed in his formal city clothes, of Macan out on his yacht with his bottles and nets. There was as yet no unbridgeable divide between field natural history and scientific discovery. People of all classes retained the capacity for wonder, and most of the first dozen New Naturalists sold in their tens of thousands.

R.J. Berry and Orkney vole, being filmed for Granada's *Evolution* series in 1981. (*Photo: R.J. Berry/Granada TV*)

One should be careful not to generalise too far, however. The books were written by individuals and in widely differing styles. Few lend themselves to 'soundbite' quotation; you cannot readily detach a paragraph to exhibit as the essential Dudley Stamp or undiluted Yonge. They are, by and large, quietly written and not self-consciously 'literary'. To remind ourselves of the disparity of subject and style covered by this series, let us savour briefly each of the first dozen titles (apart from *Butterflies* which has had a fair airing already), which, as it happens, include most of the books that were written and published during the 1940s. Collectively they may tell us something about the whole series; individually they are as distinct in style and approach as they are in subject matter.

In terms of hardback sales, the most successful book in the whole series is *Britain's Structure and Scenery* by Dudley Stamp (though it was overtaken in Fontana paperback by *The Highlands and Islands*). 'BS & S' was used widely in schools and introductory college courses, and probably owed much of its success to Stamp's reputation as a writer of text books. And a text book is what it pretty much is, though it tells a definite story with a nice emphasis on field observation and a convincing display of the author's extraordinary breadth of knowledge. By combining the separate disciplines of physical geography, geology, soil science and botany (and personal acquaintance with every part of the British Isles), Stamp presents a synthesis that is almost *de rigueur* now, but was well ahead of its time then. Its main thrust was geological, telling (as *The Observer* put it) of 'how our landscape heaved, split, settled and was pared down into countless varieties of beauty, packed within our tiny mileage'. The book remained a classic for at least 30 years, and in the preface to the softback edition in 1984, Professor Clayton described it as still useful, despite having become out of date on some subjects. Stamp seems to have written it with his usual briskness during spare moments in 1944 and 1945. The subject and its series struck a chord with him, and Clayton judged this book 'one of the very best from this prolific author'.

British Game, published at about the same time, forms an interesting contrast: the opinionated editor behind his desk at *The Field* opposed to the urbane university teacher with his diagrams and slides. Brian Vesey-Fitzgerald could be described as a professional countryman, one of the last of the literary sportsman-naturalists, though his sympathies were as least as much with the poacher as with the gamekeeper. He saw things through a countryman's eyes, and wrote *British Game* in the same style as his articles in *The Field* and his radio *Field Fare* broadcasts. He might almost have dictated it, so lively, conversational, and prejudiced is the book, as if Fitzgerald was at your elbow, in cloth cap, pipe and shooting garb. This was an appropriate style if the book was aimed at the readers of *The Field* and *Country Life*, but it sits a little oddly within a scientific series. Fitzgerald's hero was Abel Chapman, the Victorian punt-gunner, and his natural history was that of the sportsman and gamekeeper, not the scientist, whom he seems rather to have despised. *British Game* was probably not specifically chosen to be book number 2 in the series; the title was originally listed as a 'special subject'. But Brian Vesey-Fitzgerald was a professional writer and, unlike most New Naturalist authors, he finished on time. *British Game* was a successful book and is now, oddly enough, about the commonest New Naturalist title in second-hand bookshops.

Richard Fitter, author of *London's Natural History*, in 1994. (*Photo: Anna Fitter*)

About *London's Natural History* no such reservations are needed. It was, as Vesey-Fitzgerald himself noted, 'the first real natural history of a great city, the development of a large urbanised area traced through the history of its wildlife'. The book attracted more attention at the time than its better-known companion, *Butterflies*, and has influenced countless urban ecologists ever since in its judicious blend of history and wildlife. For a first book, *London's Natural History* was remarkably confident and accomplished. Richard Fitter had become experienced at marshalling facts and figures as a writer of government reports.

He had also accumulated a very large 'database' on London wildlife as a leading member of the London Natural History Society. It was a book he was therefore all geared up to write, and the invitation from his friend and BTO colleague, James Fisher, to contribute to the series must have seemed God-given. In a recent interview, Richard Fitter acknowledged a debt to a book called *The Influence of Man on Animal Life in Scotland* by Professor James Ritchie, which gave him the idea of analysing the influence of man's various activities on wildlife – digging, traffic, refuse disposal, smoke, sport etc. – against the panoramic backcloth of Greater London. This was a much more interesting and rewarding approach than 'starting at the mammals and going down to the insects, which is another way it might have been done'. And an approach, moreover, in keeping with the new vision of ecology. It was in other ways a book of its time. Surely only a book written in wartime would have devoted one of its most fascinating sections to the influence of bombs and explosives! This section is full of the sort of particular, intriguing information that characterises the book as a whole. We are told that the *Luftwaffe* killed remarkably few birds or beasts, and even trees often escaped serious damage: 'A horse-chestnut tree in Camberwell that was stripped of almost all its leaves in July was in full bloom again in September, and at Reigate lilac bloomed again and a creeper came into full leaf after its first crop had been blown into the house.' In their preface, the editors appeared to miss the point of the book in regarding the 'progressive biological sterilisation' of London as a sad story. Charles Elton was closer to the mark in his review, encapsulating the book as 'the history of changing equilibrium between wildlife and man, and in some of the curious habitats and communities and changed habits that are to be found in the new conditions of this vast temperate desert'. *London's Natural History* should not be judged as the editor's 'gloomy reading' but as a celebration of the resilience and resourcefulness of Nature.

If Fitter's was an ecological portrait of a city, *A Country Parish* was A.W. Boyd's 'natural history of an ordinary country parish'. The book was James Fisher's idea; he had edited a scholarly edition of *The Natural History of Selborne* and wanted this tradition to be reflected in the New Naturalist library. And who better to write it than his uncle and mentor Arnold Boyd, a natural history all-rounder and one of the leading amateur ornithologists of the first half of the century? Boyd had lived all his life in Cheshire, and from 1920 in the parish of Great Budworth on the Cheshire Plain with its charming village, meres and strips of woodland known as 'bongs'. On sending the manuscript to Collins, he remarked with characteristic modesty that, 'I hope it will not severely let down the high standard of several of the earlier New Naturalist books; rather a fond wish, I fear.' It did not let them down. *A Country Parish* has always been, it seems to me, one of the underrated books of the series, partly because it was allowed to go out of print in the 1950s and is difficult to

Arnold W. Boyd (1885-1959), author of *A Country Parish*, on Hilbre Island, 1956. (*Photo: Eric Hosking*)

obtain. Boyd based the book on his long-running weekly Country Diary in the *Manchester Guardian*, some 500 entries from which were collected together as *The Country Diary of a Cheshire Man*, published by Collins in 1946 (and incidentally advertising the New Naturalist series on its dust jacket). Like *Selborne*, *A Country Parish* is an 'exploration of individuality', and like *London's Natural History* it is about man as much as about wildlife. Unlike the latter, however, *A Country Parish* keeps man and nature apart. The first half of the book is a social history, charmingly described, with loving attention to local customs and dialect (there is a whole chapter on the local Mummer's Play). The honest, down-to-earth style is characteristic of Boyd, but it says something about country life too. It is full of little stories like the following:

> 'A footpath runs along the side of Great Budworth churchyard, where there was a newly dug grave just over the wall. A practical joker, so it is said, knowing that a man would pass that way late at night, lay down in the empty grave and, as his friend approached, began to moan: "Eh! it is cowd down 'ere, eh! I am cowd." The passer-by looked into the grave ... and called out: "I don't wonder th'art cowd; they'n none covered thee up," and started to shovel sand over the imagined corpse.'
>
> *A Country Parish*, Ch. 9 Folk-lore

The second half of *A Country Parish* adheres rather too closely to the 'Country Diary' format, dealing with the birds, plants and butterflies in the old-fashioned way, species by species. That might have concerned the New Naturalist editors more than it should bother us. This is a delightful portrait of a country parish caught just before postwar change eroded the rural traditions that had made every country village an individual and special place. Perhaps its time is now, rather than then. Collins received a number of requests to reprint it during the 1960s and 1970s; one day, perhaps.

From the Brueghel-esque scenes of *A Country Parish* we pass to the Landseer landscapes of Frank Fraser Darling's *Natural History in the Highlands and Islands*. This was an important book, and one which perhaps few others at the time would dare to have attempted. That, in its original form, the book was rather a curate's egg, was not surprising, given Darling's wartime isolation in western Scotland. It is a very personal book, with elements of his natural history journals, and some of his earlier work on deer, seabirds and seals, thrown into the mix, together with the notes for his later ecological study, *West Highland Survey* (1955). Parts of the book are scrappily written while others are possessed of unusual lyric intensity:

> 'To climb to one of these alps of grass and descend again in a few hours is not enough. Take a little tent..... The only sounds breaking the silence, if you get the best of the early July weather, will be the grackle of the ptarmigan, the flute-like pipe of the ring ouzel, perhaps the plaint of a golden plover or a dotterel and the bark of the golden eagle. These are good sounds and do not disturb what is for the moment a place of peace. See how the deer, now bright-red-coated, lie at ease in the alpine grassland. Listen, if you have stalked near enough, to the sweet talking of the calves who are like happy children. Here is new herbage over which no other muzzles have grazed; the very soil has been washed by fifty inches of rain since the deer were here before, in November. Of this short nutritious grass the deer are growing their clean bone and the good condition which will help them to face the winter. Of your charity dis-

turb them not in their Arcadia.'

Natural History in the Highlands and Islands, Ch. 8 The Summits of the Hills

That particular passage has stuck in my mind ever since I first read it, and especially that last Roman injunction. There is a sweet melancholy about Fraser Darling at his best that, like the best wine, is balanced by sharp phrasing and firm scientific definition. This is good science still; only so imaginatively expressed that it is easy to imagine yourself on that mountain top among the peaceful sounds of nature. Darling had had plenty of time on his own among the hills in which to sharpen his quill.

Natural History in the Highlands and Islands received the best known and most bitterly disappointing book review in the history of the series. Fraser Darling had returned from an expedition to St Kilda with James Fisher, an idyll spoiled only by Darling's appalling seasickness, to find the latest copy of *The Scottish Naturalist* among his mail. It included a scathing three-page review of the book by Professor V.C. Wynne-Edwards that criticised not only the author's own lapses but also the New Naturalist editors for daring to claim that 'every care has been taken .... to ensure the scientific accuracy of factual statements'. As other reviews also pointed out, the botanical section of the book was indeed riddled with 'half-truths and errors'. But none of them rubbed them in with quite the same abrasive skill:

> 'The yarn handed out about the migration of the herring was the latest thing in the days of Thomas Pennant's *British Zoology* (1761-66); it has long been regarded as an illusion .... The section on salmon contains a crop of dubious generalisations .... the little bivalve *Pisidium* masquerades as a "snail" .... I cannot follow the supposed relation between irritability and humidity, even with the help of the surrealist graph.'

After a great deal more in this vein, the reviewer concluded that:

> 'It would do no good to carry this uncharitable dissection further. Clearly a book like this is exceptionally difficult to write, and most of us would not have the courage to attempt it. Fraser Darling's views on conservation I most heartily endorse; his passionate love of his chosen land, and ability to inspire it in others, I admire and respect. All human authors err (and I hope in this respect that reviewers are [not?] as inhuman as they seem); we might well have been worse off with the opposite extreme, a prosy compendium of incredible dullness, richly documented with footnotes. At least this book has warmth and personality and an infectious appreciation of the good things of life.'

This review struck Darling on the nerve since it exposed what he felt to be his weakness, his lack of formal scientific training. While it wounded him, it angered and alarmed the editors, and the fatal clause was henceforth removed from the New Naturalist credo. The criticism was justified, though the reviewer had been less than charitable in focusing on the detailed errors of fact while for the most part ignoring the grand sweep of Darling's book, his beauty of expression and the great-heartedness of his vision. As his later reviser and co-author Morton Boyd pointed out, the licence used by Darling in interpreting his theme clearly irritated some of his academic contemporaries. Vero Wynne-Edwards, normally a magnanimous man himself, later regretted the review as having done more harm than good. It did at least have the happy consequence of bringing together two of the greatest Scottish naturalists. Despite the review, they became friends and remained so until Darling's death in 1979. Wynne-Edwards travelled far to be present at the unveiling of a plaque to Frank Fraser Darling's memory at Dundonnell in July 1991. Bad reviews were one reason why the original *Natural History* was not reprinted (though they did not seem to affect sales). The errors of fact were relatively easy to put right once Morton Boyd came to revise the book in the early 1960s. Under a new title, *The Highlands and Islands* (1964), 'Darling and Boyd' went on to become one of the most successful books in the series, with a total sales in excess of 40,000 hardback and twice that in paperback, and remaining in print for 36 years.

When the botanical titles of the series were being planned, the editors envisaged two introductory plant life books that would cover the broad field from different perspectives, followed by a series of books about different habitats and groups of wild flowers. In practice the sequence of books was broken by the late appearance of the first book *Wild Flowers*, so that the first botanical volume in the series was the more specialised *British Plant Life* (1948) by W.B. Turrill. This book broke new ground by attempting to present the evolutionary biology of British plants to the general public, covering genetics, ecology, cell biology and classification; in short, the more 'biological' aspects of British plants, as opposed to the more 'open-air' plant-hunting approach of *Wild Flowers*. The original title, 'The Biology of the British Flora' described Turrill's book better than the adopted one. *British Plant Life* was the most 'difficult' title published so far, and helped to brand the series as 'books written by specialists for advanced students'. Even some specialists found it heavy going. In writing of the 'heavy demand' parts of the book would make on the general public's comprehension, Harry Godwin implied that it risked falling between two stools, neither sufficiently well-explained for the ordinary reader nor sufficiently rigorous for the advanced-course student. E.F. Warburg wondered whether the author was not attempting too much. *British Plant Life* seemed like two books crushed together: a fairly elementary one on ecology and plant history, and a difficult one on heredity and evolution. The book was worthy in intention, but sometimes clumsy in execution. Turrill had written it 'in intervals of spare time during the past three years [1943-46] under somewhat difficult conditions'. But in those days when the New Naturalists walked on water, even *British Plant Life* sold quite well at first.

The tardy appearance of *Wild Flowers*, eight years later than planned, made that book seem more of an aftermath than its intended rôle as an introduction, but it was worth the wait. The subtitle 'Botanising in Britain' expresses its main theme perfectly. The book is an invitation to go botanising with Gilmour and Walters along tracks and dales, past woodland and commons. It could, wrote an anonymous reviewer in *The Illustrated London News*, 'be read in a field, in a wood, or on top of a cliff'. It is, indeed, very readable. Charles Sinker devoured it 'at a sitting'. *The Listener* discerned differences in the co-author's styles: 'Gilmour is the more poetical companion, Walters the more prosaic; but both write in clear, straightforward, unpretentious English, carrying the bare minimum of technical jargon.' More than one reviewer picked out John Gilmour's chapter on 'How Our Flora Was Discovered' for special praise. New to many readers, this was familiar ground to Gilmour, who had written the book on British Botanists for the 'Britain in Pictures' series ten years earlier. Another Gilmour idea was the short introductory chapter on 'The Anatomy of Field Botany', which, in his hands, becomes the anatomy of a passion, an expression of the spirit of the amateur. It is typical of a bright, outdoorsy and curiously *cheerful* book, 'full of amusing detail' (said another reviewer), and as splendid an encouragement to the pursuit of flowers as has ever been written. I confess it had the desired effect on me as an undergraduate, and glancing over its pages again, I find myself yearning once more for the woods and byways.

W.H. Pearsall (1891-1964), author of *Mountains and Moorlands.* Pencil portrait by Delmar Banner, 1961. (*Photo: Freshwater Biological Association*)

Godwin summed up *Mountains and Moorlands* by W.H. Pearsall in a phrase: 'good ecology from cover to cover'. The first habitat book of the series (unless you count London as a habitat), *Mountains and Moorlands* is the classic account of upland ecology in Britain by the pioneer and master of such studies. It is possibly the most influential book about ecology in English, certainly one of the all-time classics of natural history. His colleague, A.R. Clapham, devoted two pages of a Biographical Memoir of Pearsall to the book, and clearly regarded it as a major achievement in a remarkably full and busy life. The style is plain, no prosy flourishes here, though, like Darling, Pearsall was good at conjuring up a sense of the mountain scene. A son of the cold north, Pearsall was a somewhat buttoned-up personality. His literary gifts were clarity of expression and the imagination of the big country naturalist, seeing nature simultaneously in close-up and in vistavision. Consciously or not, he planned *Mountains and Moorlands* as the companion to university field excursions, preferably his own (which were notorious for their mileage and indifference to the weather). His introduction, which captures something of the New Naturalist spirit, deserves quotation:

> 'A visitor to the British Isles usually disembarks in lowland England. He is charmed by its orderly arrangement and by its open landscapes, tamed and formed by man and mellowed by a thousand years of human history. There is another Britain, to many of us the better half, a land of mountains and moorlands and of sun and cloud .... to the biologist at least, highland Britain is of surpassing interest because in it there is shown the dependence of organism upon environment on a large scale. It includes a whole range of habitats with restricted and often much specialised faunas and floras. At times, these habitats approach the limits within which organic life is possible, and they are commonly so severe that man has avoided them. Thus we can not only study the factors affecting the distribution of plants and animals as a whole, but we can envisage something of the forces that have influenced human distribution. Moreover, in these marginal habitats, we most often see man as part of a biological system rather than as the lord of his surroundings.'
>
> *Mountains and Moorlands*, Ch.1 Introduction

Those who studied ecology at university will know this book well, and would probably agree with Clapham that 'Pearsall succeeded admirably in the task he

John Ramsbottom (1884-1974), author of *Mushrooms and Toadstools*, explains the finer points of a specimen to school children, c. 1955. (*Photo: Natural History Museum*)

set himself'. Winifred Pennington, who revised certain sections of the book in 1971, was in still greater awe of the master: 'It is a classic, and must not be touched by a lesser hand.'

Mushrooms and Toadstools, subtitled 'A Study of the Activities of Fungi', stands by itself as a contribution to natural history literature: a mycological *tour de force*, 'an Anatomy of Toadstools'. No one but John Ramsbottom, the 'master of strange learning', the man 'who had read everything mycological, met every living mycologist of note, and forgotten nothing', could have written it. Brian Vesey-Fitzgerald put it another way: 'I think that there is just about everything that anyone can want to know about fungi in this book'. Ramsbottom takes us on a journey through the strange Gothic world of fungi, pausing at places where mankind and fungi have clashed or co-operated, and scattering pertinent quotations from the botanical classics on every page. It manages to be what *The Listener* called 'comfortably humorous'. However bizarre the tale, the author usually keeps a straight face, as no doubt he did in real life when telling one of 'his fund of stories, both proper and improper'.

This book was another late arrival. Commissioned to write it in 1943, Ramsbottom found one excuse after another to put it off: his obituarist noted that 'to exasperate people by procrastination was an integral part of his character'. When he finally got around to it, he produced a sprawling epic the size of a Dickens novel. Back went the manuscript with instructions to make cuts of between a third and a half, and so, to Ramsbottom's regret, the habitat chapters became 'but shadows of their former selves'. This might have altered the balance of book by making it less of an ecological history and more of a study of the relations of fungi and man. When Ramsbottom showed every sign of fiddling about with it indefinitely, the despairing editors resorted to subterfuge. Under the pretext of wishing to read it again, the editors sent a taxi round to the Natural History Museum with instructions to secure the text at all costs. Having done so, the taxi driver took it straight to the printer. As *Mountains and Moorlands* will to Pearsall, so *Mushrooms and Toadstools* will serve forever as Ramsbottom's memorial.

Insect Natural History is yet another classic. With A.D. Imms, author of the standard, but notoriously dry, *General Textbook of Entomology*, editors and readers alike might have been agreeably surprised at its readability. And all the more so since the author was obliged to concentrate on the less appealing insects, as the more popular ones, like butterflies, moths and dragonflies, had been allocated books of their own. Even so, the insect kingdom was too vast to cover systematically in a single book. Wisely, Imms decided not to try, choosing instead to review particular aspects of insect life that were shared by all insects: senses, feeding habits, protective devices and social life. It is a very comparative book, almost a series of essays. There is a chapter on aquatic insects included for no better reason – and what could be a better reason? – than that Imms thought they were interesting. Within these general themes there are more than enough examples of the ingenious, the bizarre, the amusing and the horrifying. The book is simply written with a nicely judged balance of science and popular natural history. It has been said of *Insect Natural History* that the book could not have been written much earlier, nor much later either, for the subject soon outgrew the bounds of a single book. Twenty years after its publication, it was still being hailed as 'incomparably the best-written semi-popular account of British insect lore .... appealing to entomologists at all levels'. Technically, the weakest part of the original book was the chapter on insect flight. It was later rewritten by Imms' reviser, Professor George Varley, though the latter confessed that 'When last I had a serious thought about it, I merely succeeded in convincing myself that I could not understand how a fly worked, which was rather disappointing.'

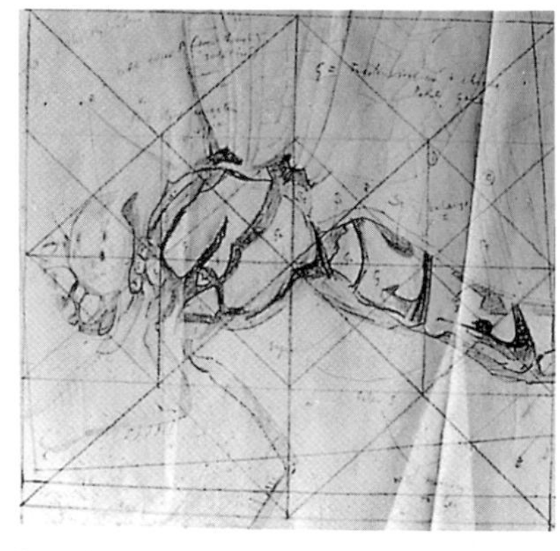

Pencil sketch of an *Aeshna* dragonfly by Clifford and Rosemary Ellis, the basis of the jacket design for *Insect Natural History*. (*Private collection*)

And so to the seaside and *The Sea Shore* by C.M. Yonge, later Sir Maurice Yonge FRS, the perfect blend of well-honed text, lovely coloured plates and fine line-drawings. Even by the standards of the previous books, *The Sea Shore* was uncommonly well received. One thing to emerge from its highly readable pages is Yonge's immersion in shore life literature in the widest possible sense, from bivalve anatomies in obscure journals to the poetry of Crabbe and Southey, and the seaside natural histories of Victorian England. Geoffrey Taylor writing in *The New Statesman* compared Yonge's style with that of Gosse in its economy and clear-sightedness. He added pertinently that there was in fact very little with which to compare *The Sea Shore* since Victorian times, a point that could be extended to not a few other New Naturalist titles. One of Yonge's literary gifts was charm, which probably came naturally because by all accounts he was a charming man – and a good lecturer, despite his reserved nature and stammer. He was one of several authors who saw the popularisation of his subject as a positive duty. As early as 1926, he had written a natural history classic, *The Seas*, with his friend (and later fellow marine knight) Frederick Russell, and thirty years later wrote the best-selling *Collins Pocket Guide to the Seashore* with John Barrett. It may be significant that, like John Ramsbottom, he got on well with young people. He was a natural-born teacher. To Morton Boyd, one of his eminent pupils, Maurice Yonge was simply 'The Master'.

The Sea Shore was begun in pain. Yonge's first wife Mattie had died from a brain tumour in 1945, leaving two small children, and he began the book to occupy his mind. It bears the most poignant dedication of all the New Naturalist books: 'In

Sir Maurice Yonge (1899-1986), author of *The Sea Shore*, at work on a bivalve c. 1979. (*Photo: Natural History Museum*)

Memory. M.J.Y. Who will walk on no more shores with me.'

The first dozen New Naturalists covered British natural history in considerable breadth and with a depth not previously found in popular natural histories. They are all to some extent concerned with ecology and behaviour, while four of them, *London's Natural History, Natural History in the Highlands and Islands, Mountains and Moorlands* and *The Sea Shore* are concerned throughout with an animal's or a plant's relationship with its environment. The broad brush treatment was probably an important ingredient of their success. It is perhaps fortunate that at this stage in the series there were no books specifically about birds, and that a separate home was created for the more specialised titles. Each book, in its way, reflected a peculiarly English (or Scottish) approach to ecology, concerned with living organisms in their natural surroundings, on the subtle and entangled influence of rocks, climate and scenery, and, above all, on intelligent observation as a means of discovery. At a time when Britain led the world in its happy combining of ecology and field natural history, of professional and amateur, these were the most influential books of their *genre*, and they captivated a whole generation of British naturalists.

The National Park and regional books

In the 1940s, National Parks were in the news. The New Naturalist editors were keen to commission natural-history-oriented guide books on individual Parks as part of a prospective series of regional volumes on 'parts of this country that have a special appeal to all naturalists and lovers of the countryside'. Their eagerness to expand the series in this direction is not surprising given the personal involvement of Stamp and Huxley in the planning of the Parks, the former as vice-chairman (and main draughtsman) of the Scott Committee, and the latter as a member of its successor committee chaired by Sir Arthur Hobhouse. Correspondence and formal contracts for several National Park books (Snowdonia, Dartmoor and The Lake District) survive dated 1946-47, which places the commissioning of these books well in advance of the National Parks themselves. Billy Collins must have been anxious to cater for the anticipated public enthusiasm for the Parks by having guide books ready and running when the Parks themselves were designated. Unfortunately no one then knew exactly where the Park boundaries would fall. It was a matter for negotiation, in which people affected by them would have their say. The authors had to guess, and in the case of Snowdonia, they guessed wrongly. Although subtitled 'The National Park of North Wales', the *Snowdonia* book covered Caernarvonshire only, whereas the Park was to include most of Merioneth as well.

On the flyleaf of *Snowdonia*, published in 1949, are listed four 'National Park

Books' then in production. These were Dartmoor, The Lake District, Pembrokeshire and The Broads (for at that time the Broads was one of the candidates for National Park status). Whether or not the Board intended eventually to produce guide books for all the National Parks is doubtful. More likely, their approach was pragmatic, commissioning books on the more popular areas first and then awaiting the public response. The books would offer a rounded portrait of the natural and social history of each region, stressing the links between man and nature, and the present with the past. In their wholeness of vision, the National Park books were of a piece with the rest of the series: each area would be given the 'New Naturalist' treatment.

That being so, the editors went about things in a rather odd way. With most titles in the series, they sensibly insisted on a policy of one author, one book, and strongly discouraged multi-author volumes. For the National Park books, on the other hand, they deliberately commissioned a team of three or more authors presumably in the belief that no single person could cover the whole field with the same authority. There would be a nominated lead author for each title, responsible for co-ordinating the efforts of the others and harmonising the different texts. It was a recipe for confusion and delay, and that is exactly what happened in nearly every case. *Snowdonia* was the only book to emerge more or less as planned. Here, Bruce Campbell was nominally the senior author, responsible for the natural history, while F.J. North, Keeper in Geology at the National Museum of Wales, tackled the rocks and Richenda Scott, an economist with a keen interest in the countryside, the social and historical background. Unfortunately, all three authors lived not in the north of Wales but in the south. Bruce Campbell made a whistle-stop tour of the region after the war, picking the brains of local naturalists as he went, but his part of the book does suggest only a superficial acquaintance with Snowdonia. Moreover, he seems to have made little attempt to edit the book, and the result was really three books inside one wrapper, with no introduction and no overview. This was not the most economical way of writing a book. *Snowdonia*, with 468 pages, is one of the longest in the series. Its size ensured it a short life in the shops for once the edition had sold out, it was impossible to reprint it at an economic price. Nor was there much incentive to do so, since the book was already badly out of date. Bruce Campbell seems to have been glad to wash his hands of it, though Frederick North, the most ardent populariser of geology in Wales, deplored its early demise.

William Condry in 1962, minutes after being invited to write *The Snowdonia National Park*. (*Photo: Eric Hosking*)

Instead, the Board decided to commission a new and shorter book covering the whole of the Snowdonia National Park, and asked the writer and naturalist William Condry to write it. The Condry book focused fairly and squarely on natural history. The first half deals in turn with rocks, scenery and wildlife in the traditional way, while the second offers a wildlife tour of the region. And a pleasant tour it is with Condry as a guide; the book is full of lively personal touches and good descriptive writing. There are a few healthy prejudices thrown in, such as Bill Condry's hatred of motor cars, a point of view evidently shared by the New Naturalist editors (and also by Elgar, who, like Condry, fled west to take refuge from the internal combustion engine). *The Snowdonia National Park* was published inside an unfortunate laminated jacket in 1966. It was a successful book, remaining in print for 15 years, and being reprinted in paperback.

Though it barely acknowledges the existence of a predecessor, the second *Snowdonia* was the first 'replacement title' in the library. Given that the editors wanted to retain a Snowdonia book in print, they had in this case no other course of action. It does, however, bring to mind the broader question of whether it is wise to replace 'out of date' titles in the series with new ones. In my opinion, it is not. The classic titles of the series have a quality that transcends the passage of time, and people will go on reading them for the same reason that one still reads *The Origin of Species* or *Tarka the Otter*. The New Naturalist library represents a tradition of British natural history that does not date, and the best of the books have a wholeness and breadth of vision that would be difficult to emulate today. We can refresh the series with contemporary insights and perspectives but the library should plough new ground, not resow last year's crop. A third Snowdonia would be a hill too far.

Of the remaining National Park books, one of the most satisfactory is *Dartmoor*, precisely because it was mainly the work of one man, Professor L.A. Harvey. *Dartmoor* had been conceived along the lines of *Snowdonia*, with different authors contributing chapters on natural history, customs and folklore and prehistoric remains. In practice, though, one of the contributors died, and another dropped out, and the two mellow chapters on the history and social customs of Dartmoor people by Douglas St Leger-Gordon dovetail into Harvey's text without discord. The photographer, E.H. Ware, worked closely with Harvey, and the marriage of illustrations and text is well above average in this book, though the colour reproduction is no better than usual. Harvey admitted that it was Ware's pictures 'which have lifted me over the more difficult passages of my own'. Furthermore, Dartmoor had been designated as one of the first National Parks in 1951, allowing the authors time to revise the book before publication and for Harvey to include a chapter on 'Dartmoor as a National Park', as well as a postscript deploring its continued use for military training. The only serious weakness was the long series of species lists at the end which, as Harvey acknowledged when revising the book, 'seem to me neither one thing nor the other. I would happily excise them.....'

Dartmoor forms an interesting contrast with its immediate predecessor in the series, *The Weald*. They are regions of similar size and both books were written by university professors. But, whereas Leslie Harvey's ecological approach is well suited to the New Naturalist series and forms the consistent wholeness of view on which to hang details of Dartmoor's bogs, moors and rivers, Sidney Wooldridge's preoccupation with 'land sculpture' works in the opposite direction. Wooldridge was a geomorphologist based at the University of London, and had used the Weald as an open air laboratory to work out how the region's scenery had evolved and formulate certain general principles of land erosion. Physical geography, therefore, forms the heart of the book; the chapters on plant and animal life, and

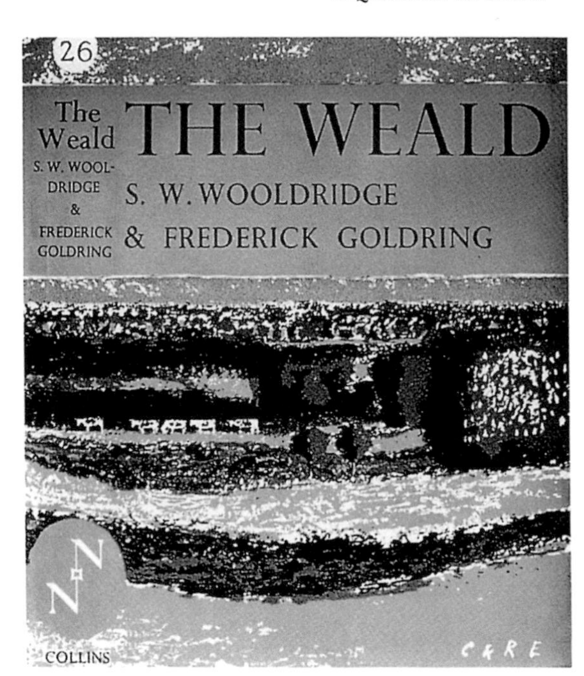

Jacket of *The Weald* (1953).

on human settlement, have a more perfunctory, knocked-on quality, and the style is as dry as the sands of Ashdown Forest (though, in person, Wooldridge was anything but dry). Geography and geology seem to resist easy popularisation. Reviewing *The Weald* within the context of the series as a whole, Cyril Connolly found it 'prickly with unfamiliar fact and theory'. The ordinary reader heads for the catalogue of fauna promised in the Appendix, 'only to find a miserable list of earthworms and woodlice'. In writing *The Weald*, Wooldridge may have had his London students foremost in mind. Its use as a university textbook and as a guide to field courses ensured *The Weald* a steady sale, but it would be hard to claim that the book has much wider appeal.

What *The Weald* and *Dartmoor* did have was an author capable of seeing the book through to publication relatively speedily and efficiently. The lack of such a *primum mobile* bedeviled production of *The Broads* and *The Lake District* and delayed their publication for years (while *Pembrokeshire* seems never to have left the ground). Both books had their origin in the 1940s, although they were not published until 1965 and 1973 respectively. To explain in detail how a book can take as long as that to write would take a whole chapter (researching this book, I compiled a four-page summary chronicling the ups and downs of *The Broads*, which I will spare the reader). The fundamental problem with *The Broads*, which was never really resolved, was that of too many cooks. Ted Ellis, the doyen of Norfolk naturalists, had the unenviable task of weaving together some dozen individual contributions, written at levels ranging from popular guidebook to university thesis. Ellis lacked the requisite ruthlessness. He admitted that things had rather 'got out of hand and my contributors would not let me edit or cut their contributions, on which they had galloped away on their own hobbyhorses'. Early on in the proceedings, the blockmakers lost all the colour plates; and then in 1952 Joyce Lambert made the startling discovery that the Broads were almost certainly artificial lakes caused by ancient turf cutting. That revelation alone meant that much of the book would have to be rewritten. Progress remained at a snail's pace throughout the 1950s, with James Fisher contributing his bit by sitting on the draft for a year or more. Another of Ted Ellis' problems might have been too much information. He had drawers, cupboards and cardboard boxes crammed full of Broadland matters scattered about his 'den' at Wheatfen Broad; James Fisher had been filled with horror at the sight of it. Eventually, after weary effort involving Stamp and Gilmour, as well as Fisher, *The Broads* was published in 1965. Inevitably, perhaps, it lacks coherence as a book, being full of bits and pieces by different authors. It might have been a better strategy to allow Ted Ellis to write his own book, though that modest and well-loved naturalist might have been reluctant to do so. *The Broads* was to become out of date very soon, through no fault of the authors, by the catastrophic pollution of the Broadland ecosystem in the 1970s. Today, when we read of 'the fantastic water-gardens' described by Ellis, and learn that the otter 'is thoroughly at home in every river and throughout the broads', the book seems older than its years.

The Lake District is a much more coherent book, indeed, technically it is possibly the best of all the regional guides, though not the easiest read. But it was not the book that had been anticipated a quarter of a century earlier. That one was to have been written by the Carlisle-based naturalist Ernest Blezard with the help of no fewer than five co-authors, who between them agreed to produce 100,000 words for delivery by 1950. Blezard's synopsis arranged the book conventionally into chapters on topography, geology, history, plant life, birds and so on. A specimen chapter on 'bygone birds' written by Blezard himself was lively and up to scratch, but the other contributions were not (though the pictures by the photographer, J.A. Jenson, are said to have been 'breathtakingly good'). The deadline passed without much further progress being made. By 1953, the Board had decided they would get a better book out of W.H. Pearsall, author of the successful *Mountains and Moorlands*. But then Pearsall could not get on with it either. He had evidently produced a first draft shortly before his sudden death in 1964, but was presumably unhappy with the result for he is said to have burnt it, intending to make a fresh start. When Winifred Pennington (Mrs T.G. Tutin) agreed to complete the book, she had only Pearsall's rough notes and a handful of second-rate slides to work from. At length she completed her task with the help of eleven 'guest writers' on special subjects, among them the old New Naturalist hands T.T. Macan, Gordon Manley and Winifred Frost. The bulk of the book (and a bulky book it was) is Pennington's, however, and admitting Pearsall as the first-named co-author was an act of generosity. The structure of *The Lake District* showed how much had changed during the past quarter of a century. Gone were the old reductionist divisions of topography, wildlife and land-use, and in their place were chapters on ecology which united the disparate elements of history, land-use and biology. In consequence, this is the most thoroughly ecological of all the regional or National Park books, and, being ahead of its time, reads remarkably freshly even today. Pearsall would undoubtedly have approved.

The last of the books to be considered here, *The Peak District*, by K.C. Edwards, was published in 1962. Like its fellows, *The Peak District* has its own particular flavour, strongly reflecting the main author's preoccupation with regional geography. Half of this rather short book is man-centred and concerned with the villages, factories, water supplies, farms, and stone quarries of the area. Feeling less sure of himself on scientific subjects, Edwards co-opted R.H. Hall, a tutor at

Nottingham University extra-mural department, to write about botany, and the aged H.H. Swinnerton to contribute two chapters on the physical landscape. While not a title to set the bookshelf on fire, *The Peak District* covers the field competently, if somewhat uncritically, with one extraordinary exception – there is hardly anything in it about animal life apart from short notes on birds and fish relegated to an appendix. This book seems to have been commissioned on Edwards' own initiative. Despite the popularity of the Peak District as a weekend venue, the New Naturalist editors had doubts about the book's saleability. Perhaps they were right, for in the event the main users of the book were university students. Fortunately, they bought it in sufficient numbers to guarantee a steady sale over nearly 20 years.

Co-authorship

About one quarter of the New Naturalist books is credited to two or more authors. In practice, however, the division between lone and multiple authorship is somewhat blurred, and co-authorship expresses different things in this series. For most titles, one person took the lead, and collaborations of equal partners are in the minority. There is little consistency in the attribution of authorship. *The Broads* and *The Peak District* have only one named author on the spine, yet the title page confirms that they were co-operative efforts. *Life in Lakes and Rivers* and *Bumblebees*, on the other hand, are credited to co-authors, but they were largely one-man efforts. The 1970 revision of *A Natural History of Man in Britain* is credited, rightly, to co-authors on the title page, but only the name of the original author, H.J. Fleure, appears on the jacket and spine. And so on.

The situation suggests a degree of muddle, but each case reflects the different circumstances in which the book came to be written. *Life in Lakes and Rivers* had been assigned to E.B. Worthington, then the director of the Freshwater Biological Association (FBA). Barton Worthington told me that 'It was I think 1944 when Julian [Huxley] got me for a discussion with Collins and Eric Hosking to see some of the latter's magnificent photos intended for a volume in the series. They suggested that I should do one on freshwater biology, and I set about a rough synopsis and drafted one or two chapters.' Fate then intervened. In 1946, Worthington was 'yanked back to Africa' to rejoin the East Africa High Commission, and it was at this point that his deputy, T.T. Macan, was brought in as a collaborator. The bulk of the book was written by Macan during Worthington's sojourn in Africa, though he drew extensively on the work of FBA colleagues (it was very much an FBA book). *Life in Lakes and Rivers* reflects Macan's unrivalled knowledge of the invertebrates of Cumbrian lakes and streams, with an ecological slant appropriate to

T.T. Macan (1910-1985), co-author of *Life in Lakes and Rivers*. (*Photo: Freshwater Biological Association*)

the series. Worthington's overall plan is probably reflected in the structure of the book, with its emphasis on different kinds of lakes and their productivity. It was an advanced book for its day and could easily have been written in the 1960s during the vogue for production ecology, in which Worthington played a leading role.

Bumblebees had a similar history in that it was commissioned at Colin Butler's suggestion, but written by John Free, whose PhD and recent postgraduate work on bumblebees Butler had supervised. Butler himself contributed the splendid photographs and helped to edit the book for publication. Without Butler, it is doubtful whether *Bumblebees* would have been commissioned at all. As always with Billy Collins, the illustrations were the key selling point of New Naturalist books, and he had been so impressed by Butler's pin-sharp close-ups of honey bees that he had asked him there and then to write the commercially successful *World of the Honeybee* on the strength of them. That book must have provided cause for optimism about a similar success for *Bumblebees*. Unfortunately, the apiarists and university students in Britain and the United States who bought *Honeybee* in large numbers were less interested in wild bumblebees. Through no fault of either author, this is now one of the rare books of the series.

The Art of Botanical Illustration was another collaboration that hovered somewhere between joint and single authorship. Its commissioning had been the occasion of a celebrated New Naturalist muddle. In about 1946, the editors had decided that a volume on botanical illustration would be 'a pleasing and useful addition to the series'. Unfortunately, while John Gilmour turned to his old Cambridge colleague, William Stearn, another editor (probably James Fisher) had quite independently approached Wilfred Blunt, the art master at Eton College. In consequence, they ended up with different authors for the same book, neither of whom knew of each other's existence. In Stearn's words:

> 'The embarrassed Editors suggested that we should co-operate. Blunt came to the Lindley Library of the Royal Horticultural Society, of which I was then the Librarian, and we agreed on joint action. He was an unmarried art master with no marking of classwork to occupy his evenings; moreover he had long holidays and ample time and money to visit public collections in Holland, France and Italy; I, on the other hand, was a busy librarian with no such facilities and already engaged in gathering material for my *Botanical Latin* and bibliographical papers. We accordingly decided that he should write the book and that I should later revise and augment it.'
>
> *The Art of Botanical Illustration*. Introduction to the 1994 edition

Thus began a collaboration and friendship that extended over several more books and lasted until Wilfred Blunt's death in 1987. The editorial blunder had a happy consequence since, as Stearn pointed out, *The Art of Botanical Illustration* was a more comprehensive and balanced book than either author would have been able to achieved alone. They complemented each other perfectly: Blunt, the craftsman and aesthete, and Stearn, the professional botanist and bibliographer.

The Shetland and Orkney books, published in 1980 and 1985 respectively, form an interesting contrast in authorship. The Shetland book was the idea of R.J. 'Sam' Berry, then a regular visitor to the islands in search of genetically isolated populations of field mice, moths and other organisms (they had already formed part of the subject matter of Berry's earlier contribution to the series, *Inheritance and Natural History*). There was no comprehensive natural history of Shetland in print and Berry wanted to write one. Feeling that he needed a resident collabora-

R.J. Berry (left) and Laughton Johnston (right), authors of *The Natural History of Shetland*, catching field mice during hay making at Quendale, Shetland c. 1972. (*Photo: R.J. Berry/Dennis Couttts, Lerwick*)

tor, he roped in Laughton Johnston, a Shetlander by birth and, at that time, the Nature Conservancy Council's representative on the islands. Surprisingly, in view of Shetland's major importance as a Mecca for wildlife and its new-found prominence at the start of the 1970s oil boom, it proved hard to interest a publisher in the book. Collins at first turned it down (the mid-1970s was a thin time in the book trade, and Shetland was a long way from London). A dowry from BP made the difference. The two co-authors set to work in about 1975. Laughton Johnston remembers writing some of it in the glass-fronted verandah of his house overlooking the west shore of Shetland, now and again looking up to gain inspiration from the cavorting wildlife below, which included divers and otters. His contribution to the book were the chapters on whales and seals, birds, lochs and burns and most of the conservation chapter (chapters 6, 7, 9, 10 and 14). Berry wrote chapters 1, 3, 5, 8 and 12, and four 'guest authors' contributed a chapter each.

The book was given a leavening of editing by Sam Berry alone, but the result was something of a potpourri and the treatment inevitably uneven. As Berry himself admitted, the book contained some 'turgid stuff' more suited to a university seminar than a popular natural history publication. He resolved to approach the complementary *Natural History of Orkney* (also made possible by an oil company grant) in a different way: 'I intend to write it as Winston Churchill used to write his books: commission individual chapters from experts, and then re-write them into a flowing whole.' This is, surely, the right decision for books of this sort. The complications and disagreements behind the scenes remained, especially as Berry did not know Orkney as well as Shetland and was therefore even more dependent on local help. But as the product of a single hand, *Orkney* has a better balance and a more even flow. Its main misfortune was to be printed in a minute

Left John Raven (1914-1980). Co-author of *Mountain Flowers*. (*Photo: Faith Raven*)
Right Max Walters c. 1965. Co-author of *Wild Flowers* and *Mountain Flowers*. (*Photo: S.M. Walters*)

typeface that strains the eyes; had its readers been treated more considerately by the publisher, they might have agreed that this is one of the better written books of recent New Naturalist times.

Three botanical books were that unusual thing, a genuine fusion of two equal parts: *Wild Flowers*, *Mountain Flowers* and *The Pollination of Flowers*. In the pollination book, the contributions of Michael Proctor and Peter Yeo are so similar in style, that the book might easily pass as the product of a single hand. But one of the strengths of *Mountain Flowers* is the contrast between Max Walters' scientific lectures and John Ravens' exuberant botanical travelogue, rather as though *British Plant Life* had married *Wild Flowers of Chalk and Limestone*. It is perhaps the most perfectly realised marriage of the amateur and professional viewpoints in the whole series, and, as in *The Art of Botanical Illustration*, it thereby gains in balance and comprehension. There were, of course, some who would have preferred undiluted Raven. John Raven was a Cambridge classics scholar, but his life's passion was for plants. He lived among them; he travelled hundreds of miles and scaled the remotest hills to find them; he cultivated them; he painted them; and, of course, he wrote about them with panache. He had a story to tell about every plant he found, and an artist's memory for the form and atmosphere of hill scenery. As a botanical writer he was, in short, extremely likeable. His survey of the Scottish hills is, as Geoffrey Grigson found, 'extremely tantalising and makes one wish to calculate miles at once on a road map. How far to Glen Doll or The Storr or the great Ben Lawers?'

It was not to be expected that people would take to Max Walters' drier, more technical style with the same degree of enthusiasm. Grigson certainly did not:

> 'The authors do separate jobs. Mr Raven (in normal English) introduces the flowers historically and emotionally. Dr Walters (in appalling professional botanist's English) explains them. Then Mr Raven takes over again, sketching the mountain flora region by region..... Persevere with Dr Walters. When he wants to say: "The heavy falls of rain and snow are important," he says... "Precipitation in the form of rain or, in winter, snow, is generally high, and this is

of the first importance." Inside this entanglement of barbed wire are facts and speculations about the way flowers live and maintain themselves at such heights and about the origins of our mountain flora.....'

Geoffrey Grigson. Review of *Mountain Flowers, The Observer*, August 1956

Max Walters retorted that 'It isn't a charge against which one can defend oneself – except to say, with respect to the particular example he chooses to drag from its context and set up as an Aunt Sally, that "precipitation" is a technical term, the use of which prevents one having to say throughout "rain, snow, hail, sleet or dew".' Reading the Walters half of *Mountain Flowers* today, one wonders what all the fuss was about: it is, of course, a scientific text, but a perfectly readable one which either avoids technical terms or carefully explains those which are used. In praising Raven at the expense of Walters, Grigson missed the significance of the blend of the amateur and professional naturalist which lies at the heart of the New Naturalist library. A travelogue without the ecological background would have been like a statue without a pediment, especially in the context of mountains where the physical environment is so dominating, and so extreme. It is the mixture of seminar room and field excursion which lifts *Mountain Flowers* above the herd. The New Naturalists were trying, gently, to guide the reader beyond mere rarity-ticking to ask why the plant was rare in the first place, and where it came from, and how it survives. In that aim, I suggest the book succeeds triumphantly. For me, at any rate, it is the apotheosis of co-authorship, and one of the best books in the series.

The evolving series

The standard of the magnificent first dozen New Naturalist titles was maintained through the second dozen and, perhaps, even the third. We cannot hope to do justice to each book here, nor even offer these books the same consideration as their predecessors. Fortunately, some of them have their place in the sun in another context, for which I must refer you to the index. Here the object is to step back from the chronological sequence of titles to pick out some highlights and see whether we can discern overall trends that have changed the nature of the series.

The early 1950s saw a pile of new titles, as authors commissioned in the mid-1940s all breasted the tape in a rush. While most of these books were written in the same spirit as their predecessors, we can note in *Wild Flowers of Chalk and Limestone* and *The Wild Orchids of Britain* an improvement in colour reproduction to match two of the most popular texts of the series. *Chalk and Limestone* is the amateur plant-finders book *par excellence*. It is still pretty useful 40 years on despite J.E. Lousley's deliberate vagueness about localities. It is easy to agree with *The Field* that it is 'a wonderfully friendly book'; more surprisingly, it was equally well reviewed in the scientific press. Harry Godwin was charmed by the author's 'obvious affection for the beauty of plants in natural surroundings'. Though plainly written – no Richard Jeffreys transports here – the book does have an air of the downs in summer and the sweet savour of thyme-scented grass. Since everything in this book is first hand, the author excels at capturing the individuality of each rare flower, and placing it into its context in the landscape. Lousley adds a couple of fairly perfunctory chapters on ecology, but he knows what most of us want and goes on to provide it in the second best botanical travelogue ever written (I'm sorry Ted, but the best is John Raven's). *Wild Orchids* is more of a herbarium curator's book, full of detailed leaf-by-leaf descriptions of the British

J.E. Lousley (1907-1976), author of *Wild Flowers of Chalk and Limestone*, at Wicken Fen, 1960. (*Photo: BSBI*)

orchid flora that relate form to function. But for all that, there is an open-air feel to this book too, helped by Robert Atkinson's lovely colour portraits. Orchids have always had a special mystique, and V.S. Summerhayes gave field botanists exactly the book they were waiting for. It judged the overlap in amateur and professional expectations perfectly.

In the same influential league were the two vertebrate titles, *British Mammals* by Leo Harrison Matthews and *The British Amphibians and Reptiles* by Malcolm Smith.

Both were the first serious books on their subject for decades, and helped to start a revival of interest in our then neglected four-legged fauna. Malcolm Smith's book is regarded today by herpetologists as a classic, and, if now out of date in some respects, it is nevertheless still read with great respect and not a little affection. Smith had helped to found the British Herpetological Society in 1947, shortly before starting work on the book, and he more than anyone is considered the father of modern herpetology. Similarly, Matthew's great (in more ways than one) work gave mammalogists what they needed and its publication influenced the formation of the Mammal Society in 1955. Both books reflect their period in their preoccupation with anatomy and traditional life-history. Smith's book was continually updated through five editions, after the death of the author in 1958, by Angus Bellairs and J.F.D. Frazer (who also wrote the replacement 'herptile' title in 1983). *Mammals*, on the other hand, had grown badly out of date by 1970, and for economic reasons it proved impossible to reset this very long book. Having already produced the necessary revisions, Matthews was naturally put out when belatedly informed of this. Eventually, however, he put the work to good use by writing a new, shorter mammal book for the series, published in 1982. *Mammals in the British Isles* indicated how much had changed in the intervening 30 years. The second *Mammals* abandons the systematic treatment of the first for a comparative one on a range of topics on the behaviour and evolution of mammals in the wild, much of it gathered in recent years by radio-tracking devices and, oddly enough, by archaeologists. It was another masterly and personal review, neatly side-stepping the fund of mammal lore already available. In one respect, though, Matthews refused to move with the times. Conservation, to him, was futile. What mattered to him was not the sentimental tinkering of well-meaning naturalists but 'the evolution of new habits and habitats and ways of co-existence between the wild mammals and man'. It was the dynamic of evolution, with the rise and fall of species, that fascinated him. His strictures about the folly of trying to preserve nature in aspic are currently unfashionable, but they deserve more serious attention and reflection than they seem to have received.

Titles of the highest quality continued to swell the series in the 1950s. There was *The Art of Botanical Illustration* (1950), 'a glittering and intoxicating gallery of flowers .... a new world of beauty' (Sacheverell Sitwell in *The Spectator*). Forty years on, it is still a unique book, still the standard work internationally, peerless in taste, scope and style. There was *Life in Lakes and Rivers* (1951), perhaps exceeded only by *Mountains and Moorlands* as a masterpiece of popular ecology. And what about *A Natural History of Man in Britain*, 'conceived as a study of changing relations between Men and Environments', a simply written but imaginative 'picturing of British life' that links the progress of the human animal with the evolving landscape he creates. It is, as one reviewer noted, the first ecology of man. Gordon Manley's *Climate and the British Scene* (1952) offers another synthesis, this time the relationship of earth and sky, confidently treading the no-man's-land between the professional meteorologist, on the one hand, and the naturalist, on the other, and always with the casual observer or the walker in the hills in mind. In the choice and expression of such titles we detect lively imaginations at work. How easy it would be, and how dull, to simply knock together books on identification, landscape and weather. All these titles, and most of their peers, involve the reader in the subject because they are about things he can see and find out for himself, yet are presented in ways that were new and surprising and, above all, interesting. How well they wrote, these Blunts and Manleys and Fleures, masters of their subject and never crushed by their complexities. They were of the great age of nature interpretation, when the enthusiasm and sense of wonder of the Victorian naturalists had been informed by new learning and new perspectives, yet their message was still comprehensible to all educated people.

Perhaps the New Naturalist ethos is seen to greatest advantage in two of the last of the originally planned titles to be published: *The Open Sea* (2 volumes 1956 and 1959) and *The World of Spiders* (1958). Alister Hardy's brief had been broad enough: 'to write about the open sea as a habitat'. As many New Naturalist authors did, he interpreted his instructions broadly and eventually rewarded the editors with a book beyond their dreams. Or rather, two books. Hardy explained to James Fisher that 'I wanted to produce as good a book as possible and went on adding, thinking you must be going to let me produce something getting on for the size of your *Fulmar*.' [touché!]. Billy Collins grumbled about the length and the 'complicated scientific names' in the first part, but Fisher had his way and after some revision *The Open Sea* was published uncut in two instalments. To Julian Huxley, this title was the embodiment of what they were all trying to do. He especially liked 'the unobtrusive personal note that comes in so often' and 'the insistence in the possibilities of amateur work .... At last it seems that the regrettable confusion between popular scientific writing and vulgar journalism is being swept away .... No book since *The Sea Shore* (and that in a different way) has given me such a boost.'

It was, indeed, as the *Yorkshire Post* put it, 'a book for all naturalists'. Much of the success of *The Open Sea* is attributable to Hardy's lucid, enthusiastic style, which he complemented perfectly with 40 beautiful watercolour drawings reminiscent of the Victorian nature books he so admired, as well as a large number of text figures, also mostly by the author. It was not surprising that *The Open Sea* had taken years of labour to complete. The two volumes represent an extraordinary achievement for any man, let alone someone like Alister Hardy who was busy with his new department at Oxford and a hundred other things. There was never any more dedicated New Naturalist than he.

W.S. Bristowe (1901-1979), author of *The World of Spiders*, dressed for an arachnological conference at Slapton Lea, Devon in 1977. (*Photo: D.R. Nellist*)

With one possible exception. *The World of Spiders* by W.S. Bristowe was published in-between the two volumes of *The Open Sea*. It, too, had been long in the making, but when the manuscript and drawings for it finally came in, they created a sensation. Fisher exclained to the board: 'This is the best NN book we've ever had!' It was something quite new in spider literature: a natural history of living spiders in their natural environment, vividly described. Bill Bristowe wrote a great deal of it in the first person since he was describing what to a large extent was his own life's work. It seemed, and still seems, incredible that one person could manage to find out so much about spiders, and not only in Britain but all over the world. It is impossible to convey Bristowe's delightful style, so full of modesty and humour and delightfully quirky anecdotes, in a few lines. But let us at least try with the following short extract which offers some indication of Bristowe's dauntless quest of the eight-legged. It concerns a dreadful-looking beast called the Daddy-Long-Legs Spider:

> '*Pholcus phalangioides* must be well-known to people who live in the South of England.... She sits unobtrusively in corners of rooms between ceilings and walls hanging motionless from a scaffolding of fine invisible threads. Her presence is not resented because she seldom moves and is regarded as an innocuous creature which may be useful in catching mosquitoes or clothes-moths. *Pholcus* did not live in my childhood home at Stoke d'Abernon, Surrey, although she thrived only ten miles further south, so the quest of an explanation inspired me to trace her distribution. This had to await the acquisition of a motor-bicycle and then, with the impudence of youth, I zig-zagged across England ostensibly seeking rooms in hotels or lodgings whose ceilings I viewed with nonchalant interest. My apologies are no doubt due to a host of hoteliers for gaining entry under false pretences, but in the result their unwitting co-operation enabled me to draw a map which showed that *Pholcus* inhabited houses coinciding with the narrow southern strip where the average temperature throughout the year exceeds 50°F. North of this strip she is normally confined to cellars ....'
>
> *The World of Spiders*, Ch. 10 The Scytodoidea. Spitting and Daddy-Long-Legs Spiders

There you are, a project within the means of everyone!

If the text of *The World of Spiders* is enticing, the illustrations are simply stunning. As if to make amends for the unappetising nature of his subject, Bristowe produced what is surely the most beautiful book in the series. The measly four-colour plates sink into insignificance compared with the magnificent gallery of some 200 line-and-wash drawings of living spiders, all of them by a leading natural history book illustrator, Arthur Smith. It is a measure of the labour of love that is *The World of Spiders*, that Bristowe paid for most of the illustrations out of his own pocket. He probably never made any money out of the title that will be his memorial for as long as people study spiders.

By sheer virtuosity, Bristowe turned a potentially specialised subject into a popular work of natural history. With some other titles published in the late 1950s and 1960s, we start to find an increased scientific or scholarly rigour, and a growing detachment between author and subject. At an early stage, the editors decided there was room in a successful best-selling series for a few more specialised (though not *highly* specialised) titles on subjects of special interest. They themselves described *An Angler's Entomology* (1952) as 'the most specialised volume in the series ... to date'. A book about mayflies written for the fly fisherman, it seemed an unlikely topic for a general natural history library. But it turned out to be yet another example of an imaginative synthesis of man and nature, symbolised on the dust jacket by a living fly and a fishing fly. In terms of sales, it did quite well in Britain, Ireland and America. The author J.R. Harris had plenty of 'river credibility' among the angling fraternity, especially in his native Ireland, and he had the same large 'captive' market that *The World of the Honeybee* had. Two later

Philip Corbet, co-author of *Dragonflies*, in 1951, when a Cambridge undergraduate. (*Photo: Philip Corbet*)

specialised titles, *Insect Migration* (1958) and *The Folklore of Birds* (1958) represented a greater financial risk. The latter was a throwback to the optimistic days of the 1940s. The author, Edward Armstrong, had a contract, wanted to write the book and was supported in doing so by James Fisher. By 1958, however, book buying habits had changed and such a title was no longer likely to have wide appeal. Nor did it: *The Folklore of Birds* had the shortest shelf life of any mainstream New Naturalist book. Both it and *Insect Migration* are impeccably scholarly, but there is no denying that they are tough going compared with *Butterflies* or *Mountains and Moorlands*.

The changing face of natural history can be traced in a well-known and much-praised title published in 1960, *Dragonflies*. While Bill Bristowe wrote about spiders in the same spirit as Fabre or White, much of the material in *Dragonflies* is based on PhD theses (as was John Free's on *Bumblebees*). The more technical style was recognised by T.T. Macan who thought that 'Perhaps more than any other in the series [*Dragonflies*] merits the words "new" and "naturalist".' By this he meant the study of the living animal in its natural setting, of which Philip Corbet and Norman Moore were keen exponents. At one point, in the book Moore follows a particular dragonfly around for hours, and when it finally goes to sleep, so does Moore, in a sleeping bag a short distance away. The material was really no more 'difficult' than Tinbergen's *The World of the Herring Gull*, but at that time the majority of naturalists were unfamiliar with the British dragonflies. *Dragonflies* was too specialised a title for the intended market. A generation later more naturalists would have been ready for it, but by that time the book had retreated into the better class of second-hand bookshop.

This trend deepened with time. *Weeds and Aliens* was engaging, but an awful lot of it was concerned with the small print of seed weight, root length and germination period. It had a ready made market with the Botanical Society of the British Isles and 'there were cries of grief' when *Weeds and Aliens* was dropped from the list in 1972. *The Common Lands of England and Wales* (1963), solidly based (too solidly based) on a recent Royal Commission report was useful but scarcely exciting, despite a contribution from the normally entertaining W.G. Hoskins. *Grass and Grasslands* (1966) was presumably aimed at agricultural colleges, for there is hardly any natural history in it. Peter Yeo remembers John Gilmour 'tearing his hair out' over the manuscript, which he had had to re-write in places. The original concept of a book about wild grasses had been scuppered when C.E. Hubbard's book, *Grasses*, appeared in Penguin in 1954. Gilmour decided to go instead for a predominantly agricultural book, 'a natural history of pasture', but in the event he must have wished he had abandoned the subject altogether.

The more technical style of the New Naturalists from the 1960s onwards can be exemplified by the series of books published by scientists working for the then Nature Conservancy at Monks Wood Field Station: *Pesticides and Pollution* (1967), *Man and Birds* (1971) and *Hedges* (1974). At that time, Monks Wood was a remarkable institution, almost an apotheosis of the new natural history, dedicated to habitat surveys, ways of managing nature reserves and examining the effects of pesticides on wildlife. It was, nevertheless, like any other professional institute, a place where experts spent a great deal of time in each others company, talking their own language. It was natural that their work was tailored to the critical standards of peer groups rather than elementary students or the general public. Many of its staff were gifted field naturalists, but they brought to their professional subjects the due rigour of experimental science, with its cargo of data, theory and analysis.

Pesticides and Pollution is one of the most controversial titles in the series; and also one of the shortest. It was hailed by the editors as 'a calm book, and a deeply thought-out book, and patently a balanced book', a kind of antidote to Rachel Carson's best-selling *Silent Spring*. For some, it was a bit *too* calm and balanced, especially in its apparent defence of the much maligned DDT. Writing in the scientific press, G.R. Sagar was also 'disappointed to find the ecological aspects of the subject so scattered .... the reader may emerge knowing rather more about the chemicals than the ecology'. Others have found Mellanby's treatment rather sketchy and, on some topics, superficial. It seems to me, though, that on stylistic grounds at least, *Pesticides* is admirable: crisp, well argued, tough talking when it needs to be and somehow managing to be simultaneously punchy and dispassionate. It is a personal book like all the best books in the series; it was also for a few

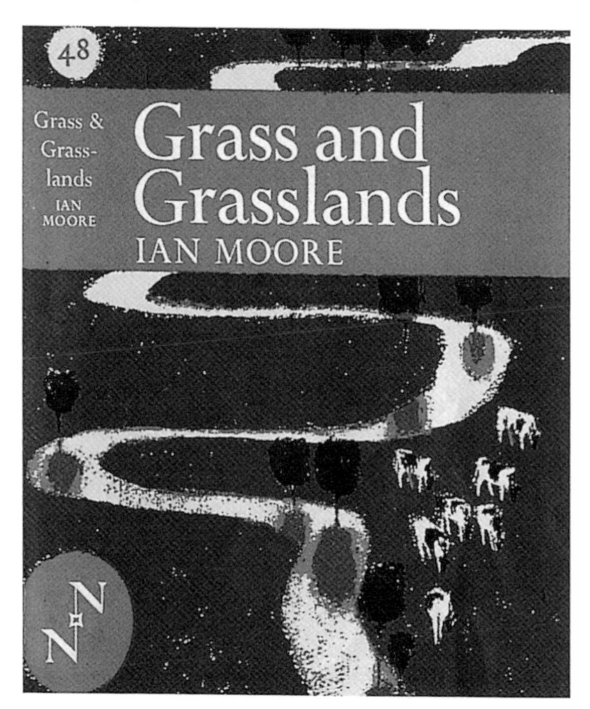

Jacket of *Grass and Grasslands* (1966).

Max Hooper, co-author of *Hedges*, in 1988. (*Photo: Institute of Terrestrial Ecology*)

Ian Newton, author of *Finches* in 1972. (*Photo: Pamela Harrison*)

years the standard work on a popular subject and therefore sold quite well, especially in the student's paperback edition. Devoid of fussy textual references and completely jargon free, it points in a direction that the series might usefully have taken.

Unfortunately, some of the succeeding titles were more like *Man and Birds*. Not that R.K. Murton's book is a bad book by any means. It covers the subject very thoroughly, the material seems well organised and what he has to say is interesting. But it is a prime specimen of the kind of scientific writing Geoffrey Grigson and Cyril Connolly had complained about. It is laden with the sort of 'Scientific Correctness' which decrees that it is better to say, 'adult birds mostly experience accidental mortality' than 'adult birds usually die by accident'. We notice the advance of graphs and data, and of what Klopfer and MacArthur did in 1960. The book would do very well for ecology courses at Monks Wood and at universities all round the world, but it was not the popular natural history envisaged by Fisher and Collins.

Peter Yeo, co-author of *The Pollination of Flowers*, in 1977. (*Photo: Peter Yeo*)

Even so, scientific writing need not be turgid, and even the most difficult, mathematics-based science can be made exciting in skilled hands. There is a festive, bird-nesting spirit in parts of *Hedges*, which is what one would expect from such excellent all-round naturalists as Ernest Pollard, Max Hooper and Norman Moore. *Finches* was justly praised as 'a model of clear and logical writing'. *Pollination* is firmly rooted in English amateur botany. But titles like *Ants* (1977) and *Pedigree: Words from Nature* (1973) which need a light, humorous touch to make them palatable, are positively Germanic in their ponderous tones. I have tried hard with *Ants*, but after a few pages found myself praying for Fabre. In a long review, Maurice Richardson summed up exactly what was wrong with the series at this time.

> '....The important thing when you go to the ant is to stick to the facts. This you can rely on M.V. Brian to do .... Much of the text is based on his own observations. His writing is lucid, on the dry side. He rarely hazards a metaphor and any hint of anthropomorphism is alien to him. When describing worker ants dragging big prey to their nests, he does not compare them to a gang of labourers on the lump [sic]; he gives you exact, experimentally-derived figures for their horse power: "Single workers of *Myrmica rubra* developed 0.8×10^{-6} horse power and of *Formica lugubris* 3.2×10^{-6} horse power. When this [measurement] was tried for pairs of ants it proved difficult. However in thirty per cent of cases the second ant contributed nothing at all".'
>
> *Maurice Richardson*, review in *The Times Literary Supplement*, 1977

By this time, Billy Collins, James Fisher and Julian Huxley were no more, and the most active replacement editor, Kenneth Mellanby, might have been more tolerant of this sort of thing than they. The last bestseller of the series, *British Birds of Prey* (1976), was, perhaps significantly, written by an amateur naturalist (though with 'amateurs' like Leslie Brown one needs to use quotation marks). That book deserved its success, and even survived the critical and not always friendly attention of British ornithologists (the expatriate Brown had been expecting their reviews to be 'bad and ungenerous, if not actually libellous'). *Birds of Prey* was a legitimate New Naturalist title, but one has more reservations about the flocks of birds that were entering the series at this time. Eight out of 20 titles published between 1971 and 1985 were bird books, whereas among the first 40 titles, only two were wholly devoted to birds. There was some justification for this in that the editors had decided to replace the unsuccessful bird monographs with books on families of birds in the hope that they would sell better. There were also many knowledgeable and experienced ornithologists ready to contribute to the series, whereas in other areas suitable authors were harder to find. Even so, there were far too many Thrushes and Tits, Waders and War-

Eric Simms, author of a record four volumes in the New Naturalist series, photographed in 1971, at the time of the first of them, *Woodland Birds*. (*Photo: Eric Simms/Stanmore Studios*)

blers for the health of the series, especially with several other publishers competing in the same market and publishing the same sort of books. Over-specialisation was a common accusation made against the series then, and understandably.

We could note one more unfortunate trend in the later books. Topical, scientific books date more quickly than broadly-based natural history. We have noted this with *Pesticides and Pollution*, and it was even more evident in *Nature Conservation in Britain* (1969), the classic instance of a title that was pensionable at birth. Moreover, while most of the earlier titles went through several editions when revision was at least theoretically possible, (*Highlands and Islands, Mountains and Moorlands, A Natural History of Man in Britain*, and *Life in Lakes and Rivers* were all given a new lease of life by this means) most of the later ones were not afforded that luxury. One near-guarantee of inbuilt obsolescence was to employ an elderly author. That was the trouble with *Fossils* (1960). The octogenarian H.H. Swinnerton had written a readable, well-illustrated book, and one of the few to concentrate on British fossils and relate them to the chronological sequence. But, as a colleague later pointed out, 'the contents seem older than the book'. *Fossils* might have been written in the 1940s or even the 1930s: few of the references cited are of later date than 1940. The same was true to a lesser extent of *Man and the Land* (1955), Dudley Stamp's sequel to *Britain's Structure and Scenery* on the evolution of the landscape. Stamp wrote it with his customary efficiency, probably relying on his general knowledge of social and economic geography, built up over previous decades. His prehistoric chapters looked rather stale even in 1955 and from a modern standpoint read almost like myth. In an unfavourable review of the book, Geoffrey Grigson noted various 'whoppers' and claimed to have discovered others. The book was a good candidate for early oblivion, but it caught on in school and college libraries and remained in print as a sort of literary coelacanth until 1983.

It would be unfair to place Herbert Edlin's well-known book, *Trees, Woods and Man* (1956), in the same bracket, but it, too, was very much a product of its time. In its dismissal of indigenous trees and woods as virtually worthless and its espousal of conifer forestry during its most ruthless phase, Edlin was doing no more than voicing current practice. He himself loved trees and traditional crafts, but as an employee of the Forestry Commission he may not have felt a complete independence of opinion. Today, of course, attitudes have changed, and a New Naturalist woodland book would now be concerned far more with native ancient woods than with parks and plantations. The best-known sentence in *Trees, Woods and Man* concerns elm: 'We need not fear that these splendid trees will cease to stand as landmarks along our hedgerows.' Shortly after that was written, the elms were all gone.

This review threatens to turn uncharitable. If I have criticised some aspects of the later books of the series, it is only to stress the very high standards of the earlier ones and on the whole, its maintenance through rapidly changing educational and commercial circumstances. The scientific pedigree of the New Naturalist series is well-nigh impeccable. On the question of style, the books vary from pedestrian to brilliant. In terms of readability, the series has, I am happy to suggest, shown distinct signs of improvement in recent years, witness *Ferns, Caves and Cave Life*, and *Wild and Garden Plants*. Perhaps, after a decade or two of 'Scientific Correctness', we are witnessing a revival of an older literary tradition and a reversion to the popular appeal of the founding New Naturalist authors.

7

Naturalists All

As one might expect, the authors of the New Naturalist series come from a wide variety of backgrounds and vary enormously in terms of personality, outlook and career. The one thing they have in common, of course, is that they wrote a book for our series. In doing so, and bearing in mind that the majority of them were asked to do so, they were expressing no superficial aspect of their outlook. Reading about them – and by the nature of things, more is available about the dead than the living – it is clear that natural history meant a great deal to these people. In many cases, it lay at the root of things, whether they became professional biologists or remained as gifted amateurs. They were 'new' naturalists in the 1940s sense of embracing the recent advances of science to illuminate the workings of nature; but in another sense their approach was rather old-fashioned. Their contributions to science, though considerable, were almost entirely non-technological. Nearly all of these authors were first and foremost masters of field study, based on observation and simple experiment. They were not so much interested in data sets and theoretical modelling as in the relationships and behaviour of living wild animals and plants, or with the shaping of the scenery. The other matter that we can usefully notice at the start, because it is less common among scientists today, is that they were good communicators and could write in plain, everyday language without over-simplifying; and, in some cases, they wrote extraordinarily well. The series was blessed, in its first two decades at least, with writers who seemed to find the right level intuitively. It may not have been possible to do so on such a broad range of subject matter at any other period. These people were exceptional not only in their personal qualities but in their experiences.

We can note two broad 'generations' of authors: those who pioneered new academic disciplines, new ways of studying nature, and later became founders of institutions for nature conservation and field study that are still with us today; and a younger 'generation' who were, or are, often members of staff of those selfsame institutions. I owe my own professional background in the Nature Conservancy to the work of Max Nicholson, Julian Huxley, Dudley Stamp and other writers in the series, and have been fortunate enough to have been a colleague of Norman Moore, Morton Boyd, Niall Campbell and Colin Tubbs. In this chapter, as elsewhere in this book, I have devoted more space to the earlier than the later generation, partly for reasons of space but also because the New Naturalist ethos is rooted in those now rather distant days. Many of the older writers were among the founding fathers of ecology and modern field study. Though they also founded new institutions, they rarely became institutionalised themselves.

There are more than 100 authors in the New Naturalist series, and this chapter has to be selective. I have teased out four strands from a tangled skein: childhoods (because an interest in natural history is nearly always rooted in childhood experiences); what I have called the 'moss-gatherers' of the series, that is to say those who stayed in the same place for much of their working lives, relative to those in the third strand whom I have classed as the 'world adventurers' (arbitrary classes of course, but Linnaeus might have sympathised with that problem). Finally, I have included a postscript about Frank Fraser Darling and J. Morton Boyd who, between them, represent a Scottish tradition of natural history that I might have neglected elsewhere.

The sources for this chapter are legion, but I outline the main ones at the end of the book.

New Naturalist childhood

When reading about the lives of the 'first generation' New Naturalists from a gradually acquired collection of cuttings and off-prints, I was surprised at how few of them had the advantage of a privileged background. Relatively few went to Eton, Harrow or other top public schools, and many attended grammar school. They were self-made men, whose fortune was not in their pockets but in their genes (pun unintended). The 'typical' New Naturalist author came from a solid middle-class background around the turn of the century, often in business or one of the professions. A few, the Fords, Salisburys and Pearsalls, were scions of ancient country families (in Cumberland, Hertfordshire and Worcestershire respectively). More frequently, their background was in the prosperous urban middle-class of the late Victorian era. S.W. Wooldridge, A.D. Imms and C.B. Williams were the sons of bank managers. Alister Hardy's father was an architect, Harrison Matthews' a manufacturing chemist, Macan's and the Campbells' were career officers in the army. W.B. Turrill's father seems to have stepped straight out of a novel by Thomas Hardy as provision merchant and sometime Mayor of Woodstock. Perhaps surprisingly, in only a minority of cases did the New Naturalists belong to families of high academic achievement. Huxley is, of course, the best known example, but Maurice Yonge, James Fisher and W.H. Pearsall were the sons of headmasters and Kenneth Mellanby might have owed much of his irrepressible self-confidence to his family background, a small galaxy of scientific professors and distinguished medical men. Men like John Raven, Mellanby and E.J. Salisbury must have grown up surrounded by brilliant, successful people. John Russell might have had a less comfortable upbringing in that his schoolmaster father is said to have displayed 'an independence of judgement that led to frequent conflict with his employers and to consequent changes in employment'. He ended up as a Unitarian lay minister in the Midlands.

A few had to count the pennies. Ramsbottom senior was a letter-carrier in the Manchester leather trade, which was unlikely to have been particularly well-paid. His son John, the future mycologist, earned his school fees as a pupil-teacher, which enabled him to pass the Cambridge entrance exam. J.E. Lousley, too, is said to have grown up in straightened circumstances. But while there were few rags-to-riches stories among the New Naturalists, neither were there many millionaires. The two *Grandes Dames* of the series, Miriam Rothschild and Cynthia Longfield, came from privileged backgrounds. Miriam's pedigree, daughter of one great naturalist and niece of another, is too well-known to repeat here (I refer the interested reader to Derek Wilson's book on the Rothschilds). Cynthia Longfield (1896-1991) was the last surviving member of a wealthy Anglo-Irish family which held broad estates at Cloyne in County Cork. Both ladies chose, very much against the social grain, to follow a scientific career, which in Cynthia's case took her all over the world in quest of dragonflies, supported by a modest income from a family trust. She was one of the great individualists of the series, tough and intrepid, but who seemed to retain her health and femininity throughout her adventures. Some of her sayings are remembered: 'If you're interested in the peoples of the world, you'll *just have* to get used to travelling on local transport.' and, 'I do find pangas *so* useful in the jungle, don't you?' There are not many left like her.

Bruce Campbell (1912-1993), co-author of *Snowdonia*, in c. 1975. (*Photo Eric Hosking*)

The home life of many young people at the turn of the century was strongly tinged with religion. Sometimes this could be stultifying. But for the fortunate ones, like W.H. Pearsall, H.H. Swinnerton and E.B. Ford, religion led, through an enhanced sense of wonder, to nature study. For such people, natural history had a profound moral and religious dimension. Today, the former is most often expressed as a concern for endangered species and damage to the environment. Former generations were more philosophical and less wholly secular about man and nature. John Raven's father, Canon Charles Raven wrote a biography of the French scientist-philosopher Teilhard de Chardin, whose transcendental speculations about speciation and human evolution led him into original, if theologically unorthodox, arguments that proved the existence of God. Edward Armstrong (1900-1978), Anglican priest and natural history polymath, found no contradiction between his faith and wonder at the beauty of the natural world; rather the opposite, a conviction that the first leads, hand in glove, to the second. Similar views and concomitant ethical considerations seem to have occupied the minds of several New Naturalist authors, whether Anglicans, like Alister Hardy and Sam Berry, or Methodists, like W.H. Pearsall, or Congregationalists, like Sidney Wooldridge. Bruce Campbell (1912-1993), whose religious convictions made him both a socialist and a pacifist, broadcast a sermon from the parish church at Selborne, at the beginning of National Nature Week, in 1963, on 'Conservation and Christianity'. To Campbell, man's 'power to change things overnight' carried with it the moral obligation to husband the earth's resources. In accepting that evolution was the means God had chosen to work towards his ends on earth, Christians had a special obligation to become good conservationists: 'Surely the exercise of the mind to the glory of God includes the study and appreciation of the whole of His world and the taking of thought to conserve it?' Thirty years on, one seldom hears much talk about God by conservationists, and environmental issues have taken on a secular, humanist slant. But without admitting the presence of God, even if only as a symbol, can there be a proper explanation for the naturalist's sense of wonder? And without it, what is nature writing but the repository of sickly rural longings and matter for the boring article?

Natural history is one of those vocations, like art and music, that one seems to be born into. The New Naturalist authors were no exceptions. It is said of Edward Armstrong, as no doubt it could be of many others, that 'as a young child, he became entranced by the beauty of nature and natural things and

W.B. Turrill (1890-1961), author of *British Plant Life*, holding open his book on *Knapweeds*, written with his long-term collaborator, E.B. Marsden-Jones in 1954. (*Photo: Royal Botanic Gardens, Kew*)

concluded .... that he could himself find out things which grown-ups did not know'. One might hazard an unscientific guess that many New Naturalists were clever but rather shy little boys who spent a lot of time on their own. The American writer Stephen Fox pointed out that a remarkable number of leading conservationists over there were 'only children', and perhaps there is a tendency for children to turn to nature as a compensation for less than happy human relationships. Armstrong himself declared that to study a species intensively you needed silence and concentration, which are not easily reconciled with company. Nevertheless, nature study in late Victorian and Edwardian England was a sociable, even a fashionable, activity, and many of our authors seem to have received plenty of parental encouragement. We could pick out examples almost at random. As a native of Woodstock in Oxfordshire, W.B. Turrill (1890-1961) spent his boyhood in an area rich in wild habitats and uncommon plants – Blenheim Park, Wychwood and the winding wooded banks of the lovely River Evenlode. 'Here', as his biographer C.E. Hubbard recalled, 'he spent his days searching the woods, fields, broad green lanes, ponds and watercourses for natural history specimens, drawing inspiration and delight from each fresh discovery, and studying them in greater detail over the years, with an increasing specialisation on plant life.' In this, he was encouraged by his mother, who was from a farming background and gave young William his own garden plot to grow a variety of flowers and vegetables. Turrill dedicated *British Plant Life* to the memory of his mother, 'to whom I owe my love of plants'. C.B. Williams (1889-1981) was less fortunate in his surroundings, living above the bank on a busy street corner in the middle of Liverpool, where 'there was not a living tree within a mile'. Nor were his parents at all scientifically minded. 'But to compensate for the absence of countryside he and his elder sister were given books on natural history, they kept an aquarium, and on Sunday mornings were taken to feed the seagulls from the landing stage on the River Mersey.'

At least three of the New Naturalist authors were fortunate enough to have for a father one of the foremost naturalists of the day: Canon Charles Raven; W.H. Pearsall senior, sometime Secretary of the Botanical Society (now the Botanical Society of the British Isles); and Colonel Ronald Campbell, ardent ornithologist and nest-finder *extraordinaire*. The sons took after their fathers. Charles and John Raven collaborated in a project to paint the entire British flora in watercolour; Bruce Campbell, whose earliest memory was of his father lowering him down a bank of the Itchen to look into a grey wagtail's nest, went on to write *Finding Nests* (1953) and *A Field Guide to Birds' Nests* (1972 with James Ferguson-Lees); and the

North Fen, Esthwaite Water, the scene of W.H. Pearsall's classic study of an English Lake. (*Photo: Peter Wakely/English Nature*)

two W.H.Pearsalls went fly-fishing and plant-hunting by boat, the one lowering a pronged and weighted dredger to haul up submerged plants, the other manning the oars. W.H. Pearsall the younger's classic study of Esthwaite Water was begun on one of these excursions in 1914. All in all, the impression one receives from the memoirs and obituaries is that of fairly sunny childhoods, in which every sympathy and encouragement was given to the young naturalist. No doubt that is not the whole story. Some probably went through hell at public school, especially during the austere years of the First World War. But, on the whole, the flavour is closer to *Bevis* than to *David Copperfield*.

The precocity of luminaries like Dudley Stamp (1898-1966) and H.J. Fleure (1877-1969) might owe something to childhood illness. As we have seen, Stamp had been laid up with so many illnesses of one kind or another that he received hardly any formal schooling at all. Fleure was another invalid, delicate, blind in one eye, with pleural cavities full of fluid. Possibly this was a consequence of elderly parentage. Fleure's father, accountant to the States, Guernsey, was born in 1803 and was aged 74 when Fleure *fils* was conceived (the Fleures father and son spanned a remarkable 166 years: from Napoleon to The Beatles!). At the start of a career in natural history, severe illness is not necessarily a drawback, though there is an element of make-or-break about it. For Fleure, only two pursuits were available: reading and, when he felt up to it, walking. But that was more than enough: 'Alone with his thoughts he explored Guernsey and came to know his native island intimately, to become the born field naturalist, conscious at a very early age of the wonders of the natural history of shore animals and plants,' wrote his biographer, Alice Garnett. Like Stamp, Fleure became an avid reader, and from his reading and love of nature there grew 'a desire to try to understand

Pencil portrait of A.D. Imms (1880-1949), author of *Insect Natural History*, by Paul Drury, 1945. (*Department of Zoology, Cambridge*)

more deeply all the physical and human features – past and present – of his native island, and from this to study current theories of evolution'. If, as has been suggested, enforced idleness as a child can lead to super-industry in the adult, this might explain the relentless dynamism of Stamp and Fleure in later life. Though frail and, on the face of it, rather meek, Fleure habitually pushed himself to the limit. Remarkably, he lived to the age of 92, becoming revered as a teacher and, through the medium of geography, a promoter of international understanding and peace.

A.D. Imms (1880-1949) was another frequent patient, laid up with attacks of asthma. He, too, was tougher than he looked. His Royal Society memoirist recalled that 'the boy, debarred from many of the activities of his fellows, quickly devoured and assimilated all the popular handbooks for [butterfly] collectors ... And then the chance purchase of Todd's *Encyclopaedia of Anatomy and Physiology* evoked a more scientific interest ...' The way to the *General Textbook of Entomology* and *Insect Natural History* had opened. We might note the same precocity in Derek Wragge Morley (1920-1969) who, crippled by a severe illness in his early teens, began to study the ants that crawled about his wheelchair. Natural history is a good antidote to adversity.

A few authors recalled what to them was the defining moment that made them naturalists. To Julian Huxley, it was on being confronted by a toad at the age of four: 'Out of the hawthorn hedge there hopped a fat toad. What a creature, with its warty skin, its big eyes bulging up, and its awkward movements! That comic toad helped to determine my career as a scientific naturalist.' For Max Nicholson, at about the same age, it was 'the sparrow-hawk seizing and making away with a favourite yellow chicken'. The young Alister Hardy had several experiences to choose from: the rock pools on the family holidays at Scarborough, a neighbour's collection of beetles, and the mantelpiece in their study, made of fossiliferous marble and full of the remains of marine creatures from the age of the dinosaurs.

In his book, *The Environmental Revolution* (1970), Max Nicholson devoted several lyrical pages to his own childhood experiences, generally enjoyed in solitude or in the company of just one or two like-minded friends. Real nature was more exciting, and vastly more challenging, than the books he had read or the museums he had visited: 'There was a special magic in visible change – a sudden fall of snow, a swirl of mist, an overnight sheet of hoar-frost on the grass, the shape of great clouds passing across with their shadows, the buds and catkins and new foliage of spring and the eagerly expected but ever surprising return of the swallow and the cuckoo.' The theatre of such observations and feelings was often very small: 'a chalkpit or sandpit, a little pond, a gorse thicket or the bend of a stream became a world in themselves, where time stood still'. It is in such places that naturalists are made, so long as such places exist near one's home and one can roam without fear.

Perhaps a discussion of boyhood wonder is the place to mention what to many of the authors was the first phase of their natural-history careers – collecting. The older generation of authors belonged to a time when specimen-collecting, often of the widest kind from rocks and minerals to birds' nests and feathers, was eagerly indulged in. Many of the authors of the series collected butterflies or birds' eggs in their youth, and the latter became expert nest-finders in the process. A few never quite grew out of it. Part of the antipathy between Desmond Nethersole-Thompson and Leslie Brown was due to the former's egg-collecting exploits in the 1920s and 1930s; Brown's labelling in *British Birds of Prey* of Thompson as an egg-collector in the apparent present tense nearly brought a libel action (and since Brown himself collected eggs as a boy, his suit might have stood on brittle glass). One notices a broader magpie instinct at work among the New Naturalists. Lousley, Steers and Dudley Stamp were keen stamp collectors throughout their lives. It was Dudley's influence that persuaded the Post Office to issue a set of geography stamps in 1964 (in which, incidentally, the 8*d.* stamp bears a passing resemblance to the jacket of *Dartmoor*). Steers and Pearsall were also train-spotters. Turrill collected pamphlets. Yonge amassed marine curios and illustrated books. Lousley owned the largest private herbarium in the country.

Collections of skins, seashells and microscope slides were, of course, in many cases a necessary adjunct to serious study. It is quite likely that most serious naturalists of the inter-war years were collectors of some sort. Possibly there is also some psychological impulse at work here. One sees it in a sublimated form in the obsessive 'twitching' of rare birds by some present-day birders – not to mention collectors of the New Naturalist books! Collecting, like gardening and fishing, binds us to the object of our devotion. The apparent absence of that same acquisitive sense among 'conservationists' today, except in debased forms, is rather puzzling, and it is one of the significant differences between an older generation and our own: from active field pursuits to passive spectating, from the particular to the generalised. How many amateur naturalists today study beetles, or slime-moulds, or diatoms? And if you don't know anything about beetles, how can you hope to protect them?

Before we leave New Naturalist childhood, a word or two about Christian names. The older generation of New Naturalists belonged to a period and class where people were commonly known by their surname or, among friends and colleagues, by a nickname. Alister Hardy is an interesting example. He evidently disliked his Christian name, regarding it as a bit soppy. His second name, Clavering, was there for sentiment, not use: it was his mother's maiden name. To many scientific colleagues, he was 'A.C.' or 'The Prof' (even 'Uncle Prof' to his family). His father-in-law called him 'Ali', old school chums 'Glider', and the shipmates and miners he mixed with as Acting Captain Hardy during the First World War referred to him affectionately as 'Uncle Mac', which suggests that he seemed old and wise beyond his years. To be known by one's initials was not unusual at that time. Hence, Williams was always 'CB', Macan 'TTM', Edwards 'KC', and even Winifred Frost was 'WEF'. I have never heard anyone refer to Pearsall as 'William', or, for that matter, to Tansley as 'Arthur'. It implied no antipathy. Names seem to be formed by an unspoken, perhaps subconscious, mass agreement as to what is appropriate. We have already dealt with the troublesome etymology of E.B. Ford's nickname, Henry. Equally inexplicable to me is the nickname, Sam, of Ford's spiritual successor, R.J. Berry. Sam Berry told me it had something to do with his Lancashire family background – 'Sam, Sam, pick oop tha musket'. Fraser Darling went through a period of calling himself Frank Darling, but fortunately pulled himself together and reverted to the full, fine Frank Fraser in time for the publication of *Natural History in the Highlands and Islands*. Since the 1960s, initials have gone out of fashion and abbreviated nicknames are in universal use, though, unless we except Niko (short for Nikolaas) Tinbergen, not yet on the title page of the New Naturalist books.

Teachers and moss-gatherers

In a sense, all field natural history is self-taught. James Fisher became one of the great ornithologists by going birdwatching when he was supposed to be attending anatomy classes. Max Nicholson equipped himself as a writer and journalist by reading the works of W.H. Hudson. Richard Fitter, Ted Lousley, Ronald Lockley, Ted Ellis and Bill Condry are among the best-known British naturalists, doyens of the art of field study, but they do not have a science degree between them. Yet no one, amateur or professional, knew more about docks and knotgrasses than Lousley, or shearwaters and gannets than Lockley, or field micro-fungi than Ellis. One of the special qualities of the New Naturalist library lies in its blend of the 'amateur' and 'professional', each book absorbing something of both approaches. During the 1930s, and increasingly so after the Second World War, these hitherto separate worlds were moving closer to one another. *The Sea Shore* is written by a professional in the enthusiastic spirit of the amateur. The ecological chapters in Ted Lousley's book *Wild Flowers of Chalk and Limestone* or Ian Hepburn's *Flowers of the Coast* could have been written by Pearsall or Salisbury. But very probably no one at the time could have written a more satisfactory book on either subject.

The majority of New Naturalist authors did have a scientific background, and the full-time university men generally held a doctorate in some relevant discipline. Cambridge was the *alma mater* of the majority of botanists and geographers of the series, though the three King's College scholars – Leo Harrison Matthews, Kenneth Mellanby and M.V. Brian – were, as it happens, all zoologists. John Raven read the classics at Trinity College (and later became a Fellow of King's), W.S. Bristowe science at Gonville and Caius, and Leslie Brown a postgraduate course in tropical agriculture. Oxford and London tie for second place. The Oxford men and women included A.W. Boyd, Wilfred Blunt, Ian Hepburn, V.S. Summerhayes and Margaret Brown. The Londoners, mainly at University or Imperial Colleges, were L.A. Harvey, S.W. Wooldridge, L. Dudley Stamp, E.J. Salisbury and H.R. Hewer. Relatively few of the pre-war generation held degrees from 'provincial' colleges and universities, though Yonge and H.L. Edlin studied forestry at Edinburgh, Swinnerton natural sciences at Nottingham and Pearsall botany at Manchester.

Before the Second World War, university science degrees were more broadly based than today, though, unless one was fortunate in one's tutor, the biological part was still heavily centred on anatomy and physiology. Geology, genetics and animal behaviour were not much in vogue then; indeed, New Naturalist authors were among the pioneers in those very subjects. The student specialising in botany, zoology or geology in his final year would first have received a thorough grounding in chemistry, physics and mathematics. This gave him or her a broad scientific outlook to add to the public school or grammar school education, which was likely to have been weighted towards English and the classics. What is notable is that several future authors switched to botany or zoology in mid-term, having originally read some other subject. James Fisher, Bill Bristowe and Leo Harrison Matthews all started off by studying medicine, before deciding that zoology was more their line. Pearsall switched from chemistry to botany at Manchester. Even so, they might not have received much stimulus for field study from the arid biology classes of the 1920s and 1930s. This they gained from extra-curricular activity: university expeditions to Spitzbergen and the Amazon, special courses at the marine laboratory at Plymouth, or from meeting people like Julian Huxley.

Many of those born before 1930 had to endure a major disruption in their studies – war service – whether in the First or the Second World War, or sometimes both. Alister Hardy, for example, had no sooner begun his first term at Oxford than he was busy training in the Officers' Training Corps to gain a commission. His eyesight was not up to Western Front standards, and he spent the war in command of a bicycle battalion of Durham miners, digging trenches on the east coast, and later in the camouflage school, where his knowledge of animal camouflage came in useful (as did Clifford Ellis' during the 1939-45 war). He did not return to Oxford until 1919, nearly five years after his first term. Some authors distinguished themselves during the war. A.W. Boyd was awarded the Military Cross for his bravery at Gallipoli. John Ramsbottom, serving with the Army Medical Corps in Salonika, was mentioned three times in despatches and appointed an OBE. Military service sometimes took people to remote foreign parts for the first time in their lives, with lasting results in the case of W.B. Turrill who, as a result of his service in Macedonia, began a long association with the Balkans culminating in his magnum opus, *The plant life of the Balkan Peninsula* (1929). At least two of our authors were permanently disabled: Boyd lost an eye, though that did not seem to hinder his birdwatching, and Pearsall was knocked about and permanently deafened by artillery shells (ours more than theirs). Such men probably returned from war traumatised, but with new perspectives and holding new ideals. This is expressed most clearly in the Royal Society memoir of Maurice Yonge (1899-1986). Before joining up, Yonge had intended to become a journalist, and after demobilisation had applied to Lincoln College, Oxford, to read modern history. It took a while for him to discover that his true aims had changed:

> 'Those that survived the carnage believed themselves committed to bettering the world for which so many of their generation had paid such a great sacrifice. Many sought a more meaningful life, other than quiet scholarly retreat. Maurice also rebelled against his self-centred solitude and began to look at life with eyes opened by the War and the need to help his country in a more practical way. He reasoned that primary production was what was needed urgently..... He burned his Oxford boats, sold his history books and persuaded his father to let him study forestry at the University of Edinburgh.'
>
> *B. Morton. Charles Maurice Yonge 1899-1986. Biographical Memoir of the Royal Society* (1992), **38**:382

'A more practical way' – practicality was the keynote of pioneer ecology: the subject became an applied science almost from the start. Some of the New Naturalists worked on problems of food production, others on tropical hygiene, agriculture or fisheries. They studied species of economic importance – herring, whales, rabbits, trout, shipworm and oysters. The Bureau of Animal Population, founded by Charles Elton, saw its glory days during the Second World War, when its studies on how animal populations are regulated suddenly took on great significance in the effort to boost agricultural production at the expense of rooks, rabbits and wood pigeons (it had already achieved a spectacular success in getting rid of the muskrat). Plant ecology was more concerned with the processes and dynamics of vegetation, as developed in the inter-war years by Arthur Tansley, W.H. Pearsall and E.J. Salisbury. Their work was ignored by agriculture departments, however, and its most significant application lay not in food production and forestry, but in the development of nature conservation during the 1940s and 1950s. Pearsall's lake studies did have one early consequence, however, in the founding of a British centre for freshwater research, the Freshwater Biological Association, in 1929.

While a few New Naturalists, like T.T. Macan and C.B. Williams, were engaged in full-time research, the majority of those with university posts had a full teaching load. This was especially so of those, not a few, who held a chair in a newly established department. A university professorship may sound very grand, but during the 1920s and 1930s 'The Prof' might find himself teaching much of the syllabus himself, with only a couple of assistant lecturers, with limited office facilities and little or no secretarial help; quite likely he even did his own typing. A description of H.J. Fleure's study reminds us of the conditions in which some of these books were written:

> '... he prepared all papers in his own immaculate handwriting. His study was always a large room with abundant floorspace. Involvement with so many co-authors and publishers, and with so much editorial work, at times would have disconcerted many; but he evolved his own simple methods of control despite the lack of mechanical aids, filing cabinets, card indexes and the like. The floor ... would be covered in piles of manuscripts, typescripts, and galley proofs etc. When asked where a particular item was, he would walk slowly around the room, finally to alight on one pile and to say that it "was about an inch below the middle of that pile"; and, sure enough, there it would be.'
>
> *Alice Garnett.* Herbert John Fleure 1877-1969. *Biographical Memoir of Fellows of the Royal Society* (1970),16:269

One of those piles might have been the manuscript of *A Natural History of Man in Britain*!

Among the New Naturalists were some of the most influential teachers of field study the century has known. Fleure himself ranks very high among these. He brought to his classes a philosophy: that Man was part of Nature, and that living organisms and their environment are inseparable, and so must be studied together. Half a century on, he is revered as a father of British geography, but he also taught zoology, geology, history and anthropology, in a unifying synthesis of interlocking subjects. His influence was worldwide. W.H. Pearsall (1891-1964) was another great teacher with a large number of disciples. Like Fleure, he had the knack of getting students to think for themselves. His was a versatile mind, bubbling with ideas, and he retained a boyish enthusiasm to the end of his days. Macan noted that 'A detached lay observer watching Pearsall would have gained no inkling that it was unusual for a botanist to be capable of discussing five lines of research in three disciplines.' Derek Ratcliffe, who was less of a devotee, suggests that part of Pearsall's method was his mastery of what Stephen Potter called 'lifemanship': 'He had a great capacity for making *ex cathedra* pronouncements and making other people into supplicants for knowledge at the feet of the master ... He was said to come out with outrageous statements to see the effect they had on listeners, but I am sure they came to believe some of his more fanciful ideas.' E.B. Worthington recalled that Pearsall was notably stubborn at defending his ideas, right or wrong. And that he used his deafness constructively at meetings.

S.W. Wooldridge (1900-1963), author of *The Weald.*

Geology and physical geography lend themselves to field work, perhaps because these subjects, stony in more senses than one, only really become alive on the field outing. One of the champions of field studies was the geomorphologist, Sidney Wooldridge (1900-1963), author of *The Weald.* Wooldridge is one of the moss-gatherers of the series. The son of a Surrey bank manager, he went up to King's College, London in 1918 to read geology and remained there for most of his career, latterly as Professor of Geography. Shortly after graduation, he helped to found the Weald Research Committee, and from then on the Weald became his outdoor laboratory where he worked out the relationship between the scenery and its rocks, and particularly the role of rivers in reshaping the landscape. These studies helped to establish geomorphology as a scientific discipline. He was also interested in the human geography of the area, especially in what influenced its settlement in ancient times. He was good at getting all this across to his students with clarity and a characteristic 'pungency of phrase', perhaps acquired in his extra-curricular capacity as a Congregationalist lay preacher. For him, the Weald fulfilled most purposes. 'The eyes of a fool are in the ends of the earth,' he quoted. Wooldridge was an Englishman to his fingertips, 'full of an almost Chestertonian zest for life', especially for cricket and Gilbert and Sullivan, chapel and natural history. His happiest hours were those spent on fieldwork in the countryside of Surrey, Sussex and Kent. He had a natural eye for the lie of the land, and for clues in the rocks and landscape, and he became a leading light of the newly founded Field Studies Council. The establishment of its second centre at Juniper Hall, Mickleham, was to a large extent his doing, providing a permanent centre for south-eastern studies to take over from the Timberscombe Guest House, run by his friend and collaborator Fred Goldring.

Another apostle of field study was Wooldridge's fellow geographer, K.C. Edwards (1904-1982) of Nottingham University. Edwards' early teaching duties were typically onerous. In the 1920s he would spend 30 hours each week teaching geography and geology to undergraduates, trainee teachers and, appropriately enough for that county, mining engineers. He must quickly have learned a certain versatility, and soon became to Nottinghamshire what Wooldridge had been to The Weald. And like Wooldridge, the field was his true classroom. He organised regular field courses not only for his own students but also for adult education courses and he welcomed students and teachers from the Continent, especially from countries where he himself had studied, notably Hungary and Luxembourg.

K.C. Edwards (1904-1982), author of the *Peak District*. (*Photo: Department of Geography, Nottingham University*)

His field camps at Easter were a popular innovation, one of Edwards' students recalling 'the lasting bonds of friendship and goodwill' forged among the tents, which helped to make Nottingham geography a notably convivial subject. 'K.C.' was much involved with rambling, youth hostels and footpath preservation societies, all of which made him a frequent treader of paths in the Peak District and established his credentials as the author of that volume in the series.

Leslie Harvey (1903-1986), the author of *Dartmoor*, was a fellow spirit. Harvey, Exeter's first and last Professor of Zoology, moved to the South-west from Edinburgh in 1930, and immediately fell under its thrall; he remained there for the rest of his life. Like Edwards, he took on a level of teaching duties that would have devastated anyone but the total dedicatee; reflecting his location, his classes varied from trainee teachers to agriculturists and meat inspectors. Like him, too, Harvey introduced field camping expeditions to the syllabus, with his wife Clare looking after the botany. After the Second World War, these courses were usually based near the sea, and Harvey became an adept at marine ecology and shore life: 'His knowledge of marine organisms', wrote a colleague, Tegwyn Harris, 'was genuinely encyclopaedic and, since he was a good biologist, his teaching of them did not stop at a mere recitation of names and classification, but embraced their broader biology, ecological relationships and behaviour.' He was, in a phrase, a New Naturalist. Almost exactly the same wording has been used in praise of Pearsall and E.J. Salisbury, and is equally applicable to other authors in the series. Harvey retired to the scene of one of Exeter's popular marine courses in the Scillies, to a house 'overlooking the wild headland of Peninnis and the weed-blanketed shores of Porthcressa'. There he remained, with his rock garden and his crossword puzzles, and there he died, suddenly as we would all wish, in 1986.

Local field study is *par excellence* the province of the moss-gatherer. World adventurers like Darwin may acquire great insights into the processes of nature, but the observance of detail, the grain of the wood, is the hallmark of one of England's gifts to the world: the local naturalist. There can be few who lived up to that tradition more successfully than Ted Ellis (1909-1986), whose name will always be linked with his adopted county, Norfolk. Ted Ellis' biography is aptly sub-titled 'The People's Naturalist': his enthusiasm for nature was always a hallmark of his regular writings for the *Eastern Daily Press*, and later, radio broadcasting and television. He was a born communicator, and had that increasingly rare ability to describe exactly what he saw, with the aid of a first-rate pair of eyes and a detailed nature diary which he maintained throughout his life. Though he was, strictly speaking, a 'professional', as the keeper of natural history

Ted Ellis (1909-1986) author of *The Broads* at Wheatfen Broad in 1982. (*Photo: Peter Loughran*)

Ian Hepburn (1902-1974), author of *Flowers of the Coast*. (*Photo: The Old Oundelian*)

at Norwich Castle Museum, in Ted Ellis there is the spirit of the amateur English naturalist at its best. It is there in the fleeting detail of weather, light and sound in his nature notes, in his untidy study stuffed full of diaries and specimens and books and microscope slides and a thousand other things, in the steady trail of visitors to his marshman's cottage in the wilds of Wheatfen Broad. He was an authority on fungi, especially rusts and smuts, but always remained a natural history all-rounder, interested in everything. He was offered tempting inducements to move, to the Imperial Mycological Institute in London, or the wardenship of the Field Studies Centre at Flatford Mill. But he stayed where he was, and is remembered not only as a naturalist but as a Man of Norfolk and a great local character. The reedy wilderness where he lived with his wife and family is now a nature reserve, run by the Ted Ellis Trust, and open on every day of the year.

One would never know from reading *Flowers of the Coast* (1952) that Ian Hepburn (1902-1974) was an amateur botanist, and in fact taught chemistry at Oundle School. It is unlikely that Hepburn, the most modest of men, volunteered for the job of writing that book. Perhaps he was talked into it by James Fisher, whose father was headmaster of Oundle at the time that Hepburn was housemaster. His friend, Ioan Thomas, remembers them going through the list of new boys, with Kenneth Fisher taking all the good games players for School House while Hepburn took all the able boys for his Laxton House. The two often went off birdwatching together, and each year 'Hep' used to count the herons at Titchmarsh and Milton Park. At Oundle, Ian Hepburn is remembered more for his support of school music than for his natural history, and evidently few knew that he was on the Council of the British Ecological Society and was one of our leading amateur botanists. He was by all accounts a delightful man and a born

V.S. Summerhayes (1897-1974), author of *Wild Orchids of Britain*, working on African orchids at Kew. (*Photo: Royal Botanic Gardens, Kew*)

teacher. He must have loved Oundle, remaining there for 39 years until his retirement in 1964. Botanists remember him as a mainstay of local naturalist trust activities in Northamptonshire and, later, Cambridgeshire. On excursions, he always had a copy of the McClintock and Fitter *Pocket Guide to Wild Flowers* in his pocket which he referred to as 'his crib'. 'If you're not interested in anything,' he would warn the boys in his House, 'you're going to be a pretty boring person, aren't you?' He dedicated *Flowers of the Coast* to his musician wife Phyllis, 'who loves the sea but is sometimes uncertain of her botany'.

Perhaps it is not quite right to include the three 'men of Kew', E.J. Salisbury, W.B. Turrill and V.S. Summerhayes among the moss-gatherers, for they all travelled widely, collecting plants and giving lectures, but one associates them with the stable and relatively tranquil worlds of the botanical library and the herbarium. Turrill and Summerhayes were quiet men who spent most of their careers at Kew. Victor Summerhayes (1897-1974) was the orchid specialist, assiduously building up Kew's collections of African Orchidaceae, bottling and labelling each plant himself. By training, however, he was not a systematist but an ecologist, and in 1921 took part in the Oxford University expedition to Spitsbergen, resulting in two well-known papers by Summerhayes and Charles Elton on animal populations and food-chains. His ecological outlook is evident in *The Wild Orchids of Britain*, with its detailed first-hand descriptions of habitats and his interest in evolution at work among the marsh orchids. Before the Second World War, he got about the country looking at orchids in Edgar Milne-Redhead's tiny BSA 3-wheeler; and afterwards he was often a passenger in his fellow orchidophile Donald Young's open MG. Typically, he based the book on his own studies in the field, not the dried up or pickled specimens back at base. Milne-Redhead recalled him as 'a small man with slight physique, but mentally always very much alive. He had a wonderful habit of gesticulating when, in conversation, he was describing the features of a particular orchid flower which he imagined before him enormously magnified.' He always wore a dilapidated hat when botanising and insisted on pronouncing the word species as 'svecies' for no reason that anyone could gather.

William Turrill had joined the Royal Botanic Garden at Kew as a temporary assistant at the age of 18, and obtained his degree at evening class. He was honoured in later life as one of the great systematists, helping to turn plant taxonomy from a dusty, herbarium-based science into a modern experimental one, taking in the latest advances in evolutionary and cell biology and biochemistry. But amateur botanists remember him as one of themselves, never happier than on field collecting expeditions and in his own garden. He and E. Marsden-Jones formed one of the great amateur-professional partnerships, carrying out together the extensive investivations of knapweeds and bladder campions which were published in book form by the Ray Society in 1954 and 1957 respectively. Turrill was a natural hoarder, accumulating a fine reference library and collections of pamphlets and cuttings. He hated administration, though he does not seem to have suffered from it unduly. His was exactly the life a would-be professional botanist might dream about, but today, alas, he or she would have to go on dreaming.

Sir Edward Salisbury (1886-1978), author of *Weeds and Aliens*. (*Photo: Royal Botanic Gardens, Kew*)

For 13 years during and after the Second World War, Turrill's and Summerhayes' boss was Sir Edward Salisbury (1886-1978), the author of *Weeds and Aliens*, and much else besides. Salisbury's directorship at Kew was dominated by the recovery and restoration effort after the war. There was too much to do with too little money, and Salisbury sat on far too many advisory boards, governing bodies, councils of learned societies and the like to give the Royal Botanic Gardens his full and undivided attention. Nor was he, in any case, much good at administration. Like Dudley Stamp and James Fisher, he tended to take on too much. Edward Salisbury had enormous self-esteem. He was an incessant talker and was rather obviously 'very pleased with Edward Salisbury'. David Streeter remembers him as 'a nineteenth century character' in his dark clothes, old-fashioned collars and spats. Miriam Rothschild recalls the brilliance of his intellect; he always had a story to tell about every plant and he shone on field courses at Box Hill and elsewhere. It was typical of Salisbury that, when asked to contribute a book on some elevated aspect of British botany, he replied that he would much rather do one about weeds. At the heart of Salisbury was his 'passionate interest in the living plant' whether in the wild or in a garden, and by conveying that passion well he became a gifted populariser. His studies of woodland and coastal vegetation, written seventy years ago, are still worth reading today (for they remind us of things since forgotten or overlooked), as are his books *The Living Garden* (1935) and *Downs and Dunes* (1952). He was a rare example of a phyto-geographer who retained his interest in the living individual plant. As *Weeds and Aliens* constantly reminds us, plants have function as well as form, and Salisbury's weeds behave like little seed-bombs ticking over, adaptable, aggressive and quick to exploit an opportunity. Salisbury's papers and books are usually illustrated by his own drawings and diagrams which have an endearingly home-made quality – some of his characteristic labels look like newspaper cuttings. They emphasise as much as anything that nearly all of Salisbury's work lay well within reach of the gifted amateur. He was the last survivor among the founders of plant ecology in Britain, perhaps the last great 'professional amateur', and he always retained the breadth of vision he had shared with F.W. Oliver and Arthur Tansley.

World adventurers

James Fisher once described his friend Leo Harrison Matthews (1901-1986) as 'an old-fashioned naturalist'. He meant it as a compliment. In Fisher's opinion, *British Mammals* was the best book ever written on the subject, precisely because it ignored contemporary whims and fashions: he 'pays just as much attention to the ancient and just now rather demodé subjects of anatomy, physiology and adaptation as to animal populations and animal behaviour, which are the fashionable things to be interested in'. He also, added Fisher, loved travel and adventure. 'Old-fashioned' in Fisher-ese meant the opposite of over-specialised. It meant seeing nature in the round, to pursue the romantic quest of the fulmar and the albatross and to have fun doing it (fun, that is, in the intellectual as well as the physical sense).

Due attention to traditional anatomy is exactly what one would have expected from Harrison Matthews, who was noted for relating form to function. From his student days, when he found a Barbastelle while bat-hunting among the rafters of King's College Chapel, he had been fascinated by the hearing apparatus of bats. His published work on British bats was concerned mainly with their reproductive organs and sexual cycles (another obsession), but his interest in bat communication led during the Second World War to his involvement in the development of aircraft radar. As Matthews had discovered, the battle between an aeroplane and ground radar rather resembles that between a bat and a moth. (Nature has invented countermeasures in which moths can detect the sonar of bats in time to take violent evasive countermeasures. Unfortunately, a moth can do what an aircraft cannot and snap shut its wings to plummet unexpectedly to the ground). One can trace a similar quirky interest with the external bits and pieces of mammals in many of Matthews' papers and notes written before and after the war: the ears of newly born seals, the genital anatomy and physiology of gibbons, extra nipples in chimpanzees and an inter-sexual brown rat. The 'catch-phrase' of his breezy, refreshing lectures at Bristol University was: 'I don't know what it's for – and I don't think anyone else does either.'

Leo Harrison Matthews (1901-1986) taken on the occasion of his 80th birthday. (*Photo: Marcus Matthews*)

Matthews led a full life by any standards. Apart from the war-time interlude, it could be divided broadly into his early travel years, which were probably the most enjoyable in his life, his years as scientific director at London Zoo and his busy retirement, spent writing books. He was 'the last of the great travelling naturalists'. His university students noted something in his manner and bearing that suggested a long acquaintance with the sea and ships. He was already a fairly seasoned mariner when in 1924, shortly after graduation, he joined an expedition to South Georgia aboard Captain Scott's old ship, *Discovery*. He remained in the South Seas for four years, among the world's wildest land and seascapes, studying elephant seals, sea-birds and whales. His *Wandering Albatross* (1951) recalls those days of adventure, and displays Matthews' prowess as a storyteller. The book was about real albatrosses, but in another sense the wandering albatross was Matthews himself. Characteristically, he dedicated it 'to all bird lovers, particularly those that love them piping hot with bread sauce'. One can see why James Fisher liked him.

Matthews belonged to the last great age of world discovery, when promising young naturalists began their careers with a major ocean voyage in the same spirit as Sir Joseph Banks and Charles Darwin. We should remind ourselves that these expeditions were not the perfunctory things of today, of no more than a month or two's duration, but major commitments that could last several years and yield results that would take several more years to analyse and write up. They could be, and often were, a crossroads in life, a test of field skills and of character, of a broadening of horizons and a change in perspectives. It is safe to say that those who embarked on these journeys while the going was good, during the 1920s and 1930s, returned home different people.

Take Alister Hardy, Matthews' fellow *Discovery* voyager. A few years older than Matthews, Hardy (1896-1985) had made a name for himself as a clever and hard-working marine biologist of distinctly outdoor bent. Like E.B. Ford, he had had the good fortune to run into Julian Huxley at Oxford 'who introduced him to the excitement of new ideas and discoveries'. His subsequent work on the food of the North Sea herring and his invention of a plankton detector had attracted attention, as had the fact that Hardy was a good team player and a good sailor. In 1926, he was invited to join the *Discovery* expedition as Chief Zoologist. The expedition's science was a cross between Matthews-style form-and-function zoology and the kind of animal ecology then being developed by Charles Elton, which related an animal to its food supply. Hardy's scientists were given the pleasant task of measuring and disembowelling whales brought back by the catchers. Although Hardy himself spent some time aboard a whaler, and later wrote a disturbing account of the killing and cruelty he witnessed, his main preoccupation was with the mystery of the whale's food. This led him into the world of marine plankton. His team were the first to track down and sample the vast shoals of krill that abounded in the whaling grounds. 'Imagine our excitement and joy,' wrote Hardy in *Great Waters*, 'as we saw [the] nets rising to the surface as if aglow with fire – the blue-green fire of phosphorescence; each tow-net bucket was full and the sides of each net were plastered thick with krill, all glowing brilliantly.'

Plankton was henceforward Hardy's subject in all its biological complexity and physical demand: the tow-nets, detectors, sample bottles and other laboratory clutter pitted against the roughest seas of the world. From the pursuit of the world's largest animals to some of the smallest, it was, as Hardy himself remarked, 'a challenge to modern oceanographic methods'. Yet by modern reckoning, the science was simple and straightforward enough. It was Victorian methods of identification and distribution allied to the new interest in *relationships:* with ocean currents and depths, water of different salinity and temperature, and the teeming, ever-hungry world of grazers and predators. Hardy was a master of another Victorian skill: painting from life. Wherever he went, but particularly on the *Discovery* adventure, he maintained a detailed journal, crammed with delicate, beautifully realised watercolour sketches. He put them to good use when, 30 years later, he finally found the leisure to write a biographical account of the *Discovery* days, *Great Waters*, which rivals *The Open Sea* as his masterpiece. In writing, as in life, Alister Hardy was a likeable, engaging personality, who takes the reader into his confidence in the most natural and unaffected way. Perhaps his literary secret (and he would have denied he had one) was simply to be himself. It is departing from our theme a little, but let us pause to listen to his friend, J.R. Lucas, recall the essential Alister Hardy:

> 'He was not a man to push his views either in college meetings or at dinner, but when he was persuaded to speak, he would hold his colleagues rapt with his accounts of sailing round the Horn, ballooning over England in his youth, and nearly drowning himself in his search for plankton. He could also convey his sense of excitement and inspiration in expounding his own ideas about man's evolution and man's relation with the Almighty.....
>
> 'Alister had a strong sense of fun. On the occasion of a special college dinner, he could be persuaded, without too much difficulty, to sing one of the songs of his youth, recalled from the days of the Gaiety Theatre. Some of the older fellows remembered him dancing a hornpipe. I remember him joining in the spirit of the occasion when an electric shock machine was used to send a current through the whole governing body, although he later assured me that I had misremembered him applying the electrodes to a medical fellow's face to make his ears waggle. On another occasion he read to a college essay society a brilliant spoof paper on why he believed in mermaids. At the time of the Queen's coronation he designed and constructed a hot-air balloon in one corner of the quad: work ceased in the College for a fortnight, while the paper was glued together, and heating mechanisms devised. Coronation day itself being wet and cold, every electric fire in the College was commandeered to provide extra lift, but in vain: nevertheless at a second attempt the balloon took off: it sailed across Oxford before coming to rest in the University Parks.
>
> 'A fellow once asked him how many things were named after him. There was a boat in Hong Kong, an octopus, a squid, an island in the Antarctic and – he would add, lowering his voice in a tone of comic embarrassment – two worms.'
>
> *J.R. Lucas*. Quoted in Alister Clavering Hardy. *Biographical Memoir of Fellows of the Royal Society* (1986), **32**:257

The third great ocean traveller among the New Naturalists was Maurice Yonge, a friend, as it happened, of both Matthews and Hardy. Yonge's pioneering work on the feeding and digestion of oysters and shipworms had brought him early recognition as a brilliant and original marine biologist. At the age of 26, he had written a natural history classic, *The Seas*, with his lifelong friend F.S. Russell. That book contained a chapter about coral reefs, and, since neither of its young authors had ever set eyes on a living coral, they tossed a coin over who should write it. Yonge lost the toss and wrote the chapter (it followed another Yonge chapter concerning shipworms, entitled Boring Life). This was to have unlooked-for consequences, for a year or two later, Yonge's apparent expertise on corals helped him to be chosen to lead the Great Barrier Reef expedition of 1928-29. For more than a year, Yonge and his colleagues, who included his wife Mattie and the young physiographer J.A. Steers, lived in huts on the reef itself, surrounded by a natural aquarium and attended by aborigines. It was another stupendous adventure, the first modern scientific investigation on one of the great natural wonders of the world. On his return in December 1929, Yonge wrote a book about his experiences, *A year on the Great Barrier Reef*, as well as a shower of papers about reef corals, giant clams and the ingenious stomachs of snails. Though already a biological pioneer and an obvious high flyer, it was the Barrier expedition that made him. From 1930, Maurice Yonge consolidated a worldwide reputation as an authority on marine invertebrates. Like Hardy and Matthews, though in a different way, he was a great communicator, clear-thinking, enthusiastic and good-humoured, and he brought to his lectures a touch of the romance of the sea. He learned the art of teaching the hard way, overcoming shyness and a stammer, and a punishing work schedule. Like most professors of small, newly established biology departments, he bore much of the teaching load himself. He, too, developed the popular touch and his books are always highly readable. Success made him confident: a pretty, clever wife, sparkling academic achievement and a professorial chair (at Bristol) at the age of 32. He was not a field naturalist born and bred like Matthews and Hardy, rather the other way round, a biologist who could also view his charges with the eyes of the Victorian naturalists, whose works he admired and collected.

Ecology is a science that thrives on travel and a variety of landscape and seascape. The first generation of New Naturalist authors were blessed with the advantages of the steam age and a worldwide trade empire still largely at peace. Some of them were true commonwealth naturalists who spent much of their early careers abroad. As with her armed forces and civil service, Britain's responsibilities in agriculture and forestry were imperial, and her scientists were as much concerned with tsetse fly and locust control as with home-grown concerns (indeed, possibly more so). The career of C.B. Williams (1889-1981), the author of *Insect Migration* is a case in point. 'C.B.' (as he was always known, having had the misfortune to be christened Carrington Bonser) was a born naturalist, but in his day biology usually meant medicine or agriculture. His big break came in his third year at Cambridge when William Bateson, the genetics pioneer, perhaps noting C.B.'s fondness for rearing caterpillars, offered him a research studentship in entomology. Then the First World War intervened, and the unfortunate Williams found himself engaged 'in that unheroic branch of warfare, the examination of the stools of dysentery patients'. As the story goes, one day he was called from his grisly duties by a stranger from the Colonial Office who asked him to go to the West Indies to look into the case of a blightful froghopper that was devouring Imperial sugar cane. C.B. went, and spent the next six years among the sugar plantations of Trinidad and Central America. There he witnessed for the first time the spectacular mass migration of a butterfly, the cloudless sulphur (*Phoebis eubule*) which, unbeknown to him until that moment, was to form his life's work. Butterfly migration started as a hobby aside from his main work as a travelling empire entomologist, investigating boll worm in Egypt or locusts in Tanganyika. But, by 1930, he knew enough to write *The migration of butterflies*, which immediately established him as the world authority (or, more accurately, the only authority) on the subject. His New Naturalist book *Insect Migration* (1958) is its natural continuation, with the benefit of thirty years additional experience in places as far apart as West Africa, South America and the Pyrenees. It is also a great deal more readable than the original.

Back from his travels in the Colonial Service, C.B. joined the staff at Rothamsted in 1932 as head of entomology under Sir John Russell. It was a timely move, since Russell had recently employed the great the statistician R.A: Fisher to design

C.B. Williams (1889-1981), author of *Insect Migration*, in c. 1949. (*Photo: Rothamsted Experimental Station*)

Sir John Russell (1872-1965), author of *The World of Soil*. (*Photo: Rothamsted Experimental Station*)

field trials that dragged the study of insect activity out of temperature-controlled cabinets and back outdoors. Noting C.B.'s particular talents, Russell was happy to give him his head in working out the relationship between insect activity and weather. This study occupied the rest of C.B.'s career. Essentially, it was Alister Hardy and plankton all over again, if for ocean currents one substitutes wind and air – though C.B. Williams went well beyond Hardy in developing quantitative methods, which he summed up in his magnum opus, *Patterns in the balance of nature* (1964), written at the age of 75. Despite the formidably mathematical nature of his learning, C.B. was at heart an amateur naturalist, though of the sort that is interested primarily in causes and principles. On retiring in 1955 to Kincraig in the Spey Valley, he immediately set up light and suction traps and a private weather station to monitor the movement of moths, blackflies and other insects. Those who have worked a Rothamsted light trap are using a device pioneered by C.B. Williams in the 1950s.

Malcolm Smith (1875-1958), revered author of *The Amphibians and Reptiles of Britain* (1950), was a rolling stone of a different kind. One of the elder statesmen of the series, Smith was at the peak of his medical career at the outbreak of the First World War and by 1925 had retired at the age of 50 to devote more time to reptiles and amphibians. Like C.B. Williams, his boyhood was steeped in natural history, and he was often found with some toad or snake in one of his pockets. But in his day, and for long afterwards, there was no obvious career to be made from natural history. Of Smith and his age it was said that 'The dissecting-room and the lecture of the medical school furnished the only regular training for the naturalist, while he found in the medical profession the likeliest means of earning his bread.' And so Malcolm Smith qualified as a physician. That did at least offer opportunities to study snakes and crocodiles in tropical countries.

As it happened, it also offered to Smith the great adventure of his life as physician to the Court of Siam in the 'The King and I' days, before western influences finally swept away the ancient rites and mysteries of the Orient. Smith tells the story as only he could in *A Physician at the Court of Siam* (1947), written 30 years after the events he describes. His duties were unusual – among other things his presence was required at public executions. But he was not much interested in telling readers about himself. Instead, the book is about the customs and vivid personalities of the Court, for whom Smith had an obvious affection, though he tells his story with a certain detachment. He was blessed with an excellent memory, extending to the tiniest incidental detail, and had acute powers of observation. And while he was slicing open sickly Siamese courtiers, he was using these skills to study, collect and classify lizards and snakes. His first major work, *Monograph of the Sea Snakes* (1926), was based on the large collection of these animals gathered during his years in South-East Asia. He presumably made plenty of money as Court physician, for he paid collectors to hunt for his snakes and, after his retirement at the age of 50, had saved enough to support his herpetological work for the rest of his life. Reptiles and amphibians have some unfathomed capacity to inspire intense devotion among the select, to whom Malcolm Smith is a sort of patron saint. Though primarily a museum-based systematist, he also enjoyed a day in the field. British species only really occupied him at the beginning and the end of his life. His disciple (and later reviser) Angus Bellairs recalled their reptile-collecting expeditions to the Dorset heaths: 'He was an expert at catching snakes and lizards, pouncing on them with a sudden, darting movement. He caught the first smooth snake which I ever saw in the wild. He had an ingenious method of capturing adders, picking them up with one of the side pieces of his spectacles.'

Malcolm Smith (1875-1958), author of *The British Amphibians and Reptiles*, holding his 'adder-catching spectacles', 1953. (*Photo: Natural History Museum*)

While most of the long-distance travellers of the series went to the tropics and the South Seas, at least one, the meteorologist Gordon Manley (1902-1980) sought the cold regions of the earth, on mountain tops and in the Arctic. Manley, who listed as his hobby in *Who's Who* 'travel among mountains', was the pioneer of weather studies in the British uplands. Always an advocate of a 'hands-on' approach to weather recording, Manley is perhaps best remembered for his heroic days in a hut near the summit of Great Dun Fell in the North Pennines, 'personally experiencing the worst of the weather conditions of which he writes'. This remote place, which later became part of Moor House National Nature Reserve, is the coldest part of England. Manley showed that its weather has more in common with the arctic regions than Cornwall or Kent. At that time his hut on the treeless wastes was the only high-level recording station in Britain (the one on

Gordon Manley (1902-1980), author of *Climate and the British Scene*. (*Photo: Department of Geography, Royal Holloway*)

Ben Nevis having closed long before). Gordon Manley was interested in a phenomenon there called the 'helm wind', first described by a local vicar more than 200 years earlier, which produces gusts of exceptional strength on the lee side of the hill. Such was his devotion that he used to camp in the hut for spells, including the worst winter conditions, when he might find himself completely cut off for days at a time. Although this work had to cease in 1941, Manley's unique run of climatic data has helped to establish Great Dun Fell as one of the leading weather stations, which is now, through automation, monitoring atmospheric pollution. Manley was one of the most gifted writers of the series; his book *Climate and the British Scene* reveals a man of wide learning, always ready with an apposite quotation from literature or from the ancient classics of meteorology.

There have been many more recent world travellers in the New Naturalist library. Leslie Brown (1917-1980) led an adventurous life as an agriculturalist and, later, a travelling ecological consultant in East Africa, and was full of stories of encounters with not always friendly animals and natives. T.T. Macan (1910-1985), though normally thought of as a fairly sedentary freshwater biologist, cut his teeth on a great voyage to survey the Indian Ocean. Macan's main job was to sort out and preserve the catches and take care of the apparatus, but he also took the opportunity to study starfish, later basing part of his PhD thesis on them. Nor is the spirit of global travel by any means dead, though the circumstances have changed. Among the recent authors of the series, Philip Chapman has participated in no fewer than 11 expeditions to tropical caves to film wildlife documentaries and carry out ecological research. Chris Page is another seasoned expeditionary who has travelled to remote places all over the world in search of ferns, fern relatives and conifers, by air, land and sea. The world may be a smaller place than it was in the 1920s and 1930s, but parts of it are still little known ecologically, and British naturalists continue to exert an influence out of all proportion to their number.

Fraser Darling and Morton Boyd: 60 years of Scottish natural history

Britain, as W.H. Pearsall pointed out, is physically two countries, highland and lowland, almost as different in their biological aspects as Norway and Holland. From the point of view of naturalists and conservation bodies, it is a pity that this difference is not reflected in the political map, for the exploration and conservation of nature takes on very different forms in each 'country', the one densely populated with a patchwork landscape, the other of wide open spaces in which man is often outnumbered by sheep and deer. The different scales produce different ways of thinking. The naturalists of lowland England are used to

Sir Frank Fraser Darling (1903-1979), co-author of *The Highlands and Islands*, in the early 1970s. (*Photo: Eric Hosking*)

regarding nature in terms of penny packets and are becoming increasingly willing to take on vested interests over a precious few yards of turf or ancient woodland. That landscape is reflected in New Naturalist titles like *Wild Flowers of Chalk and Limestone*, *Hedges* and *Common Lands in England and Wales*. For the other, to Pearsall 'the better half', one's thoughts are drawn by the distant serrated horizon to the essential wholeness of nature, not a chequerboard partitioned by hedges, walls and roads but an obvious entity. Here the only realistic way to reconcile man and nature is to think strategically, to regard nature not as a kind of crop that is grown on 'reserves' but as the landscape itself, a living, growing resource capable of use or abuse. The tragedy of recent times is that nature conservation in the Highlands and Islands has been perceived of as socially divisive, when it should be a force for unity through common sense. The wholeness of vision that is the special gift of ecology, and which was expressed with the most acute perception by Frank Fraser Darling, has since become circumscribed, rather like the once open moors and the mires have been when planted with mile after mile of tax-avoiders' spruce.

Is Fraser Darling still a name to conjure with, as it once was by naturalists in the 1940s and 1950s, and by the fledgling environmental movement in the early 1970s? He is scarcely a suitable figurehead for our contemporary conformist culture. As a philosopher-ecologist, a romantic, something of a misfit, a prophet and, in the end, a guru, Darling fitted in better with the 1960s ethos than the present day. His life was not a triumphant parade from one thing to the next, like, say, Max Nicholson's or Peter Scott's, and he failed in his two greatest crusades: the application of ecological principles to land-use in western Scotland and the campaign for National Parks for Scotland. He was an introspective, sometimes melancholy man with a habitual expression on his jowly face suggestive of some inner torment. He embodied an interesting set of contradictions: a hermit with a taste for jade, Persian carpets and fine claret; a great ecologist who hardly ever published work in a scientific journal; a loner who was three times married with four children; a man who loved Scotland, but took refuge in the Home Counties and in America.

He was a Scot by naturalisation, not birth. The derivation of his name is unusual. Darling was his mother's maiden name. The absent father is said to have been a South African army captain called Frank Moss, and the 'Fraser' was acquired later from the maiden name of his first wife, Maria Fraser. He dropped it when she divorced him to become plain Frank Darling, but readopted it later as a hallmark of Scottishness just in time to write his New Naturalist book. He was greatly attached to his mother throughout her life, but although his childhood seems to have been happy, it was lacking in parental control. He ran away on his big flat feet at the age of 15, and was evidently a somewhat unruly boy. What he did acquire, with the help of an inspirational English teacher, was a love of the English language and literature. He also found he liked animals. He went on to train in agriculture, and after a few fruitless years as an agriculturist on Buckingham County Council, went north to Edinburgh to study for a PhD on the genetics of the Scottish Blackface sheep. In 1930, he was appointed Chief Officer of the Imperial Bureau of Animal Genetics, but this was an office job and Darling hankered after the freedom of the wild. After several attempts, he landed a Leverhulme Fellowship to study the ecology and behaviour of red deer in Wester Ross which set the pattern for the next 20 years. Darling chucked in the office job and packed his cases. From now on, he made a precarious living in some of the wildest parts of Scotland on whatever income he could scrape together from grants and from his writings. He had a theory that if one had the courage and endurance to live close to wild animals on their own ground, they would 'unmask' themselves. This primitive ethic drove his work: he wished to study nature face-to-face, with all sophistry stripped away. From the red deer of An Teallach, he moved on to the Summer Isles studying the social structure of gulls and other shore birds, and later to Lunga and North Rona for a different type of social animal, the grey seal. The work, which combined ecological studies with animal behaviour, was original both as a synthesis and in the long hours of animal watching it required. Unfortunately, it did not receive the recognition it deserved, partly because of Darling's reluctance to write for the scientific press. In the meantime, in his friend James Fisher's words, 'he worked his small farm on Tanera in the Summer Isles in such fashion as to show that it was possible and reasonable to raise considerably the stock-carrying capacity of the West Highlands, and to grow a large amount of human food under crofting conditions'. He was fast becoming the New Naturalist version of Thoreau.

The seal study was brought up short by the outbreak of war. A pacifist, 'embittered by the political climate of the day', Darling found himself marooned on his Tanera croft. He farmed and wrote more books, *Island Years*, *Crofting Agriculture* and *Island Farm*, rather as Ronald Lockley had done on Skokholm in the 1930s. Darling's books were widely read and admired, especially *A Herd of Red Deer* (1937) and *A Naturalist on Rona* (1940), but they were regarded as too 'popular' to gain the approval of the austere scientific establishment; even the more 'biological' *Bird Flocks and the Breeding Cycle* (1938) fell between two stools, neither popular enough on the one hand nor scientific enough on the other. His researches on deer and sea-birds were, however, accepted as a doctoral thesis, and earned him a DSc. Darling was always acutely aware of, and inhibited by, his supposed lack of scientific credentials. He was naturally deductive and had an instinctive *rapport* with his subjects, but he was not particularly numerate and therefore did not present 'data' in the approved way.

In 1944, Darling persuaded the Development Commission to fund an idea he had been nursing while working his croft: a social and biological investigation of the West Highland area which would be a simultaneous study of 'man as part of natural history, and natural history as a large part of man's environment'. Darling saw the West Highland landscape as 'a devastated terrain', degraded by centuries of human folly, and, in *West Highland Survey*, he discusses how the present situation came about and makes suggestions for improvement from his joint perspective as a crofter and an ecologist. In its book form, the survey is pithily written, full of Darlingesque phrases like:

> 'Devastation has not quite reached its uttermost lengths but it is quite certain that present trends in land use will lead to it and the country will then be rather less productive than Baffin Land.'
>
> 'This ecological continuum .... would yield more to the nation than the subsidised devastation, rendered the more macabre by imposed mechanical industries ....'

He had written *Natural History of the Highlands and Islands* during the first two years of this six-year survey, and was probably consciously trying out in that book some of the arguments that he later incorporated into the survey. He was, then and later, strongly influenced by the views of Aldo Leopold, an American philosopher-ecologist, who combined scientific knowledge with a breadth of vision and a mastery of the telling phrase. But the conservationist ethic of *West Highland Survey* was too advanced for the conservatives of the Scottish establishment. Its influence was confined to the already converted, though it was widely quarried as a source of raw material. In similar vein, Darling's powerful advocacy, as a member of the 1945 Ramsay Committee, of a National Park system for Scotland fell on stony ground. He achieved far more with the more receptive political establishments of North America and East Africa than he ever did in his own adopted country.

The West Highland Survey was concluded in 1950, though not published until 1955 (and then, ironically enough, in southern England). Despite these disappointments, the success of British ecologists in establishing the first 'biological service', the Nature Conservancy, in 1949 should have inaugurated a time of glorious fulfilment for Darling. That he was to be disappointed in this too was mostly his own fault. He had already rocked enough establishment boats not to be appointed the Conservancy's first Scottish director, as he might reasonably have expected. He was a proud and stubborn man, as James Fisher and others discovered when they wanted him to modify some aspect of his books. They all failed. He also took on far too much and dissipated his talent. After directing the Nature Conservancy's red deer group for years, he never got round to writing up its report. He neglected his duties as a senior lecturer in ecology and conservation at Edinburgh in favour of more glamorous engagements in Rhodesia, Alaska and the American West. When the inevitable denouement came and Edinburgh University dispensed with his services in 1959, most people would have resigned themselves for early retirement. Darling, though, had a great admirer in the USA in the person of Fairfield Osborne, who invited him to become Vice-President of the Conservation Foundation, Washington D.C., a post he held until 1972. 'In the States, I was listened to,' remarked Darling later. 'Here [in Scotland], I'm less than the dust beneath the wheel of the chariot.' It was only with Darling gone, and now becoming a leading player on the world stage, that British scientists started to appreciate his contributions to ecology and nature conservation.

Not long after Darling's removal to the New World he began to give thought to a revised edition of *Natural History in the Highlands and Islands*, whose original reception we noted in the previous chapter. The idea was that the mistakes of the first edition would be corrected and the book brought fully up to date with all the significant events of the previous 17 years. 'Looking around for a suitable victim,' wrote Darling, 'I did not take long to come back to the figure I first thought of: Dr John Morton Boyd.' Morton Boyd was at that time the Nature Conservancy's Regional Officer in the Western Highlands. As the latter recalled later, 'I realised that to say "Yes" would probably land me with the whole work (as it did), but to say "No" would be to lose the chance of a lifetime. Since its publication in 1947, I had used the book continuously, knew it well and admired it both as a working tool and also a source of enjoyment.'

Morton Boyd was as an undergraduate in zoology at Glasgow University when he first met Fraser Darling. Boyd was already an avid reader of what might be called the school of romantic biology – Aldo Leopold, Seton Gordon and Darling himself – and in deciding a subject for his Honours project, he turned to the latter Master for advice. Boyd had read in Darling's *Natural History* about competition between different species of snails on the sand dunes of the Hebrides, and proposed to base his study on it. 'The reply was brief: inter-specific competition in snails was not at all suitable for an Honours study.' Instead, Darling suggested studying the different communities of animals and plants from the shoreline to the heathland interior. As a base, he recommended Tiree in the Inner Hebrides. In dashing off this laconic reply, Darling could hardly have known what he had started: 'Fraser Darling's advice contained the germ of my life's work in ecology and conservation.' As a young ecologist of promise, and supported by his tutor, Maurice Yonge, Boyd obtained a Nature Conservancy studentship to study the fauna of soil derived from Hebridean shell-sand. He married his wife Winifred in 1954 and they made their second home in Tiree, 'putting down roots at Balephuil-Sanderling, our cottage on the ocean's edge at the meeting point of land and sea and sky, our place of solace and peace between heaven and earth'. There followed for Boyd a golden decade when, perhaps in conscious emulation of Fraser Darling's exploits, he studied not only the earthworms beneath his feet, but also the grey seals of Ronay (the first person to do so since Darling) and the gannets and Soay sheep of St Kilda. His open-air laboratory was the maze of green islands and foaming water in one of the most romantic land and seascapes in Europe. Perhaps the climax to the St Kilda years was the ascent of the sea-girt pinnacles of Stac an Armin and Stac Lee by Morton Boyd and Dick Balharry in May 1969 (see Plate 16), the first time man had trodden on this most inaccessible part of Britain since the departure of the last native St Kildans in 1930.

Boyd reveres Fraser Darling as a mentor, and has written about him in detail in *Fraser Darling's Islands* and *Fraser Darling in Africa: A Rhino in the Whistling Thorn*, but he stops well short of hero-worship. He told me that it was not so much Darling's science that had struck a chord with him as his style, the directness of his approach to nature, his searching powers of introspection and the many loose

This is the original colour sketch for *Butterflies* painted by Clifford and Rosemary Ellis in late 1944 at twice the size of the printed jacket using gouache on watercolour paper. Note the hand-lettered title and the prototype New Naturalist emblem on the spine.

Plate 2

Lithographic designs produced by Clifford and Rosemary Ellis during the 1930s for novels published by Jonathan Cape Ltd (above) and posters advertising social evenings at London Zoo and an international cricket match at The Oval.

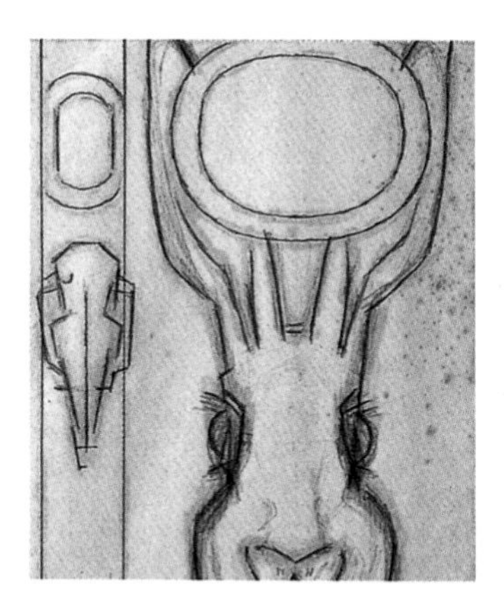

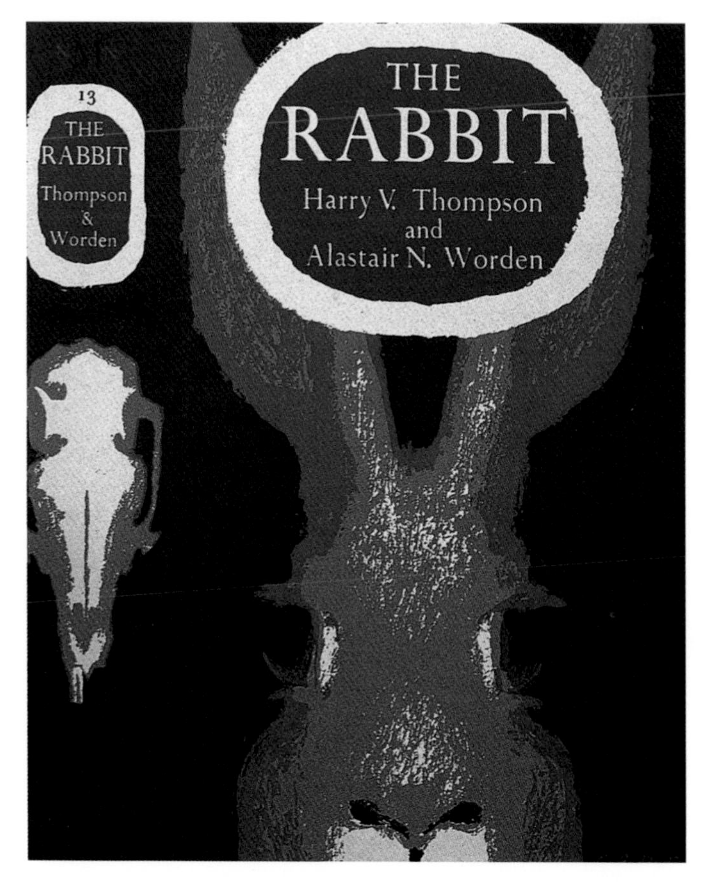

The evolution of a New Naturalist dustjacket. Starting with pencil sketches from life and material gleaned from other sources, the Ellises worked out a design, made a colour sketch and finally produced the finished artwork for the printer, in this case a perfect realisation of *The Rabbit* (1956).

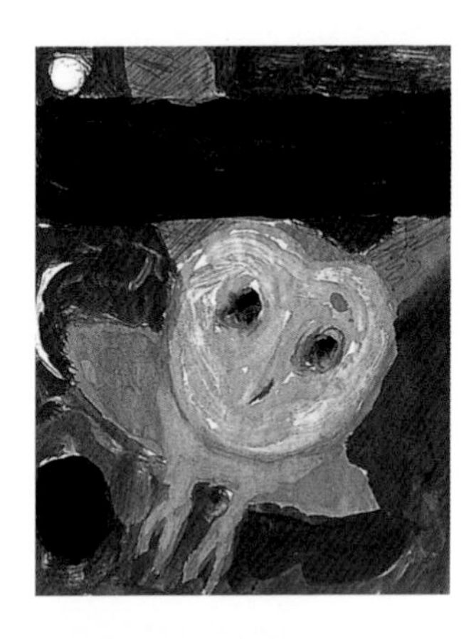

Sketches and printer's proof for *The Folklore of Birds* (1958).

Colour sketches for *The Herring Gull's World* and *Insect Natural History* and the actual artwork for *The Yellow Wagtail*, drawn to scale in three colours.

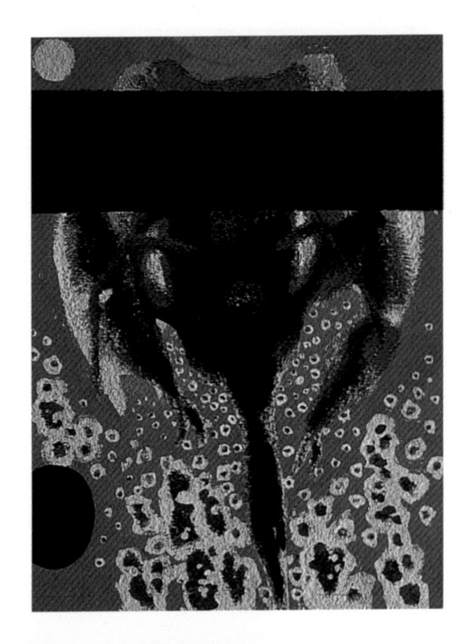

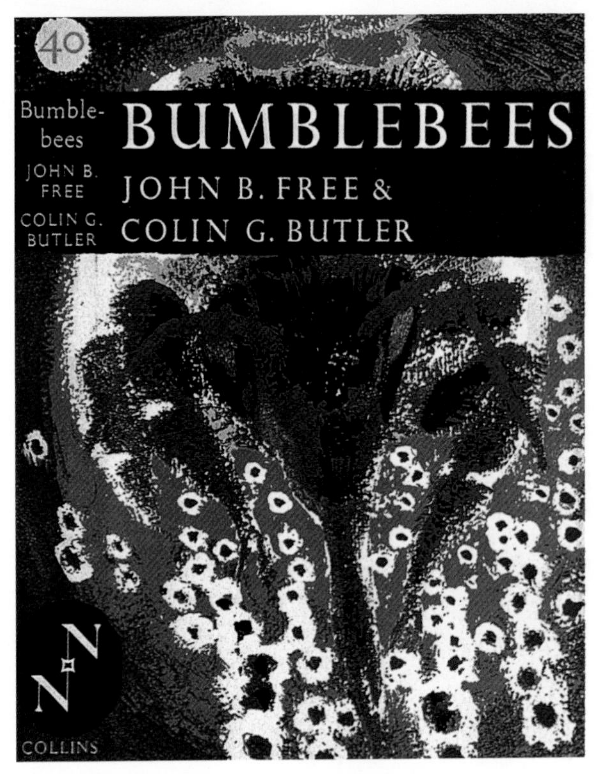

The evolution of the *Bumblebees* jacket through colour sketches to the printer's proof.

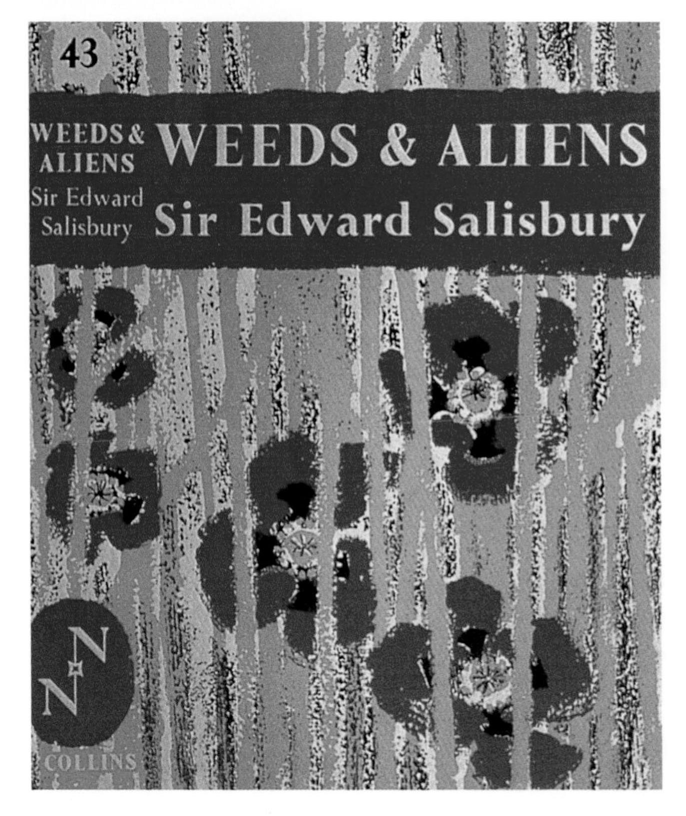

The first design for *Weeds and Aliens* (1961) – a thornapple in flower and fruit – and its subsequent replacement by red poppies in a wheat field.

Plate 8

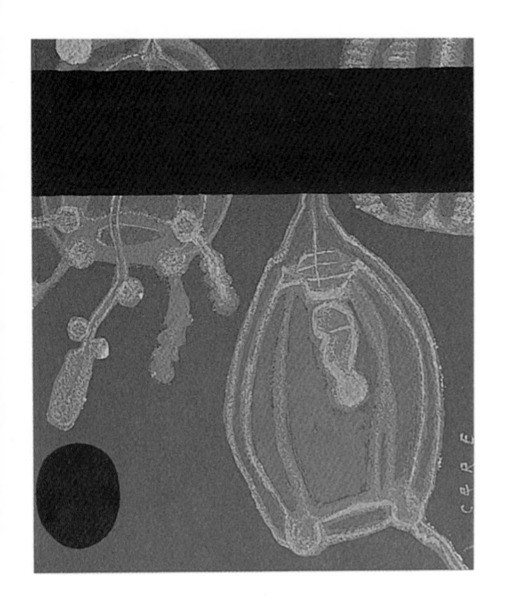

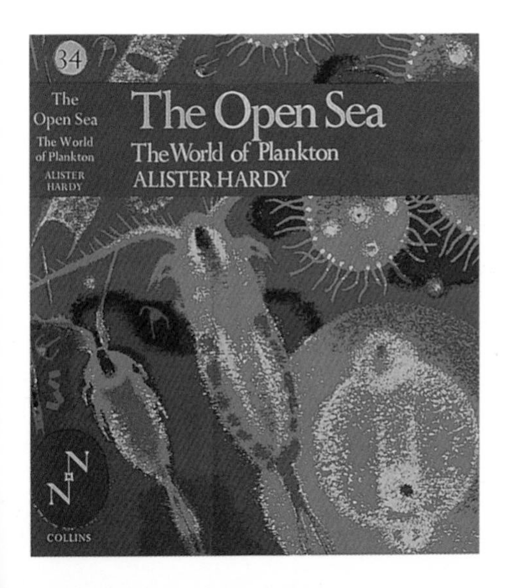

The Open Sea: (1) The World of Plankton. The original artwork of planktonic life, the seascape substituted at the request of the publishers and the printer's proof for a never-used jacket. Which one is most effective at projecting the book?

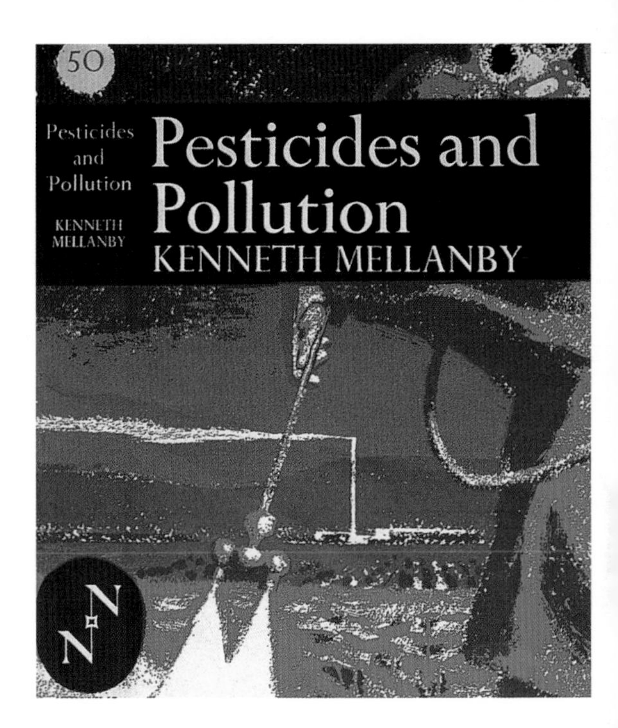

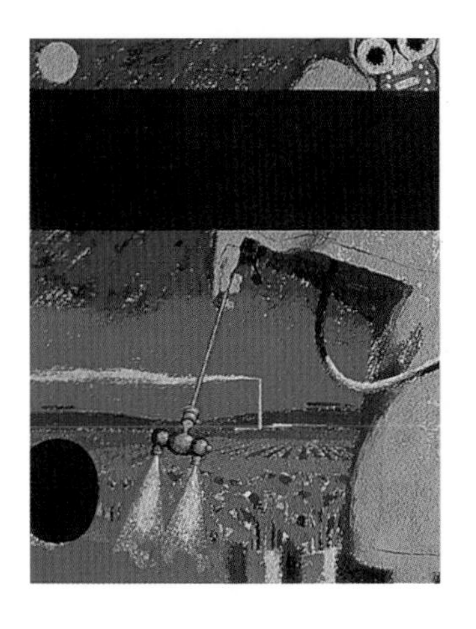

Original artwork for *Pesticides and Pollution* (1967) and *Ants* (1977) and the modified designs actually used.

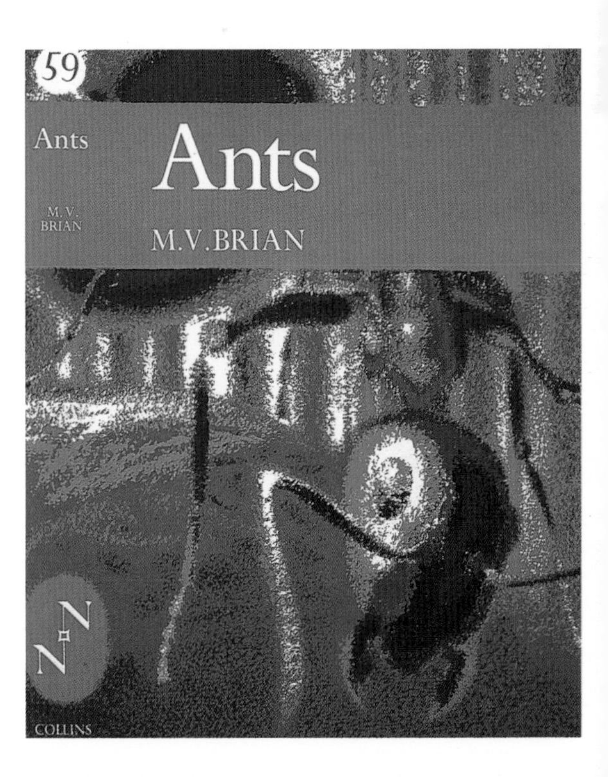

The original designs by Clifford and Rosemary Ellis for *Wild Orchids of Britain* (1951), *Fossils* (1960) and *Man and Birds* (1971); and the unused design for a reprint of *Mushrooms and Toadstools.*

Colour sketch and printed jackets of *Trees, Woods and Man* (1956) and *Lords and Ladies* (1960).

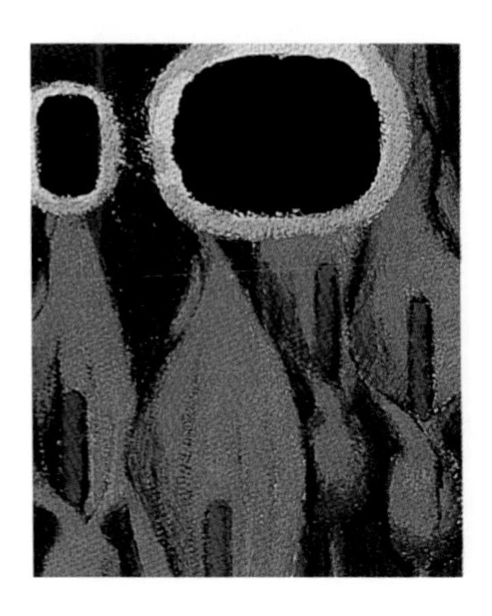

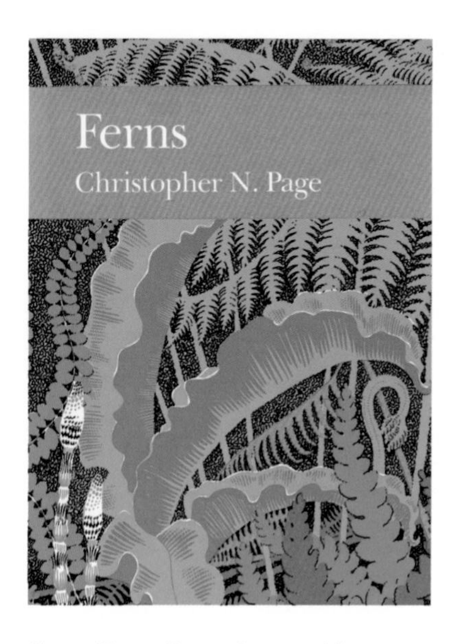

Ferns: Four flat colours with some overprinting.

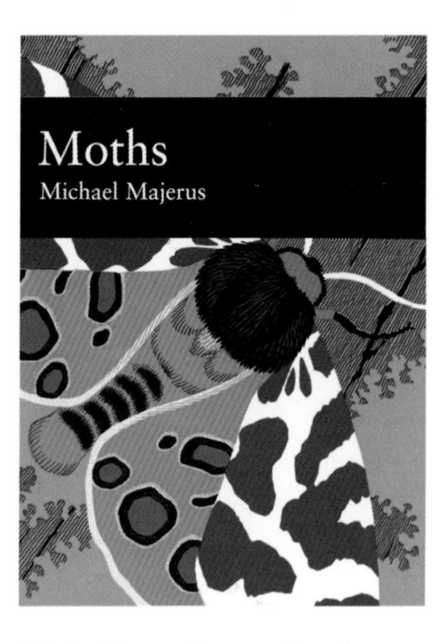

Moths: Four colours using tints.

Jacket design techniques.

Seashore: four colours using black lino block and tints.

Northumberland: a full 11-colour linocut print. On the printed jacket the white county border accidentally slipped a little to the right.

(a) The initial pencil sketch was chaotic with too many images.

(b) An idea using a map of the British Isles was tried and discarded.

Evolution of the *Nature Conservation* jacket.

(c) Instead a stretch of countryside was introduced.

(d) The finished jacket. The harvest mouse was changed at the last moment. The flying lapwing survived from the earliest sketch, its boldly patterned head working well on the spine.

First ideas and abandoned designs. The first idea for a jacket often contains the germ of the eventual design. Some need little change, while others were radically revised.

In the original *Lakeland* design the peregrine was too dominating and the right side was rather blank.

The first design for *Amphibians and Reptiles* combined several images that didn't integrate well. The adder on the spine survived on the final design.

On the original design for *British Bats* the large bat came out 'looking as if it had been pinned to a dissecting board'. A more lively bat was substituted.

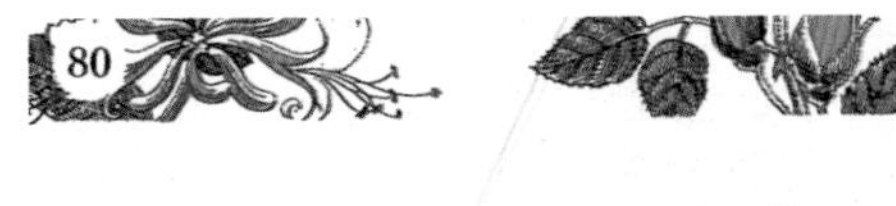

Printer's colour separations for the jacket of *Wild and Garden Plants* (1993).

A New Naturalist in his element. J Morton Boyd ascending Stac an Armin, St Kilda in 1969. Photograph by Dick Balharry.

leads he generously left dangling for a trained ecologist to pick up. Darling himself compared the energy, physical and intellectual, of the young Boyd with his own 'natural contemplative idleness'. While Darling was a visionary, Morton Boyd is a romantic, a passionate man with a deeply felt spirituality, which he expresses without inhibition. His life has taken on extraordinary parallels with that of Darling. Both had their island years (each lasting about 10 years), followed by extensive world travel on behalf of conservation causes, Darling in the 1950s, Boyd a decade later. But, while Darling became the vice-president of an independent foundation, Boyd became the Scottish director of a public institution, the Nature Conservancy (later the NCC), with a much more onerous administrative burden. I suspect he enjoyed his travels abroad more than his time behind the director's desk in Edinburgh. 'I know your job is not like mine,' Morton once wrote in one of his friendly Christmas letters to the troops, from somewhere in the South Pacific. He could say that again! He, too, suddenly seems like a figure from a lost age: a traveller, poet, painter and mystic, someone more at home on some guano-spattered rock or palm-fringed atoll than in an office chairing meetings and dictating memos. It was not that he lacked the qualities for the latter. His personal rule was marked by passion, paternalism and sometimes prejudice. I for one remember him with affection, especially in view of what happened later (i.e. growing fractiousness and division, resulting in the abolition of the NCC and its replacement by a separate body under the thumb of the Scottish Office). Morton Boyd must have missed the outdoor island life. He supported the publication of detailed symposia on the natural history of Shetland and Orkney, to which Sam Berry contributed and later used as a source for his New Naturalist books on the northern isles. Later on, Boyd organised and edited symposia on the Inner and the Outer Hebrides, published by the Royal Society of Edinburgh. For his beloved Hebrides, Boyd decided to write a readable, single-author book himself in a style similar to that of the early New Naturalist classics by Darling, Yonge, Pearsall and others. He was unquestionably the best qualified person to do so, though the book is also a testimony to the work of the Nature Conservancy team that Boyd had led. Morton told me he had aimed to provide both a personal book and a view of things in the round; the particular word he used was 'valency', the combining power of separate elements. *The Hebrides* is a three-parter: the first, with a strong central story line, takes the reader from coast to mountain-top; the second is more of a synthesis of species and islands; and in the third he overviews the all-important human dimension, including the many famous naturalists, from Martin Martin to James Fisher, who had made the Hebrides their Mecca. After a period of ill-health, Morton decided to share the work with his son Ian, a chip off

J. Morton Boyd, co-author of *The Highlands and Islands* and *The Hebrides*, in 1990.

the old block, currently in charge of seal research for the British Antarctic Survey. *The Hebrides* was published in 1990, and is a fitting testimonial to the 60 years of Scottish natural history spanned and exemplified by Fraser Darling and Morton Boyd.

Morton Boyd retired from the NCC in 1985, and is now a consultant on countryside and forestry issues. He was appointed a CBE in 1987, and is a pillar of the Scottish scientific and religious establishment. Frank Fraser Darling returned to work in England in the late 1960s, and, perhaps to his surprise, was greeted as an elder statesman. It was a remarkable about-turn by a scientific community that had, in Kenneth Mellanby's words, 'just discovered the meaning of the word "conservation"'. British Society has its ways of telling you that you have arrived. Darling was invited to deliver a memorable series of Reith Lectures under the title 'Wilderness and Plenty'. The following year he collected a knighthood, received a clutch of honorary degrees and was invited to join the first Royal Commission on environmental pollution. And, of course, he was in great demand as a speaker, especially by student bodies who saw him, with some justice, as one of the great latter-day prophets. By then, though, his health, hitherto robust, was failing and in his last years he was unable to travel far. On 20 September 1976, he wrote to Billy Collins about a new revision of *The Highlands and Islands*, mentioning that he himself was 'waning. Some quality of judgement remains, but the punch has gone.' As an afterthought, he asked: 'was *New Naturalist* your original idea? If so, it was brilliant and in these thirty years has made it possible for a number of gifted men and women to use their minds towards a wholeness of conception... The present vivid interest in natural history among a broad band of folk owes much to the *New Naturalist* series.' Billy Collins never read this letter. He died the day after Darling had sent it. Darling himself died at his home in Forres on 22 October 1979, aged 76. But:

'Still the blood is strong, the heart is Highland,
And we in dreams behold the Hebrides.'

Table 7. The age of each New Naturalist author on the publication of their book (arranged in approximate chronological order)

The best-known portraits of New Naturalist authors are mainly of men and women of mature years, and nearing the end of their public careers. This can give the impression that most of these books were written by elderly people. In fact their ages at this time ranged from their early thirties to their eighties, and an average age would be mid to late forties – in mid, not late, career. The average age has not changed significantly over the past 50 years.

E.B. Ford	44 (*Butterflies*)	D. N'sole-Th'pson	42	J.D. Summers-Smith	42
Brian Vesey-Fitzgerald	45	Miriam Rothschild	43	K.C. Edwards	58
Richard Fitter	32	J.R. Harris	42	J. Morton Boyd	39 (*Highlands and Iss*)
L. Dudley Stamp	48 (*Structure & Scenery*)	Ian Hepburn	50	A.E. Ellis	56
John Gilmour	48	J.A. Steers	53	William Condry	48 (*Snowdonia*)
Max Walters	33 (*Wild Flowers*)	James Fisher	40 (*The Fulmar*)	Ian Moore	61
Frank Fraser Darling	44	Derek Wragge Morley	33	R.K. Murton	33 (*The Wood-Pigeon*)
A.D. Imms	66	E.A. Armstrong	54 (*The Wren*)	Kenneth Mellanby	58 (*Pesticides & Poll'n*)
Ernest Neal	37	Sidney Wooldridge	52	Eric Simms	50 (*Woodland Birds*)
W.B. Turrill	57	L.A. Harvey	50	Michael Proctor	44
A.W. Boyd	66	Ronald Lockley	50	Peter Yeo	44
W.H. Pearsall	58	Colin Butler	40	Ian Newton	32
C.M. Yonge	49	Monica Shorten	31	M.V. Brian	58
Bruce Campbell	37	Niko Tinbergen	46	Leslie Brown	58
F.J. North	60	H.L. Edlin	39	R.J. Berry	43 (*Inheritance*)
Wilfred Blunt	48	John Raven	41	Christopher Perrins	43
John Buxton	37	Alister Hardy	60	W.G. Hale	45
Stuart Smith	42	Sir John Russell	84	Nigel Webb	44
T.T. Macan	41	C.B. Williams	68	Colin Tubbs	49
John Ramsbottom	67	W.S. Bristowe	47	Christopher Page	45
J.E. Lousley	42	Guy Mountfort	52	Niall Campbell	67
E.M. Nicholson	46	John Free	31	Peter Maitland	54
H.J. Fleure	73	Cynthia Longfield	63	Brian Davis	58
V.S. Summerhayes	53	Norman Moore	37	Philip Chapman	43
Malcolm Smith	74	H.H. Swinnerton	84	Michael Majerus	39
L. Harrison Matthews	50	C.T. Prime	41	Peter Marren	44
Gordon Manley	50	Sir Edward Salisbury	75		

8

The Pursuit of a Species: the Monographs

The trouble with the new natural history was that it was all too easy to lose sight of individual birds, beasts and insects in all the attention being devoted to their habitat. Although the ecologists of Tansley's and Elton's generation were in general good field naturalists, their subject had, by the 1960s, begun to creep back indoors again. That was the ecology I learned, a curiously abstract subject seemingly obsessed with checks and balances, inputs and outputs, models and statistics. The lab man had taken over again. It is harder to avoid the sunshine when you study a particular species, or a group of related species. There, at least, field study remains a respectable science, in which the amateur can engage on equal or more than equal terms with the full-time scientist. To get to know an animal in its natural habitat needs time, sometimes a great deal of patience, and as much craft and stealth as the observer can summon up. The New Naturalist monographs were written by pioneers in this field, by people who had chosen a particular animal as their own and dedicated themselves to finding out where the animal lived, how it behaved and what its needs were. Of *The Fulmar* (1952), which was and remains one of the most detailed bird studies ever published, James Fisher freely admitted that his book reflected a personal obsession:

> 'I have been haunted by the fulmar for half my life; and have needed no spur to explore its history, and uncover its mysteries, save the ghost-grey bird itself, and green islands in grey seas ..... Since 1933 I have lived no summer season without a sight of at least some of the great cliffs and fulmar colonies of Spitsbergen, Iceland, Shetland, Orkney, St Kilda ..... At one time or another I have seen every Scottish fulmar colony, and most of those in England and Wales .....'
>
> *James Fisher.* Author's Preface, *The Fulmar* (1952)

Fisher wrote this book as a *tour de force*, and probably intended it to be a pace-setter for future bird studies. But he was engagingly candid about his real reasons for writing it. 'I have written this book not because I have thought it "useful" to do so, but because I like fulmars and everything to do with them.' And in a line that might well have jolted Billy Collins' shaggy eyebrows, he added that every kind of bird 'is worthy of monography', all 8,000 of them.

About half of the New Naturalist monographs were written by amateurs in the best sense of the word: by people who did what they did for the love of doing it. In 1947, when the first half-dozen of them were being written, virtually nobody was paid to study the natural history of animals or birds in the wild. For most species, surprisingly little was yet known. There was, of course, plenty of anecdotal information about foxes or badgers or eagles, but very few species had been studied scientifically. Much of what was said about such species was myth, or at least untypical. The first generation of animal ecologists had tended to cut their scientific teeth in far-away places – the Arctic, the Barrier Reef or the southern oceans. Home-based natural history was the province of amateur naturalists, including sportsmen and country curates. For a long time, serious birding was open only to people, usually wealthy people, with plenty of leisure, and access to grouse moors, pheasant coverts and wildfowling marshes. From the late 1920s onwards, the scene began to change as Britain's large body of amateur bird-watchers organised themselves and undertook local and national surveys of breeding birds. Much of this work focused on particular species, notably the great crested grebe, heron and lapwing. The surveys were aimed at finding out the distribution and, as far as possible, the actual numbers of these birds in Britain. They did, however, also encourage some birdwatchers to look at other aspects of their subject, notably courtship and nesting behaviour, the variations of bird song and the aggressive way in which some birds defended their territory. A few began to specialise in a particular species, building up a detailed picture of their lives through hours of watching and volumes of birdwatching diaries. Julian Huxley led the way to scientific birding through his classic study of the great crested grebe, but by the late 1930s he had many disciples: Fisher on the Atlantic fulmar, Nethersole-Thompson on the greenshank and the crossbill, Arnold Boyd on finches and swallows, and Ronald Lockley on the Manx shearwater and puffin, to name but four.

Wartime interrupted the tranquil way of life that allowed such studies, but in a way war encouraged birdwatching. It is often said that war consists of five per cent wild excitement and fear, and 95 percent boredom. Stationary soldiers and, still more, prisoners of war need a hobby. Arnold Boyd found the trenches of the First World War very handy for birdwatching, during the quieter moments at

Jacket of *The Fulmar* (1952).

least. During the Second, John Buxton watched redstarts building their nest through the barbed wire of a prisoner-of-war camp in Germany. Meanwhile, back on the Home Front, non-combatants restricted by petrol rationing to their home ground, also began to study the birds on their doorstep. Most of the New Naturalist bird monographs were based on a particular study area, often a very small one. The extent of Edward Armstrong's 'Wren Wood' was precisely four acres, and Stuart Smith's yellow wagtails lived in a vegetable plot by the railway in a Manchester suburb.

The New Naturalist monographs were not altogether 'new', but there were not many predecessors. The first fully fledged bird monograph ever published was probably *The Grouse in health and disease* (1911) attributed to the Lovat Committee, though much of the work was by Edward Wilson – who might have contributed to the New Naturalist series in his old age had he not accompanied Scott to the South Pole. Another pioneer work was *The Gannet, a Bird with a History* (1913) by John Henry Gurney, a book which influenced James Fisher when he came to write *The Fulmar*. These were, however, the exception rather than the rule. More influential for the majority of British birdwatchers in the 1940s were the classic how-to-do-it books by Max Nicholson, Julian Huxley, Stuart Smith and James Fisher. They helped to transform the hobby of birdwatching into the science of ornithology, as symbolised by the founding of the British Trust for Ornithology in 1932, and the Edward Grey Institute at Oxford in 1938. By the Second World War, the stage had been set for anyone with time and intellectual resources to become a player.

Nor was the idea of natural history monographs confined to the bird world. The Biological Flora project, which encouraged the study of individual species of British plants, had got underway as early as 1928, though it did not gather speed until after the war. British mammals, on the other hand, had been neglected by all but a handful of intrepid amateurs. Their time did not come until the 1950s with the founding of the Mammal Society (which was itself partly the product of the surge of interest caused by Ernest Neal's and Harrison Matthews' New Naturalist books).

What *was* new was the production of a uniform series of books, each devoted to a single species and using the latest scientific techniques. To any publisher before 1945, the notion would have smacked of commercial disaster. Without the confidence engendered by the success of the first New Naturalist books, it is doubtful whether anyone, even Collins, would have considered publishing such 'specialist books', let alone a series of 22 on subjects ranging from fulmars to fleas, rabbits to wrens. The first monograph was published in 1948, but they were already the subject of animated discussion in early 1946 when only two New Naturalist books had yet appeared. The minutes of the relevant Editorial Board meetings have not survived, but the private papers of some of these most closely involved reveal in broad terms how the idea developed over the next few years. The story begins with *The Badger*.

The making of a natural history classic

In considering the origins of the monographs, we must recall that 1945 and 1946 were the *anni mirabile* of the New Naturalist library, when the public appetite for such books proved greater than Billy Collins had anticipated, even at his most optimistic. The first four books had all sold out within their first year, and reprinting them brought their total print-run to some 40,000 copies each.

Billy Collins had succeeded in launching a flagship series of books, burnishing the name of his house in natural history circles *and* making a profit. Their success was attributed, rightly or wrongly, to the novel use of photography, especially colour photography. We can assume that Collins was at that time receptive to an expansion of the series, especially if the new books were accompanied by an impressive collection of photographs. Both Huxley and Fisher were strong advocates of what the latter called 'monography', and Fisher had already begun work on a book about the fulmar. If 40,000 members of the public were willing to pay a guinea for a book about butterflies or gamebirds, would not, say, a quarter of their number be interested in paying 10*s*. for a book about the cabbage white or the partridge .... or the fulmar?

The first New Naturalist monograph, *The Badger*, had its origins, like the library as a whole, in a London restaurant. Over lunch with Billy Collins, Leo Harrison Matthews, who was writing the mammal book, mentioned some remarkable work on badgers being carried out by a schoolmaster he knew at Rendcomb College in the Cotswolds. Matthews had been out badger watching with him recently, had admired his photographs of wild badgers at play and been much impressed with the detailed portrait of the badger community he was piecing together with the help of his pupils. The schoolmaster's name was Ernest Neal. Collins liked badgers, and once photographs were mentioned needed no further encouragement. He wrote to Neal soon afterwards:

> 'When I was lunching with Dr Harrison Matthews this week he was talking about you and told me of the photographs you had taken and also about the book you are writing on Badgers.
>
> 'As you may know we are now specialising in nature books, and if you care to send us up a selection of your photographs we would be very interested. We also would very much like to see your manuscript on Badgers when you are ready to send it.'
>
> *W.A.R. Collins*. Unpublished letter to Ernest Neal, 25 January 1946

Collins evidently passed the matter on to the editorial board, for a fortnight later Ernest Neal received a further letter, this time from Julian Huxley.

> 'Your name has been given to us as being possibly interested in doing a monograph in our New Naturalist Monograph Series on the Badger. I hope this may prove possible, as I think this will be a very interesting subject.
>
> 'I am getting Messrs Adprint to send you a specimen contract and details of the series, to show you the length, business arrangements etc. We are publishing two different types of monograph, one with 8 colour plates and the other type with only one colour plate (as frontispiece). Perhaps you could let me have your views on this. You could also have a reasonable number of black and white plates, in either case.'
>
> *Julian Huxley*. Unpublished letter to Ernest Neal, 7 February 1946

Huxley went on to ask Neal to send in a synopsis of his book, and, if possible, a specimen chapter. Neal responded with several chapters, though the promised contract did not in fact arrive until April 1947, when the book was nearly finished. In the meantime, Matthews had sent a friendly letter of his own, emphasising that 'Collins are anxious to publish this for you' and that the terms were not bad.

Ernest Neal with badger cub, 1953. (*Photo: Ernest Neal*)

The Badger took about a year to write. Ernest Neal had begun it as science master at Rendcomb College, and completed it shortly after his appointment to Taunton School, which, fortunately, was also well placed for watching badgers. The success of *The Badger* was crucial to the future of the monographs. It was to some extent a pilot for the new series. If it flopped, the future of the other titles was questionable: after all, who would purchase a 'greenshank' if no one bought a 'badger'? As it turned out, *The Badger* was so successful that it raised false expectations for the other books; no other New Naturalist monograph sold as well. Some 20,000 hardback copies were printed and sold in five editions over the next 30 years, and the rights were purchased by Penguin Books in 1958 for an equally large paperback edition in the Pelican series. Even so, after the first year the book was a steady seller rather than a best seller. It was most remarkable not for its sales, but for what it did. For Ernest Neal, it led to further work on badgers resulting in a doctorate, to many new academic friendships and to a lifetime's devotion to wild badgers and their conservation. In a wider sense, the book stirred up a great deal of interest, and played its part in the founding of The Mammal Society and the last of the English county wildlife trusts, Somerset (for his part in which Neal was later appointed an MBE). As for Billy Collins, *The Badger* was his favourite book in the whole series. The Board minutes record his comments on more than one new monograph title: 'quite good, readable and full of facts – but not as good as *The Badger*'.

What makes a natural history classic? In the case of *The Badger*, it is probably the fact that most of the book consists of personal observation. Neal was writing about an animal that everyone had heard of, but relatively few had seen and about which remarkably little was known. Neal had uncovered some, but by no means all, of the facts about the badger's secretive life, and wrote about them with warmth and simplicity. His is an exceptionally well-mannered book, taking care to inform the reader exactly how the author carried out his studies and how he had managed to photograph them with the cumbersome apparatus available at the time. It was encouraging, too, that this was all based on simple field observation within the capabilities of any sufficiently dedicated amateur naturalist (though for watching several sett entrances at once, a band of sixth form helpers was undoubtedly an advantage). Another virtue of *The Badger* was the direct way in which it involves the reader from the opening page. The reader is at the writer's elbow as the sun sinks below the horizon, the wind drops and the first 'low, quavering, yelping sound' is heard from the depths of the earth. Neal's evocation of one night's expedition set an example followed by many of the later titles, engaging the reader's interest by taking him on a badger watch, rather than by inconsiderately hurling facts at him.

As one might expect, it was the reviews in the scientific press that were the most dispassionate. J.S. Watson, in the *Journal of Animal Ecology*, admitted that the book 'adds appreciably to our knowledge of the social life and habits of the badger' but that it was not (and why should it have been?) 'a critical analysis of the badger's ecology, for which, however, the data available are still inadequate'. Other reviewers responded more warmly. To Brian Vesey-Fitzgerald, who had become a faithful reviewer of New Naturalist titles, 'Mr Neal's book has thrilled me as few books have done in the past twenty years; it has held my attention for every line of the page; it has made me think ..... He has not made the mistake common to modern scientists of believing that truth to be truth must be frigid. "Cold facts" is a favourite catchphrase of the scientist. How pleasant to have the facts with some warmth for a change! How much more human!'

Turning cold facts into warm facts might be a suitable phrase to sum up the New Naturalist ethos as a whole (while not sharing Fitzgerald's antipathy to 'modern scientists' *tout court*). Since 1948, the facts about badgers have, of course, been added to and elaborated to a considerable degree. But the warmth of *The Badger* remains: it is still a good read, and surely that is the quality we look for first in a book. As a scientific milestone, its importance lies in the effect it had on Neal himself and on his disciples and successors. In a recent eulogy, from today's standpoint, Pat Morris explained why this book has been so influential.

> 'Today, with the benefit of hindsight and a wider perspective, I can appreciate *The Badger* better, not so much for what it contained as what it did. Looking at it now, this small volume seems pretty ordinary stuff – lots of people could write as much. But that's not the point. We can easily write about badgers now because of what Neal triggered off years ago.
>
> 'Today, we have a countrywide assortment of badger groups, there are lectures and meetings attended by hundreds of people, and lots of books and scientific research. All this had to start somewhere, and few would dispute that its seeds were planted between the green buckram covers of Ernest Neal's classic ..... Neal showed later authors the way, and succeeded so well that (to my mind at least) none of Collins' subsequent books have been better. That series spawned other popular mammal monographs, and there is now a significant body of British mammal literature – whereas before 1948 there was almost nothing ..... Ernest Neal's book was a powerful catalyst.'
>
> *Pat Morris. BBC Wildlife* 'A Natural Classic', July 1993

Monography

There are 22 New Naturalist 'monographs', beginning with one underground mammal, *The Badger*, in 1948 and ending with another, *The Mole*, in 1971. They were not planned as a 22-volume set, nor are those the only titles that were considered for publication in the series. As with the senior series, the monographs started slowly, built up speed in a cascade of titles in the early 1950s, and then decelerated gradually. The titles came out in no particular order, nor were they necessarily concerned with the most popular or accessible subjects. The editorial board commissioned not so much subjects as authors. There was never any particular line drawn as to which species were acceptable and which were not, and a series that could include, say, the hawfinch, might include almost any breeding bird. Nor was it entirely clear where the dividing line lay between main series subjects and the monographs. *The World of the Honeybee* in the main series is, after all, concerned with a single species. And why place *Ants* among the monographs but *Dragonflies* in the main series? *The British Amphibians and Reptiles* was originally listed among the monographs, but rightly finished among the mainstream titles. *Fleas, Flukes and Cuckoos* made the opposite journey. It is difficult in fact to find much logic behind some of these decisions, but if there is any it probably has more to do with sales than species. There was a large enough market of beekeepers to make the prospective sales of a *Honeybee* look attractive, but the world of *The Hawfinch* is likely to be smaller. The monographs were not expected to sell in the same numbers as the main series; their editions were smaller, and they were not promoted as much. Only *The Badger, The Herring Gull's World* and the two fish books did well in the shops. Most of the others, through no fault of their authors, were commercial failures.

The editor most responsible for the monographs was James Fisher, who took over much of the correspondence after Huxley's appointment to the first scientific directorship of UNESCO in 1946. It was a task which this great advocate of monography must have regarded with relish. In the 1940s and 50s, Fisher's contribution to the growth of the New Naturalist library was enormous, and nowhere more so than with the monographs. He read and commented on the majority of manuscripts with perfectionist zeal, and in one or two cases even re-wrote entire chapters himself. Without his energy, knowledge and boundless enthusiasm, it is open to doubt whether the monographs would ever have taken off. And when Fisher died in 1970, the monographs died too.

The surviving record suggests that, while some titles might have been contracted for as early as 1946, the monographs did not gather steam until the following year. By February 1947, it was time to start thinking about design

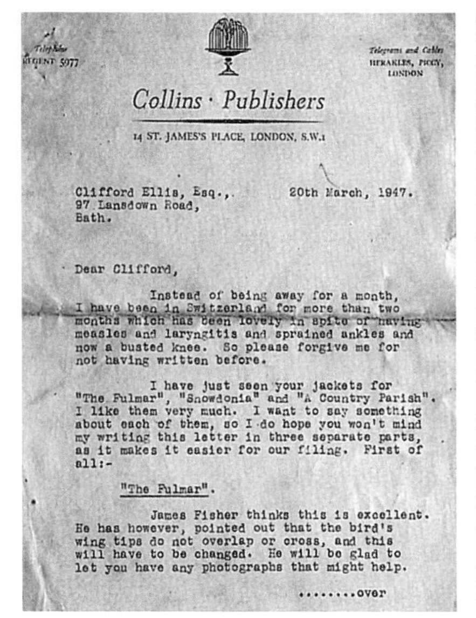

Telephone
REGENT 5977

Telegrams and Cables
HERAKLES, PICCY,
LONDON

Collins · Publishers

14 ST. JAMES'S PLACE, LONDON, S.W.1

Clifford Ellis, Esq.,
97 Lansdown Road,
Bath.

20th March, 1947.

Dear Clifford,

Instead of being away for a month, I have been in Switzerland for more than two months which has been lovely in spite of having measles and laryngitis and sprained ankles and now a busted knee. So please forgive me for not having written before.

I have just seen your jackets for "The Fulmar", "Snowdonia" and "A Country Parish". I like them very much. I want to say something about each of them, so I do hope you won't mind my writing this letter in three separate parts, as it makes it easier for our filing. First of all:-

"The Fulmar".

James Fisher thinks this is excellent. He has however, pointed out that the bird's wing tips do not overlap or cross, and this will have to be changed. He will be glad to let you have any photographs that might help.

........over

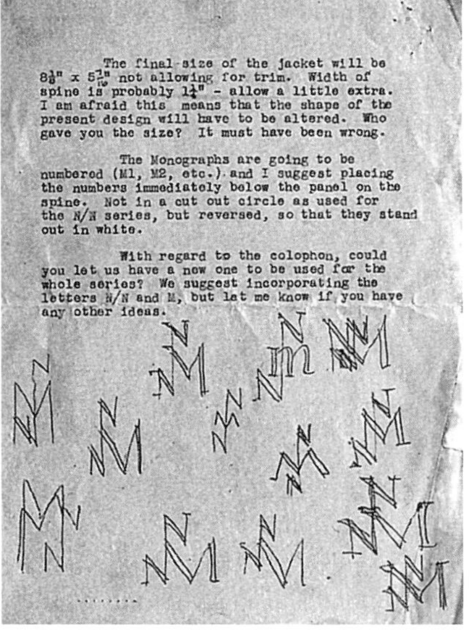

The final size of the jacket will be 8 1/8" x 5 7/16" not allowing for trim. Width of spine is probably 1¼" - allow a little extra. I am afraid this means that the shape of the present design will have to be altered. Who gave you the size? It must have been wrong.

The Monographs are going to be numbered (M1, M2, etc.) and I suggest placing the numbers immediately below the panel on the spine. Not in a cut out circle as used for the N/N series, but reversed, so that they stand out in white.

With regard to the colophon, could you let us have a new one to be used for the whole series? We suggest incorporating the letters N/N and M, but let me know if you have any other ideas.

Letter from Ruth Atkinson to Clifford Ellis, with the latter's sketches for a Monograph colophon. (*Private collection*)

matters. Fisher wrote to Clifford Ellis to inform him that 'We are extending the series to include a series of monographs. One of the first of these is to be 'The Fulmar' which I'm in the middle of writing myself'. The Ellises were asked to produce a new design for the dust jackets, and also to devise a special colophon. Judging from the pencilled doodles at the bottom of this letter (reproduced here) Clifford Ellis started work straight away, perhaps over the breakfast table. He decided to keep the basic attributes of the main series jacket designs, but to replace the title band with coloured ovals on the front and on the spine. In place of the conjoined Ns of the main titles, he devised a new monogram – NMN – placing it at the top of the spine. On the title page, the 'M' was added as a kind of badge at the point where the main series colophons bore sticklebacks and caterpillars. On *Redstart* and *Yellow Wagtail* the 'M' on the spine was lettered by hand.

The monographs were supposed to be smaller and slimmer than the main series – about 50,000 words compared with 80,000 and with 8 × 5½ inch pages as opposed to 8¾ × 6. They would have fewer colour photographs, and would sell for about half the price. Inevitably there were exceptions. When they came to be written, *The Fulmar* and *The Wren* were each well over 100,000 words, and they had to be printed in the larger size.[7] The monographs also included three 'special volumes' in plain jackets, two of which were also printed in main series size. For these reasons, the monographs do not make such a satisfactory set as the main series, and the bigger books stick out like battlements when they are arranged on the shelf in numerical order. It is a great pity, too, that the Ellises were not allowed to design all the jackets. Those three blank spines of *Fleas, Flukes and Cuckoos, Mumps, Measles and Mosaics* and *Birds of the London Area*, which have had New Naturalist collectors cursing ever since, were cost-cutting expedients.

The titles which Huxley or Fisher had commissioned in 1946 and 1947 took at least a year to complete. The first past the post was *The Yellow Wagtail*, which was ready by spring 1948, but not published until 1950 (as number 4 in the series, the numbered sequence of the monographs not always corresponding with the order in which they were written). Possibly Collins had decided to wait and see how *The Badger* fared before printing further titles. *The Yellow Wagtail* and *The Redstart* were published together in March 1950. They appeared at a bad time for the book trade, with rising costs and plummeting sales. Neither book sold well, though they were favourably reviewed. Ten years later, there were still large stocks of *The Redstart* and *The Yellow Wagtail* lying about unbound in the Collins Glasgow warehouse. At one point, the publishers considered remaindering them, having failed to persuade the Readers Union to accept a batch, and in 1962 took the decision to pulp 1,000 uncut quires of *Wagtail.* These books remained technically in print until 1972, but they seem to have disappeared from the shops long before that and I suspect they were quietly remaindered.

It was soon clear that a public which bought *Honeybees* and *Badgers* in commercially viable numbers, did not entertain the same enthusiasm for bird titles. The bird books that sold were identification guides about the British and European species generally, not those on individual birds. The experience of *The Redstart* and *The Yellow Wagtail* suggested that even an edition as small as 6,000 was to over-estimate the market rather wildly (one has to remember that even the RSPB had no more than a few thousand members in 1950). The sales of *The Greenshank* were unlikely to be as high, and even fewer people were likely to fork out 35*s.* (soon raised to 42*s.*) for 500 pages on the fulmar. The choice of four rather unfamiliar birds to blaze a trail was unfortunate. Perhaps if the series had

Jacket of *The Wren* (1955).

included David Lack's book on the robin, or if the books had been about waders and sea-birds, rather than greenshanks or fulmars, they would have been more successful. Who knows? We need not mourn unduly. Though they were not a bookselling sensation, these early bird monographs were important for the same reason that *The Badger* was important. They were remarkable books in their day, and they reached and impressed the market they were written for.

There were considerably more ructions over the next three books, *The Wren* (1952), *Fleas, Flukes and Cuckoos* (1952) and *Ants* (1953). Billy Collins seems to have been willing to countenance making a loss over *The Fulmar* – it was Fisher's reward for being a good editor. But when Collins heard about the length of *The Wren* and saw the setting costs of another ornithological marathon, he made his displeasure known. The time lag between the completion of *The Wren* (1951) and its eventual publication (1955) suggests a considerable wrangle behind the boardroom door at Grafton Street. Edward Armstrong had written a scholarly and meticulous book, covering every aspect of wrens, and above all their curious nuptial life, with nests built by the cock bird and polygamous marriages. The text was liberally sprinkled with references and in general tone it amounted to 'almost a treatise', as Richard Fitter put it in his review for *Nature*. He added that 'if it were necessary to standardise monographs on birds, this would make a very satisfactory book on which to base the standard'. Satisfactory in academic terms, that is.

The editorial board received the 125,000 word manuscript in May 1951. No formal contract survives – there may not have been one – and Armstrong had evidently not felt constrained by considerations of length. Dudley Stamp asked in some bewilderment: 'Who will *read* monographs of this length?' Fisher and Huxley were asked to look into the matter, and to recommend cuts. They were further encouraged to do so by what was described as a 'strong letter' from Billy Collins. In this delicate undertaking, Huxley wrote to Armstrong's friend, the distinguished ornithologist W.H. Thorpe, requesting his help in reducing the book to more saleable proportions. Indignantly, Thorpe refused, offering his opinion that Armstrong had been badly treated, and adding that he had advised him to withdraw the manuscript. James Fisher said that for his part he was prepared to accept the manuscript as it was, and if necessary forgo his editor's royalty. And Huxley, though he thought the book too detailed 'on comparative matters', agreed that the problem was not literary nor scientific, but commercial. Various ruses were used, such as increasing the number of lines per page, to reduce the pagination from 378 to 302, and Armstrong agreed to make various changes that entailed retyping the whole manuscript. But even so, this was not one of Billy Collins' favourite books. In a memo to his editors shortly after publication he wrote:

> 'I was appalled when I saw the setting costs of this book was £750... Even if we sell out all our first edition, we will still show a loss of £160. The present loss must be much higher. There are some very bad features of this book, and there must be an absolute rule that we never get in this position again with a New Naturalist title .... I think practically every bird monograph has shown a big loss.'
>
> *W.A.R. Collins*. Internal memo to editors, 27 July 1955

The publication of *The Wren* coincided with a newspaper strike. Macmillan took on a few hundred of the small edition of 3,000 for the US market, but since they refused to reorder after the book went out of print in 1964, there was no possibility of Collins reprinting it. *The Wren* was respectfully reviewed, and remained the standard world monograph on our smallest bird for many years. Today, and for many years past, *The Wren* has become a scarce book and a gap on many shelves. It has moreover a tendency to auto-destruct with big rusty damp-stains spreading over the Ellises exquisite jacket design. Perhaps it is just an unlucky title.

The sad story of *Ants*

We have reached monograph number 8, *Ants*, which, as many readers will know already, is the book that was supposedly withdrawn from sale after receiving bad reviews. I have heard other stories about it, some of them certainly incorrect. What are the facts? *Ants* was contracted for in 1947, completed in July 1952, and published on 8 June 1953. Its author, Derek Wragge Morley, was a remarkable young man. As a boy he developed a lifetime's interest in ants, studying their behaviour in his parents' garden and publishing his first research paper at the age of 16. A year later, he chaired a session for the International Congress of Entomology in Berlin, reading two of the contributing papers. The story is still told of the stir created by the bespectacled, shock-haired schoolboy limping up to the lectern. In 1942, Morley was made the youngest-ever Fellow of the Linnaean Society. This precocity was achieved in spite of – perhaps because of – a debilitating illness he suffered while in his early teens. It was while he was recovering from this illness that Morley began to study ants.

Ants were one of Julian Huxley's kaleidoscopic interests; he published a book about them in 1930. In his preface, Morley tells how 'Huxley answered the letters of a pestering schoolboy [i.e. himself], and encouraged his childish experiments.' It was probably Huxley, then, who spurred Morley on to write a book for the New Naturalist series. He should have known better. Morley had a genuine enthusiasm for ants, but not nearly enough experience to write a serious book about them, and especially not for a series that included books of the weight of *The Fulmar* and *Fleas, Flukes and Cuckoos*. His tragedy was that he might well have written a good book in time.

The first draft was no good. As Morley put it, James Fisher had made some 'candid criticisms which made me tear up my first manuscript. How right he was!' The revised script passed muster, however, and it is doubtful whether anyone but an expert in ants would have found much wrong with it. F.T. Smith, Collins' principal editor, had pronounced it 'a very good book' on the whole, but, of course, he was not an ant expert. Unfortunately Huxley, who would have spotted the mistakes, had recently suffered a nervous breakdown, was absent at the crucial time and probably never read the manuscript. By the time he resumed attendance at the Editorial Board in September 1952, *Ants* was already being set in type. The Ellises designed a handsome jacket for it, and the book contained some good professional photographs by Raymond Kleboe of *Picture Post*, where Morley was science editor, and some excellent close-ups by Ronald Startup. A reviewer for *The Naturalist* found little to complain about, praising its easy style, even though 'sometimes he simplifies his writing too much, and writes as though his readers were still in the second form'. All seemed set fair until, shortly after publication, the editors received a letter from Dr M.V. Brian at Glasgow University criticising the book for numerous errors of fact. The minutes of the editorial board record that Morley drafted 'a good restrained reply' which was read and approved by Huxley and Fisher, who were of the view 'that personal issues beyond science had been involved in Brian's review'. But worse was to follow, with a very adverse review in the *Entomologist's Monthly Magazine*. It began mildly:

> 'The New Naturalist volumes have set such a high standard of excellence that we accept almost complacently Mr Wragge Morley's opening remark that his contribution to the series is the first comprehensive work to be published on British ants for over twenty-five years. The chapter headings are indeed full of promise and it seems that the author is prepared to cover his subject widely.....
>
> 'Why then, with so much offered, does Mr Wragge Morley so soon begin to disappoint us? In an introductory chapter we learn about the ants he found as a boy in his parents' garden. Here is much general information on foraging habits, mating flights, colony founding and other aspects of ant behaviour. But the material is assembled in a haphazard sort of way and thrust without order at the reader. And all the time we get the impression, almost too sharply, that the author is relying on a very personal interpretation of what he has observed. Before long, however, we feel a twinge of uneasiness, a first flicker of real doubt. For on page ten he describes how one fine day in March, 1934, he witnessed a marriage flight of the Black Lawn ant with hundreds of males and females rising into the air at dusk. Now this particular species – *Lasius niger* (L.) – mates normally in the afternoon of sultry days towards the end of summer. But Mr Wragge Morley saw the event in spring and not only in his garden, but round about as well. And yet he does not trouble to tell us that this was something unusual and quite exceptional.'

The review goes on to list more factual mistakes, before concluding damningly:

> 'If Mr Wragge Morley's book fails, it is not because he lacks enthusiasm. But enthusiasm without discipline is not enough to carry a work of this kind to

success. Right from the start he underrates our intelligence. The pills we are offered are so often too big to swallow and we resent all along the author's irresponsibility, his carelessness, and the absence of anything like a critical approach to his subject.'

Review in *Entomologist's Monthly Magazine*, November 1953, p. xliii

A copy of this review seems to have been sent to Collins. Morley had not endeared himself to the editors by writing another book about ants, *The World of Ants* for Penguin, without informing them. The Board minutes state only that James Fisher told Morley that the book would not be reprinted, but that the editors would not commission anyone else to write on ants 'without telling him and giving him the opportunity to write another book'. The record leaves it open to doubt whether *Ants* was in fact withdrawn from sale as is widely believed. Out of some 3,900 copies printed, some 3,651 were sold, leaving only 244 to be accounted for, all of which could have been swallowed up in samples, review copies, bookseller's returns and so on. An alternative reading of the evidence would be that *Ants* simply sold out, and was not reprinted. Although it was in the shops for less than a year, *Ants* is not a rare book, as it would be if a large proportion had been returned and pulped. It was, and still is, quite a popular book among myrmecophiles. My guess is that Collins simply decided not to advertise it.

The bad reviews in the entomological press must have been a bitter disappointment to Wragge Morley personally, and he does seem to have ceased to publish scientific papers on ants from about this time. However, his decision to switch from insect research to science journalism had been made several years earlier. From 1952 to 1962, Wragge Morley was the science editor of the Financial Times and later worked as a science and technical consultant to Hambros Bank and on the board of a firm of compressor manufacturers. He died in January 1969, leaving a wife and four children. Ironically his book sold more copies than did the main series *Ants* by M.V. Brian, published a quarter century later.

The special volumes

The monograph series includes three 'special volumes' which are not really monographs at all but books which were regarded as too limited in appeal to form part of the main series. One of these books, *Fleas, Flukes and Cuckoos* (1952) is a science classic. The other two, *Mumps, Measles and Mosaics* (1954) and *Birds of the London Area* (1957) are half-forgotten, and are usually among the very last titles to come to the attention of the New Naturalist collector. The idea behind the first two seems to have been Julian Huxley's and they were commissioned soon after the war. Miriam Rothschild began to write *Fleas, Flukes* (its working title was *The Parasites of Birds*) in 1947, and a contract was sent to Kenneth Smith for the book about viruses which became *Mumps, Measles and Mosaics* in the following year.

The circumstances in which the first chapters of *Fleas, Flukes and Cuckoos* were written – a desolate hotel in the middle of the rubble of Calais, while a storm brewed up in the Channel – must be enjoyed and read in Miriam Rothschild's own words at the end of this chapter. Although this book is credited to co-authors, the whole of the draft was in fact written by Miriam in her distinctive style, while her colleague, Theresa Clay, contributed advice and information, especially on her special subject, bird-lice, as well as blue-pencilling the manuscript. The title was Huxley's. The 'This, That and the Other' style of title was briefly in vogue in the early 1950s, and publishers saw it as a useful way of jazzing up less appealing subject matter. The authors had completed the book by September 1949, but it was not published until May 1952. Billy Collins had in the meantime decided that the title was unlikely to make much money and no longer wanted to publish it. Unfortunately, the authors had proceeded without a formal contract and therefore had no legal redress. The editors were incensed at Collins' unilateral decision, especially Julian Huxley who had read the manuscript and regarded it as a very good book. He threatened to resign and take the rest of the board with him. Collins then agreed to look into ways of publishing the book without losing money over it, but that he still refused to commit himself is suggested by a note by Raleigh Trevelyan at the end of 1950 saying 'that is just possible that we may not even do the book, the subject matter being somewhat specialised and the cost of production rather high in relation to likely sales'.

Collins agreed to publish the book only on the toughest of terms that would indemnify his firm against the expected loss. He insisted that the authors and publishers should in effect change places, so that the former would take on the financial risks, while Collins would take a percentage of the sales – in effect a royalty. The authors were to pay for all the illustrations themselves. And the first edition was limited to a mere 2,000 copies. The meanest aspect of these negotiations concerned the jacket design. By some crossing of purposes, two illustrated jackets were designed for it, one commissioned and paid for by Miriam Rothschild, and the other (featuring a cuckoo) designed by Clifford and Rosemary Ellis. But Collins decided that, as a 'special volume', the book should be issued in plain wrappers, like a university textbook. Having resisted publication at all points, he was not minded to give it a pretty cover.

Billy Collins completely misjudged *Fleas, Flukes and Cuckoos*, which is a triumph of style over content. Miriam Rothschild presented the bizarre world of bird parasites unsqueamishly but in a wonderfully clear and accessible way. Some of the regular book reviewers in the daily papers were enthralled by it. Guy Ramsay picked on its marvellous quotability. Peter Quennell in the *Daily Mail* found 'scarcely a dull page. There is a surprise – sometimes a nightmarish surprise – in nearly every paragraph.' His colleague on the *Sunday Times*, Raymond Mortimer, agreed: 'I have gone through this book with my eyes popping out of my head, so amusing, amazing, appalling are the habits here uncovered.' *The Spectator* went so far as to declare that *Fleas, Flukes* would set a new direction in natural history writing. It was a rare example (very rare at that time) of a book that managed to be a contribution to literature as well as science.

With so many good reviews behind it, the first edition sold out within weeks. Wrong-footed, Collins was slow to reprint it, and by the time the book was back on sale, the feast had grown cold, the rave reviews forgotten. Even so, the reprint had also sold out by 1956, helped by the demand from university biology courses, and a second reprint was ordered at Huxley's insistence. There was also a Reader's Union edition and two paperback versions, one in the Pelican series, the other published by Arrow and on sale in Ireland. *Fleas, Flukes and Cuckoos* did in fact sell better than the majority of the monographs, even though the Irish paperback was a travesty of a book, full of misprints that the authors were not given an opportunity to correct. Given more considerate treatment by the publishers, it would have done better still.

Mumps, Measles and Mosaics was planned as a semi-popular book about viruses. The editors had argued about the title for some time, toying with names like

Kenneth Smith (*left*) (1892-1981) and Roy Markham (*right*) (1916-1979), authors of *Mumps, Measles and Mosaics* (1954). (*Photos: John Innes Institute*)

'Beyond Bacteria' and variations on 'The Invisible World of ....' before coming up with the one adopted. Mumps, measles and mosaics are all, of course, diseases caused by viruses. Raleigh Trevelyan sent a copy to Miriam Rothschild with a note: 'As we have rather pinched your title, I thought you would like to have a special complimentary copy. You will be very pleased to know, I am sure, that there is a howling misprint, but I leave you to find out where it is.' If *Fleas, Flukes* was an experimental title, the much shorter *Mumps, Measles* charted completely new waters as a semi-popular work. The authors (for the elderly Kenneth Smith had invited his younger colleague Roy Markham to write the book with him) were both leading experts on viruses and Fellows of the Royal Society. Smith had in fact written the standard textbooks on both plant and insect viruses and was a world authority on the subject. Even so, the editors decided the book might need a push since it dealt with organisms (if, indeed, viruses *were* organisms) that were not only invisible but strangers to British naturalists. As the responsible editor, John Gilmour compiled a publicity hand-out, full of 'amazing facts': 'In the USA every person catches, on average, 2½ colds per year. In the 1918-19 'flu epidemic half the entire population of the world was attacked. Clothes moths have been successfully controlled experimentally by infecting them with a virus. Viruses are far and away the smallest living things – about a ten millionth of an inch across' and so on over a couple of pages. James Fisher wrote to all the main agricultural colleges, hoping to get the book adopted by them as a textbook: 'as a general introduction to the viruses we believe it to be unique, and its agricultural slant is self-evident'.

Mumps, Measles and Mosaics was published on 1 February 1954. It was widely reviewed but in marketing terms it fell between two stools. The agricultural world had its own textbooks (written, as it happened, by Kenneth Smith), and since this book dealt mainly with cultivated crop plants, there was little in it to interest the field naturalist. Nor, for that matter, the medical man: there was plenty about mosaics, but only a few paragraphs about measles and mumps. The authors had done their best to write popular science, but some reviewers, like Cyril Connolly, in the *Sunday Times*, still found it far too technical. 'It is a pity we could not have had somebody who knew more about the subject,' complained Raleigh Trevelyan about Connolly's review. 'He had obviously not read the book properly,' agreed an annoyed Kenneth Smith.

The drab grey jacket and print-run of 3,000 (including 250 which went to Praeger of New York) scarcely argued confidence in the book. The title sold out within 18 months but, since the hoped-for orders from America did not materialise, there was no question of reprinting it. More than half of the copies sold did so on subscription. Raleigh Trevelyan considered that the sales 'compare very favourably with other NN special volumes', by which he presumably meant *Fleas, Flukes and Cuckoos.* Kenneth Smith's reply was: 'I am glad you consider these favourable.'

On the face of it, *Birds of the London Area since 1900* seems an odd title for the New Naturalist library, based as it is on a sectional interest in a limited area (and the library already contained a London volume). London had in some ways a claim to being a special case, however, since, with a large and well-organised natural history society and the longest continuous history of ornithological observation in Britain, it was a pace-setter for the rest of the country. The book is based on a survey by the London Natural History Society, edited into book form by a committee chaired by its President, Richard Homes. James Fisher was keen for the book to be published in the series, providing it was based on bird habitats, and was not a mere systematic list. The problem was that it was both. It was also too long, with 55,000 words on ecology and 125,000 on individual birds. Collins agreed to publish the book so long as the Society guaranteed to purchase 400 copies from an edition of 2,000, and cut the manuscript to a more manageable 150,000 words. These negotiations were stringent enough, but when Trevelyan proposed to reduce the size still further, while raising the price and cutting the royalty, they grew acrimonious as well. *Birds of the London Area* was published belatedly in March 1957. It was out of print in little more than a year, representing a favourable subscription list followed by slow sales; in other words, most of those who bought the book did so at once. One was either *very* interested in London birds, it seemed, or not at all.

The book was significant as one of the first thoroughly documented ecological surveys of any part of Britain. It was reprinted in 1963, but not by Collins. The blocks were sold to Rupert Hart-Davis, who re-set the type and brought the book up to date. Collins made a profit of £85.

An eye for a bird

James Fisher once wrote that it was the ambition of the New Naturalist to 'capture British natural history, and to transform it into lines of type, and blocks, and coloured inks'. To avoid any impression of pomposity, he added that while this might be a worthwhile aim, it was, of course, impossible to achieve. Success is in the eye of the beholder. Perhaps the books which came nearest to doing so were the dozen or so bird monographs, and their fellow travellers: badger, squirrels and oyster. In the 1940s and 1950s, bird study usually involved nothing more elaborate than a notebook and a pair of binoculars, with nest boxes and hides sometimes used for more intimate observation. The rest was a matter of reason, deduction and speculation, the same attributes as those used by James Fisher's hero, Gilbert White. There was not the detachment that seems to have built up since between the naturalist and his object of study. The authors of these books made it clear that they enjoyed their work, and loved their birds (or badgers or squirrels) without ever descending into the sentimental language of the previous generation.

Every one of the bird monographs was based on years of personal work. They all reviewed the sometimes scant available literature, but personal experience is usually to the fore and it is what lends these books their lively quality. Some of the authors were genuine pioneers. No one anywhere in the world had got to know greenshanks or herring gulls or even house sparrows so well. Their work had in some cases (notably the heron, the greenshank and the fulmar) started in the 1930s, long before the New Naturalist series existed. *The Yellow Wagtail* was based on eight years of more or less continuous observation, *The Fulmar* on something nearer fifteen. It was the apparent simplicity and straightforwardness of these studies that encouraged others to follow in their footsteps. Today, of course, there are many people who could (and often do) write such books. One would not now go to the New Naturalist monographs for new information. What we do still find in these now classic books is the excitement that goes with original discovery, and to bask in the glow of that golden postwar decade.

Rather than review each book in detail, I will sketch in a little of the background to the books, and the circumstances in which each was written: for it was circumstance more often than not that governed the choice of species. James Fisher had loved the cold grey seas and green islands of the North Atlantic since

Stuart Smith (1906-1963), author of *The Yellow Wagtail*, noting the reaction of a nesting willow warbler to a stuffed cuckoo. (*Photo: Eric Hosking*)

taking part as junior ornithologist in the Oxford expedition to Spitzbergen in 1933. The fulmar epitomised this world, and Fisher usually managed to find the means to visit its remote haunts each summer. Other naturalists found themselves marooned by wartime petrol rationing and chose birds that lived close to home. Desmond Nethersole-Thompson was exceptional in that he moved his own home to live among his beloved greenshanks. Denis Summers-Smith chose the house sparrow not so much for convenience as out of his admiration for a bird that had proved itself uniquely adaptable to man. We may be certain that there was something in each chosen bird which appealed at a more human level than mere scientific objectivity. It is not hard to imagine why Stuart Smith, for example, was so entranced by the yellow wagtail 'so lovely to look upon, and so full of a dainty and buoyant airiness of stance and flight', marvelling at the brilliant colours of the returning cocks 'before they were dulled by the soot and dirt of Manchester'. Smith was a fibre research chemist for the association of cotton industries in Manchester, living in the dormitory suburb of Gatley, close to the black waters of the River Mersey. At only a field's distance from his house was a ten-acre plot of land wedged between a railway line and the river, where the wagtails would return each spring to nest among the strips of 'potatoes, cabbages, sprouts and other vegetables' grown for the Manchester market place. Smith longed for their appearance each spring, and from 1941 began the study which resulted in *The Yellow Wagtail* nine years later.

In the same corner of England another Lancashire ornithologist, Frank A. Lowe, was studying the heron. By another coincidence, Lowe was, like Smith, a manufacturing chemist, which in his case was a family business. Well-known locally as the wildlife correspondent of the *Bolton Evening News*, Lowe's love affair with the grey heron dated back to the late 1920s and the national heron enquiry, among the first national censuses of a British bird. For many years he had watched a particular heronry at Dam Wood, a private mixed wood on the Scarisbrick Estate. Herons had nested there for at least a century, and seen great changes to the surrounding land – entirely, from their point of view, for the worse. In 1937, Lowe had constructed a hide on a platform at the top of a beech tree seventy-feet high, a task not eased by the local sparrows who made a habit of stealing the tie-strings for nesting material. During that spring and summer, he spent 100 hours there between February and July, at all hours of day and night. In 1948, he rebuilt the hide, this time spending 300 hours there. When he came to write *The Heron* (and the 1948 vigil seems to have been done for that purpose), Lowe was to include fascinating material from history and folklore, as well as a detailed survey of the heron in Britain and Ireland, but it was the herons of Dam Wood that were at the core of the book. Among its most memorable passages is the all-too-brief account of the rarely witnessed tree-top world, of the tree sparrows tucking their jute-fibre nests into the masonry of the heronry itself, of the loud snapping of the beech buds as they burst open in the May sunshine and the adventures of the young herons, now 'engaged in vigorous exercises or preening', now 'uttering low growling notes, quite unlike any sound produced by their parents' and the occasional crash as some over-venturesome fledgling fell off its branch and through the canopy.

The Reverend Edward A. Armstrong's book on *The Wren* was based on a site at the opposite end of the country, a tiny wood in the western suburbs of Cambridge. Armstrong had moved there in late 1943 from the smoky streets of Leeds 'where there had been few opportunities to watch birds'. He was to be based at

Cambridge for the rest of his life. In the opening lines of *The Wren*, Armstrong described how:

> 'Darkness was falling on a November evening in 1943 and the bombers were roaring off into the gloom when, happening to look out of my study window, I saw a small bird alight on the trellis outside and then fly up into the ivy on the wall. A couple of evenings later the wren was there again. Evidently he came regularly to these sleeping quarters. My interest was again captured by a bird which had fascinated me as a boy. Here was a species about which I should like to know more.'
>
> *The Wren*. Ch. 1 Studying the Wren

An ideal study area, a private 4-acre wood with a shallow, reedy pond, excavated in the previous century by skating enthusiasts, lay less than a mile from his home. The wood was maintained as a nature sanctuary and it was in these 'ordinary' but undisturbed surroundings that Armstrong decided 'to concentrate on studying the "life and conversation", as Gilbert White would have put it, of a single species'.

While the wren study was partly the product of wartime restrictions, that of the redstart was the war itself. The 28-year-old John Buxton had joined the Commandos soon after the outbreak of war and having taken part in the ill-fated Norwegian campaign in 1940, had had the ill-luck to be captured. He spent the rest of the war as a prisoner in Germany. And so it came about that,

> 'In the summer of 1940, lying in the sun near a Bavarian river, I saw a family of redstarts, unconcerned in the affairs of our skeletal multitude, going about their ways in cherry and chestnut trees. I made no notes then (for I had no paper), but when the next spring came, and with it, on a day of snow, the first returning redstarts, I determined that these birds should be my study for most of the hours I might spend out of doors.'
>
> *The Redstart*. Ch. 1 Introduction

In some ways, prisoner-of-war camps offer rather good opportunities for birdwatching. It is hard to imagine any other circumstances in which so many intelligent, active people would have so much spare time on their hands, nor so much incentive to find a distracting pastime. Buxton's best prison camp was at Eichstätt in Bavaria, at 'an old German barracks in the [well-wooded] valley of the River Altmühl', not far from the limestone quarry where the remains of the earliest bird, *Archaeopteryx*, had been discovered. It was in the springtime of 1943 that Buxton, with the help of fellow prisoners, 'was able to gather together a mass of notes on one pair of redstarts, covering eight hundred and fifty hours'. That represented the equivalent of a 65-hour working week continued for three months, an incredible feat. Possibly nothing like it had ever been done before; and certainly the Eichstätt redstarts were the best-studied pair of redstarts that had ever lived. This particular camp became a kind of university of field natural history. Besides John Buxton, there was Peter Conder, who studied crested larks and goldfinches, George Waterston (wrynecks), Richard Purchon (swallows) and John Barrett (chaffinches). So popular did birdwatching (and butterfly watching) become that Buxton used to produce a weekly report for the canteen notice board. He even managed to get in touch with a famous German ornithologist, Erwin Stresemann, who sent him materials allowing the Eichstätt birders to catch and ring birds in the camp. John Buxton had studied redstarts in England and

Norway before the war, and was to do so after his release in 1945, but it was the mass of notes obtained on that single pair in spring 1943 that formed the factual basis for *The Redstart*, published in 1950.

In terms of single-mindedness, Desmond Nethersole-Thompson, the author of *The Greenshank* (1951) must surely be in a class of his own. Thompson devoted his entire life to bird study; everything else (apart from politics) was subordinated to it. His early career in teaching (history and the classics) served only one purpose: to fund his weekend egg-collecting forays in the spring and early summer. These raids, which took him to many of the wilder corners of England and made him unpopular in preservationist circles, did at least make Thompson a master of the art of finding nests, a respectable and necessary skill for any field ornithologist. Though he eventually gave up egg-collecting, he maintained to the end of his days that to study breeding birds you needed to have 'a predatory hunger for the nest'. Thompson was a purist; he cut no corners and he probably spent more hours watching birds than any ornithologist alive. He threw himself into writing, as into egg-collecting, fieldcraft and local politics, with great energy – no naturalist rivals him for staccato excitement, reflected in his unique style which must set some sort of record for exclamation marks: 'What a glorious challenge the Spey Valley offers! Look up! Stand and watch. Did you ever see more lovely birds? Don't go away. Listen! Watch the crossbills courting. You will love every moment!'

The Greenshank, the first of Thompson's keenly collected monographs on Highland birds, is relatively quiet in tone but it still reads with a freshness and vitality that the cloak of scholarship cannot conceal for long. The book was begun, in Thompson's mind at least, as early as 1932 with his first visit to the Scottish Highlands at the age of 24. By his own account, he had gone there to watch some of the hill birds, presumably steal their eggs, and then write up his adventures for the *Oologists Record*. It was that most beautiful of all the waders, the greenshank, that made him decide to go native. That epiphany is best read in Thompson's own words:

> 'In 1932 I had saved enough from my miserly salary as a schoolmaster to go north. It was a grand investment! In the Spey Valley I found my first crested tit's nest and stayed with an ancient gamekeeper who had been guide to Harvie-Brown, F.C. Selous and John Millais. After my first night in his cottage I killed thirty-nine fleas! Then I trained it to Sutherland where James McNicol, an outstanding naturalist-keeper, showed me my first greenshank's nest. After hunting greenshanks on the flows of Strath Helmsdale I knew that I had lost the first round..... I have never learnt so much as in those crowded weeks. In 1933 I was back north. I watched greenshanks in Rothiemurchus and dotterels in the Cairngorms. I was now completely hooked! The Highlands were the only place for me. In 1934 I was back for good. Perhaps I was the original Counterdrifter.'
>
> *Highland Birds*. (1971)

The turning point had been the late spring of 1933 when Thompson met his first wife and fellow enthusiast, Carrie:

> 'It was a summer of great heat. The air in the valley was close and stifling, yet wind and mist sometimes made work on the high ground impossible. There were days, therefore, which we spent in hunting greenshanks in the vast forest clearings that lay in the shadow of the hills: greenshanks exchanging duties at

> the nest, greenshanks flighting nestwards after feeding, greenshanks singing passionately under a hot June sky. This was a good introduction to the unwritten story that I now contemplated'.
>
> *The Greenshank.* Ch. 1 Early Beginnings

Desmond and Carrie moved into a bothy in Rothiemurchus, which was soon 'full of rusty filing cabinets'. In 1940, he had obtained a Leverhulme research fellowship with the backing of Julian Huxley, and with it was able to devote long hours to watching nesting greenshanks. That the story took so long to put down on paper was probably due to Thompson's other obsession, socialist local politics. This is of no interest to us, but it embroiled him in the immediate post-war years and reduced even his ornithology to the level of a mere hobby. Thompson had put off the task for several years while he fought landowners and Tories on the Inverness County Council. Fortunately, his memory of the days in wet heather watching Old Glory, Castle, Myrtle and the other Speyside greenshanks was still fresh. *The Greenshank* was his first book. Derek Ratcliffe, a close friend of 'Tommy', believes he never wrote a better. It is one of the best reads of the series, and by the end of it his greenshanks seem as individual as neighbours.

Greenshanks are among the most difficult of British birds to study. Their haunts are remote, and the well-camouflaged nest is one of the blue ribands of nest-finding. That was doubtless part of the attraction. Few resident birds offer more of a challenge, but the hawfinch is probably one of them. Nesting greenshanks can become tolerant of human intruders, but to watch hawfinches you need to melt into the woodland background and spend a lot of time in hides. To Guy Mountfort, who had recently co-authored one of the best-selling bird books in the English language, the *Collins Field Guide to Birds of Britain & Europe*, it was the extraordinary elusiveness of this bulky, colourful, and in terms of numbers, not uncommon finch that was the attraction: a bird that would stretch its watcher's fieldcraft to the limit. As Mountfort noted in his foreword to *The Hawfinch*, 'When one of Britain's leading ornithologists cheerfully confessed to me that he had "never laid eyes on a live Hawfinch" I gained the first measurement of my presumption in deciding to write a book devoted exclusively to this species.' In doing so he rarely found Nethersole-Thompson's opportunities to spend long hours observing particular individuals. Mountfort had had an exceptionally busy and peripatetic war as a member of the British army staff in Washington. His birdwatching time was almost equally constrained in post-war years by his directorship of a large firm of advertising agents. The hawfinch hopped in and out of his life, as in 'my first sight of the Atlas Hawfinch in Algeria and of Hawfinches feeding nonchalantly in the shell-wracked Reichswald'. It was to find out more about the hawfinch's most remarkable feature, its massive beak, that Mountfort turned to R.W. Sims at the Natural History Museum. By constructing a model of a hawfinch skull, they measured the force it could exert on cherry stones and on fresh olives flown in specially from Palestine. The experiment suggested that to crack open an olive stone, a hawfinch applies a pressure of a thousand times its own weight: at the scale of an average male human being that would represent a force of some 60 tons!

The Hawfinch is well and elegantly written. And although it draws on all the available literature, it contains, like the other monographs, some beautifully described first-hand experiences. I cannot resist quoting at least one of them, which is Mountfort at his best:

'The Hawfinch is devoted to sunbathing and will bask for long periods whenever the opportunity permits, either in the tree-tops or on the ground. It sits motionless, fluffed out like a ball, with its head sunk low between the shoulders and the flank feathers all but obscuring the wings. Sometimes the long neck is extended, feathers on end, looking like a bottle-brush. The tail, at such moments, is tightly compressed and looks ridiculously short and narrow. One bird I watched chose a sunny nook at the base of an old gnarled Beech tree and stretched out on the dry moss, lying on its side, with neck and one leg extended and eyes half closed, in positively voluptuous abandon to the sun's rays. When a distant Greenfinch sounded a mild alarm note, the Hawfinch shot up on its feet, with its neck raised to maximum extent, listened intently for a moment and then slowly sank down again and rolled over, blinking blissfully, as a cat will blink as it settles for a nap in front of a hot fire. It did not stir a muscle again for fully ten minutes. Such incidents become treasured memories.'

The Hawfinch. Ch. 3 The Winter Flock

The last bird book written in the same spirit is *The House Sparrow* (1963) by Denis Summers-Smith. He wrote the book in 1959-60, a decade later than most of the preceding books, and in some ways it shows. Although the author could incorporate more than a decade of personal fieldwork, there was by now also a substantial world literature to draw on. *The House Sparrow* is therefore a more rounded monograph than some of the others, a thorough-going review of the sparrow over nineteen chapters and 270 pages. Summers-Smith was drawn to the sparrow because of its astounding success in following human settlement over most of the globe, and adapting itself to twentieth-century human-lifestyles while still remaining true to itself as a wild bird. He began to study the bird seriously in the late 1940s, while living in Highclere. In 1953, he moved to Stockton-on-Tees

Guy Mountfort, author of *The Hawfinch*. (*Photo: Eric Hosking*)

Denis Summers-Smith, author of *The House Sparrow*. Cartoon by Euan Dunn published in *British Birds*, 1994.

to become a development engineer for ICI, whereupon his study area on the house sparrow changed from the fields and gardens of rural Hampshire to the streets of an industrial town. After a decade of intensive work censusing and ringing sparrows, inspecting nest boxes and observing captive birds in an aviary, all in his spare time, he gave up intensive sparrow watching at the close of the 1958 season and began to write the book: 'I am convinced that the attitude of the birds towards me has changed since then – now I am accepted as a normal piece of the landscape and no longer the detested pryer into their private lives, to be viewed with the utmost suspicion.'

James Fisher, who had first approached Summers-Smith about a sparrow book four years earlier, was most impressed with the manuscript, finding it both readable and scholarly. The main problem was its 'rather awkward' length: even without the full bibliography – like many of its predecessors, that had to be sacrificed to the interests of commerce – it was an uncomfortably large volume. There was a delay of more than two years before publication, during which Collins approached a succession of American publishers to consider publishing the book under a New World title, *The English Sparrow*. 'Although I cannot believe that Collins have lost interest in my monograph,' wrote Summers-Smith despairingly, in December 1962, 'the silence from your end almost makes me give up hope.' The editors reply is not filed, but only a month later *The House Sparrow* was at last in the shops. While no more successful than its peers, it sold well enough to justify a modest reprinting in 1967 and 1976. The latter represented the last printing of any New Naturalist monograph; possibly it was a response to a letter from Tony Soper telling the publishers that second-hand *House Sparrows* were changing hands at £20: 'Too much for honest labourers like me. How about a reprint?'

The House Sparrow was the first book ever devoted to this most familiar of birds, and its conclusion, that house sparrows are not only more adaptable but more intelligent than other birds, was as startling as it was convincing. That there was still a great deal of mileage in sparrow study was proved by Summers-Smith in two later books, which broadened the field to encompass sparrows worldwide. In the best tradition of British natural history, a study that began in his garden in Highclere has taken him all over the world in search of sparrows and became his life's work: Denis Summers-Smith is undoubtedly the world authority on the genus *Passer*, an astounding achievement for someone who had only weekends and holidays for fieldwork, and the evening for writing. *The House Sparrow* was a worthy successor to a series of bird monographs that succeeded in setting new standards of natural history publishing.

The scientists take over

Most of the earlier monographs were written by amateur naturalists, if there can be any meaning in the phrase at a time when nearly all naturalists had to earn their living in some other way. From *The Herring Gull's World* onwards, the 'professional scientists' started to take over, that is those people for whom bird or animal study was a more or less full-time occupation. This reflected the growing number of trained scientists in the 1950s, employed by the newly formed Nature Conservancy, by the Agriculture Ministry departments and by the Universities (especially Oxford). Inevitably, these monographs were written in a different way to their forerunners, more 'scientific' in tone, with more data and tables, more rigorous analysis of the results, and sometimes a greater distance between author and subject. Another difference is that they dealt in the main with common and economically significant species: gulls, wood pigeons, rabbits and trout, not greenshanks, hawfinches and yellow wagtails. As James Fisher put it in his foreword to *The Wood Pigeon*, 'Until the present generation of highly trained and dedicated ecologists entered the field it has been almost unfashionable to investigate the very familiar .... we suggest that there is, or until recently was, a rule that the rare and the middle-rare, the inaccessible and with-difficulty accessible, have been the great naturalist's targets. Not the common, though.'

As often happens in this series, the exceptions threaten to outnumber the rule. *Lords and Ladies* (1960), the only monograph about a plant (or, rather, two plants, for the author covered both native species of arum lily), was written by an 'amateur'. Prime was the senior biology master (later senior science master) at Whitgift School, Croydon, and, like Neal, had written a dissertation on his chosen subject for an external PhD degree. For *Lords and Ladies*, Prime did not need to resort to purple prose, for the facts were bizarre enough, and sometimes bawdy, too. A plant that was credited with all manner of magical properties, was the source of starch for stiffening Elizabethan ruffs and, though poisonous, was eaten 'to provoke Venerie', provides a merry tale in skilful hands, especially when it also has upwards of sixty folk names, most of them highly allusive. The arum lily was a good choice for a plant monograph and the book has rightly become a botanical classic. But classics do not necessarily sell well, or at least not right away. *Lords and Ladies* was published in July 1960, as the author lay ill. 'I think you have done a very nice job,' he wrote to the publishers from his hospital bed. 'I hope somebody buys it.' A few people did. Years later, Prime received a letter from a retired American professor telling him that 'this is by far the most interesting, well written book I have *ever* read'. But it had taken him five months to track down a copy. The sales offered no inducement for Collins to publish further monographs on wild flowers. This was a great pity. Prime himself wanted to write another one on thistles, and a book about British primulas had also been considered by the editors, but there seemed no way in which such books could be published economically. Part of the trouble might have been indifferent marketing. *Lords and Ladies* was reprinted by Botanical Society of the British Isles Publications in 1981 as a memorial to Cecil Prime, and it seems to have sold readily enough by advertising in the botanical society literature.

Niko Tinbergen (1907-1988), author of *The Herring Gull's World*, at Ravenglass, Cumbria. (*Photo: Dr Larry Shaffer*)

The Herring Gull's World (1953) is another book that has always stood out from the crowd. It is probably the best known of the monographs despite not really being a monograph at all. Niko Tinbergen will need no introduction for readers of this book. The great Dutch ethologist, who won the Nobel Prize for medicine and physiology in 1973 (jointly with his friend Konrad Lorenz and Karl von Frisch, the discoverer of bee 'language'), had begun to study herring gulls in the 1930s. Since there was a breeding colony on the dunes within cycling distance, Tinbergen, then an instructor at Leiden University, included gull watching within his practical course on animal behaviour. The war brought an end to all that, but with his removal to Oxford in 1949, Tinbergen continued to study gull 'language' and social behaviour, both in the wild and in captivity. He wrote *The Herring Gull's World* at the invitation of Julian Huxley soon after his arrival in Oxford, and in perfect but ever-so-slightly stilted English which forced him to express himself simply. Having read the manuscript, Huxley and Fisher realised that this was no ordinary bird book and were keen to include it in the New Naturalist series. The trouble, from the publishers' point of view, was that *The Herring Gull's World* did not fit in well with the other books and seemed a better candidate for the university press. It did not pretend to be a complete biology of the herring gull, and in those days animal behaviour was regarded as a specialised and rather arcane field. On the other hand, it was not a book that would date quickly, as Tinbergen himself pointed out in 1969, when invited to revise the text: 'The actual facts are not what sells the book; it is the type of approach to an animal that it illustrates, and as such it will not soon be out of date .... The real market is only just opening up; the book was a little ahead of its time when you decided to publish it in 1953.' The 'sales people' at Collins wanted to publish *Herring Gull* outside the series in a plain wrapper but Huxley eventually had his way. Publication was delayed by several months in order not to clash with another behavioural book by Tinbergen, which also contained a great deal about herring gulls.

The Herring Gull's World appeared to the kind of reviews that are reserved for books that are new and exciting, and touch the imagination. Writing in the *Manchester Guardian*, Arnold Boyd thought that 'for many it will present an entirely new view of bird life'. It was, perhaps, the most intimate portrait of a bird's *life* (as opposed to its biology) ever presented, of a social system that seemed 'efficient but unadaptable', and even provided an insight into the herring gull's 'mind'. A lengthy review in *Ibis* found that it 'recreates the atmosphere of the gullery with extraordinary vividness and charm, but without sentimentality'. Though Tinbergen was careful to avoid drawing comparisons with human behaviour, such a portrait inevitably holds up a mirror to ourselves. Richard Fitter recommended the book to 'politicians and their voters'.

The Herring Gull's World was not an immediate best seller. Its day came later on in the 1960s and 1970s when it became a popular standard work on animal behaviour, was published in America and translated into German, Swedish and Japanese. It is surprising that Collins did not include the book in the Fontana paperback series, since it was on the university reading lists for psychology, as well as zoology, courses. The sales, while never spectacular, exceeded all the other monographs except *The Badger*, and the book remained in print for nearly 30 years.

Lords and Ladies and *The Herring Gull's World* were books of rather specialist appeal. So too were most of the remaining monograph titles. Collins hoped that *The Salmon* and *The Trout* would find a large market among fly fishermen, while *The Rabbit, The Wood Pigeon, The Mole* and *Squirrels* were of potential interest to agricultural colleges and the more educated type of farmer. As for *Oysters*, the publishers must have hoped that others would agree with the Rev Charles Williams that 'an oyster, only regarded as a thing to be eaten, and having but a low place in the ascending series of animals, not only demands, but will richly reward,

Jacket of *The Trout* (1967).

an enlightened examination'.

The fish books did do relatively well. Some 9,150 hardback copies of *The Salmon* and 8,300 copies of *The Trout* were sold. In addition there was as an American edition of *The Salmon* and a paperback one of *The Trout.* Both were in planning many years before they were published. In about 1952, Eric Hosking had seen a remarkable film produced by J.W. Jones, a lecturer in zoology at Liverpool University, of salmon spawning in a tank. Jones was invited by the board to write a book on the strength of it, a task he accepted 'with considerable trepidation' since the world literature on the 'King of Fish' was already vast. Watching Jones' film of finger-length salmon parr dashing beneath the hen salmon to fertilise her eggs as they were extruded from the vast silvery hull, James Fisher thought of Gulliver in the land of the giants, as if the tiny hero 'had managed to consummate a union with a Brobdingnagian lady in bed with her giant lord'. Unfortunately that is probably the liveliest passage in the whole of *The Salmon*, whose text bears the signs of over-editing. Perhaps it was unfortunate that the author was not himself an angler, and, indeed, seemed to dislike angling. The last chapter on the problems caused by over-fishing and pollution is the best, but I found it a dull book.

The Trout (1967), the work of two authors, is better, if a good deal longer. It holds the record in terms of slowness of production, having taken 21 years from conception to publication. The senior author, Winifred Frost, studied trout and their food in Ireland in the 1930s under the great Rowland Southern, who taught her fly-fishing, and later in the Lake District as a senior biologist with the Freshwater Biological Association. Peggy Brown became involved with trout later, as a contribution to the war effort, and was therefore more concerned at first with their production for food than with the wild fish. After the war, she was intro-

Winifred Frost (1902-1979), co-author of *The Trout*, in her office at Ambleside. (*Photo: Freshwater Biological Association*)

duced by Frost to the delights of wild trout in their natural world of shimmering lights, 'a room with a ceiling made of mirrors except for a round sky light in the middle, through which the outside world is visible'. There is a lot about weights and measures in *The Trout*, and some tricky stuff about growth and ageing, but the book has the authentic whiff of clean river water and the fishy world within. Most of it was written at intervals in the late 1950s, and its delayed publication was not entirely the authors' fault. James Fisher had sat on the draft for a year, and Frost and Brown insisted on restoring material that Fisher had altered or removed in an attempt to improve its readability.

The third 'fishy' title, *Oysters* (1960), was, like Yonge's *The Sea Shore*, a model of good science writing and enjoyable even by those, like myself, who have suffered from eating oysters. Early in his career, Maurice Yonge had worked on the feeding mechanisms and digestion of oysters during his years at the Plymouth Laboratory, and he had a wide knowledge of oyster culture. Billy Collins made a brave attempt to launch the book in style by holding an oysters and Chablis party at his Hatchard's bookshop in Piccadilly, and decorating the window display with oyster shells. Reviewers praised Yonge's clear, graceful writing, and scientists appreciated it as the only book about oysters in any language. Even so, it was an unlikely subject for commercial success and sold only slowly. Collins reprinted it 'not so much to meet demand but because it is such a good book we would not like to see it go out of print'.

The remaining four monographs are written, to varying degrees, from the perspective of economic damage and pest control. This is least true of *Squirrels*, for whom it is clear that the author's heart was engaged as well as her head. Monica Shorten was a young Oxford graduate when, in 1943, Charles Elton asked her to do 'a little bit of checking on the distribution of the grey squirrel'. The little bit of checking turned into a research study lasting ten years, years in which the grey squirrel expanded its range at the expense of the much-loved red squirrel. Monica Shorten also studied the biology and breeding activity of wild

Monica Shorten (1923-1993), author of *Squirrels*, with a flying squirrel 'Mischa', 1956. (*Photo: A.F. Vizoso*)

and captive squirrels, and managed to nail the widespread but unwarranted belief that the greys were winning the war with the reds by killing them off or by interbreeding with them. *Squirrels* is a pert and sometimes drily humorous book: Monica was the only monograph author to include recipes on how to cook and eat her subject! Her work on squirrels for the Oxford Bureau of Animal Populations and later the 'Min of Ag & Fish' included making a film and appearances on television and on the lecture circuit, both at home and in the United States (where she co-authored another grey squirrel book in 1973). Were it not for family commitments and the weakening effects of a diabetic illness, we would have heard more of Monica Shorten. 'I'm going to study stoats and weasels now, on the quiet,' she wrote to Raleigh Trevelyan, shortly after the publication of *Squirrels*. Trevelyan pencilled a note on the letter: 'J.F. [James Fisher]. Special Vol. on Stoats & Weasels?'

The Rabbit is harder to like, partly because the authors Harry Thompson and Alastair Worden did not seem to like the rabbit much. Thompson was in charge of mammal and bird pest research at the Ministry of Agriculture. His most recent scientific papers had titles like *An Experiment to compare the efficiency of Gin Trapping, Ferreting and Cyanide Gassing* (1952), *Rabbit Repellents for Fruit Trees* (1953) and *Power Gassing of Rabbits* (1954). Worden was a veterinary consultant, specialising in animal health and nutrition. As the editors put it, they were 'entirely qualified to give a complete, calm scientific assessment', and that is what they did. Most of the research on the rabbit, carried out first by the Animal Population Bureau and later by the Ministry, had been directed at controlling rabbit numbers. The book is therefore primarily a study of the rabbit as a major crop pest, not as cuddly bunnies, an emphasis perfectly expressed by the Ellis' haunting dust jacket (see Plate 3). The original title was to have been 'Rabbits and the Rabbit Problem'. But, though it was being written while myxomatosis was in the news, publication was laggardly and so *The Rabbit* missed the bus. By this stage in the series, no one could have been very surprised when it proved to be yet another slow seller. Readers of the series might have responded more warmly to a book like Ronald Lockley's *The Private Life of the Rabbit* which was to inspire *Watership Down* and bring rabbit biology to the attention of millions.

Although the wood pigeon is almost as injurious as the rabbit, Ron Murton's book, published in 1965, treated its subject more sympathetically. More than most New Naturalist monographs, *The Wood Pigeon* is a review of the literature, but Murton had himself worked on wood pigeon biology for a decade and he was one of the most gifted animal ecologists of his generation. His work for the Agricul-

Jacket of *The Wood Pigeon* (1965).

ture Ministry had been of considerable practical value, since Murton was able to prove against internal resistance that the Ministry's war against the wood pigeon, including a vast subsidy for shotgun cartridges, had largely been a waste of money. In place of the expensive shooting campaign, Ron Murton's team recommended better crop protection and the development of stupefying baits. *The Wood Pigeon* became a standard work on pigeon biology. Despite being written quickly it reads quite well, though, being a 'serious' book, it lacks some of the liveliness that I, for one, can remember from Murton's lectures. Murton was a gifted performer as well as speaker, and in these lectures he would in effect become a pigeon, bobbing and plucking at clover heads, and peering about, and explaining what it all meant as he bobbed along the stage. His early death in 1978 was a great loss to scientific ornithology, and to the BBC.

R.K. Murton (1932-1978), author of *The Wood Pigeon* and *Man and Birds*. (*Photo: Nigel Westwood*)

At last at Number 22, we reach *The Mole*, published in 1971, during the modest revival in the series generated by the Open University. To some extent, *The Mole* represents a return to the natural history approach of the earlier monographs. Kenneth Mellanby studied moles as a hobby during any time he could spare from directing research at the Monks Wood field station. But his hobby was informed by an ecological and agricultural background. His original purpose, to see whether moles were a good indicator of soil fertility, soon broadened into studying moles for their own sake. The mole was not exactly an unknown animal, but nearly all research had been done on farmland. Mellanby was the first to study moles in a near-natural habitat – the nearby nature reserve of Monks Wood. He soon found out that (in his characteristic words) 'every statement that is commonly made about moles is wrong'. They ate less and tunnelled less than was thought, and they were more at home inside woods than in open fields. *The Mole* is no more sentimental a portrait than *The Rabbit*, but Mellanby's style is much pithier and full of infectious idiosyncrasy. He told a journalist that he normally kept a couple of moles about the place at his home, Hill Farm, which colleagues referred to as 'Mole Hill Farm'. That it was one of the more successful monographs was perhaps due to Kenneth Mellanby's scientific fame. It went on sale in America and there was a large book club edition at home. In 1976, Mellanby brought out a children's story *Talpa the Mole*, which was considerably more accurate biologically than *The Wind in the Willows*, but which, like it, seemed to appeal to children of all ages.

Yet *The Mole* was destined to be the last of the monographs. There is nothing in the Board minutes that can be construed as a decision to end the series at that point. There were several more titles in the pipeline, and more than a few others had not completed the journey, either through the death or indisposition of the author, or because the author found himself too busy to write it, or because the book had been taken up by another publisher. Among the might-have-been monographs that surface in the record from time to time are The Partridge by Douglas Middleton, The Swift by David Lack, The Crested Grebe by Kenneth Simmons, and either the Greylag Goose or The Jackdaw by Konrad Lorenz. After completing *The Fulmar* and *Sea-birds*, James Fisher had planned to write The Rook and The Gannet (he wasn't sure which one to start first), and might eventually have got round to them had he lived. The unluckiest title of all was The Fox. In the 1940s, Frances Pitt was commissioned to write the book; when she dropped out, James Fisher invited Ernest Neal to do so, evidently not quite understanding the logistics of mammal research: 'He hadn't a clue that it would mean another decade of field work before I could do it.' Three authors later, Thompson and Worden, *The Rabbit* authors, proposed themselves. 'Make the book as lively as possible,' advised Fisher, thinking of their previous book, but in the event they too found themselves too busy to get on with it. In 1968, The Fox was offered to H.G. Lloyd, and for a while it seemed to be progressing well enough for the Ellises to design 'an excellent jacket'. But it was not to be. The first draft was far too long, and when Lloyd withdrew, the editors agreed that The Fox had finally outstripped the hounds and gone to earth.

So had the monograph series. In truth it had been ailing long before 1971, and during the previous decade the flood of titles had diminished to a trickle, as had the publisher's enthusiasm. By the early 1970s, both publishers and editors had concluded that single-species titles had had their day. They decided instead to include books about related *groups* of species within the main series. At least one new title, *British Seals*, had been intended for the monograph series, and about a third of the titles published after 1971 were in effect super-monographs. The first was *Finches* (1973), followed by *British Seals*, *Ants* and *British Birds of Prey*, and in the

1980s by titles on tits, thrushes, waders and warblers. This led to charges of overspecialisation, which, when one compares the 1970s and 1980s with the early years of the series, would seem justified. But that is another story. The true, small format monographs were (mainly) critical successes but (mostly) commercial failures. The series lost its greatest advocate when James Fisher died in a car accident in 1970. Had he lived, there might have been a few more titles to come, but the commercial facts of life were against their long-term survival. On the whole, and despite a couple of duds, they brought the series nothing but credit. They spanned the period when old-style natural history became the modern science of ecology, and one can trace this progression in the series. The best titles have not dated all that much, or at least not in ways that matter. That must be why we go on reading them and why they inspire us still.

Leslie Brown (1917-1980), author of *British Birds of Prey*.

Postscript

The account of *Fleas, Flukes and Cuckoos*, reprinted below in full by kind permission of the Hon Miriam Rothschild FRS and the publishers of *The Author*, was first published in spring 1994.

MY FIRST BOOK by Miriam Rothschild

The Scientific Advisory Committee of Collins' New Naturalistic Series invited me to write a volume for them about parasites. At the time I was expecting my second child, and I decided it was the right moment to settle down and attempt to produce a popular book on this superficially unalluring subject. Would a worm living under the eyelid of a hippopotamus and feeding upon its tears have a popular appeal? Or the fluke which passed from freedom in pond water to the liver of snails, to the body cavity of a shrimp, to the gut of dragonfly larva, finally to end its peregrinations and live happily ever after under the tongue of a frog? Or fleas with the most complicated penis in the animal kingdom? Billy Collins appeared enthusiastic and advertised my book, and I was so 'green' I never bothered to ask for a contract.

This was in 1947, the year of the Fontainebleau conference at which we founded the International Union for the Protection of Nature. After the proceedings closed, I decided to return home by boat from Calais.

Calais had been flattened by aerial bombings in the war; first by the German Luftwaffe and subsequently by the British Airforce. Rebuilding had not yet commenced. As we puffed across France a gigantic storm blew up and swept the channel and the coast. On arrival at the port I found that all crossings had been cancelled.

It is impossible to imagine a more desolate and chilling scene than that of the ruined town of Calais. The ships, crowded at anchor near the quay, hid the sea and bay from view – like the backcloth of some stage scenery. Piles of shattered concrete and bricks stretched away to the horizon. The gale was blowing slates about like so much waste paper, and strands of barbed wire were sent crawling across the cracked tarmac like nervous snakes. I suddenly realised that the surrealists had had second sight – for here was a Salvador Dali landscape complete with a pair of ragged trousers flung against a sagging wall, with a dead lobster decaying in the foreground.

One pock-marked public house stood alone and forlorn in the middle of the rubble. I was offered a room. Needless to say both hot water and electricity were lacking. I was told cooking was also impossible, but under the circumstances I could be served with a meal of boiled potatoes in my room. Curiously enough a battered telephone – so the surly publican explained – was 'in order'. It took the rest of the day to get a call through to my home. My husband's voice, like a crackling whisper from outer space, implored me not to attempt a crossing the day the boats resumed service, since thrown about by high seas, I would inevitably lose our unborn child. This sounded like good sense to me so I settled down to write the Introduction to *Fleas, Flukes and Cuckoos*. The following day the storm showed no sign of abating; the windows rattled ominously and sand and powdered concrete stuttered against the panes. Fortunately, I needed to consult no reference books for the chapter on fleas, nor flukes, which I began on day three. Except for the casserole of boiled potatoes, pushed through the door by the gloomy publican, I had no contact with the outside world. On day four, I felt that the book wasn't going too badly and I might even take a stroll. On stepping outside I realised instantly that this was a mistake, since there was still too much debris flying around, but I noticed smoke coming out of the funnel of one of the anchored tugs. Was the storm on the way out? I settled down hopefully to write the chapter on Symbiosis.

About four o'clock there was a clatter on the stairs, a loud banging on my door and the publican entered with a uniformed policeman. The policeman announced I must be vaccinated. Up to that moment I had accepted the situation philosophically, without a trace of anxiety and barely a tinge of irritation. Pregnancy gives one an irrational feeling of security and confidence. But now I suddenly became alarmed. I was filled with a sense of unreality, like a bad dream. I stared at the policeman. 'Are you crazy?' I asked. 'I am an English woman delayed by the storm. I have nothing to do with the French medical service.'

'It's a special situation,' explained the policeman politely, and with an air of great importance. 'In Calais everyone, *without exception*, is being vaccinated. We have smallpox in the town.'

In 1947 it had not yet been realised that vaccination against smallpox administered in the early months of pregnancy is exceedingly dangerous for the foetus but, fortunately, I had a strong premonition that this was so. I was also afraid of dirty needles .... and determined at all costs to resist vaccination. I explained that I had not left my room since my arrival and I would not leave it until the boats were sailing for England. 'Look,' I said the policeman, 'you are world famous for your logic in France – not like we poor English. If the entire population in Calais is vaccinated, you will agree that I am completely safe in this room – especially if my host doesn't stay here too long.....' The policeman clattered away down the stairs. Altogether, I remained incarcerated for a week. The storm raged and I ate my boiled potatoes. But when I left I felt somehow – bemused by seventy hours of writing – that *Fleas and Flukes* wasn't at all a bad book.

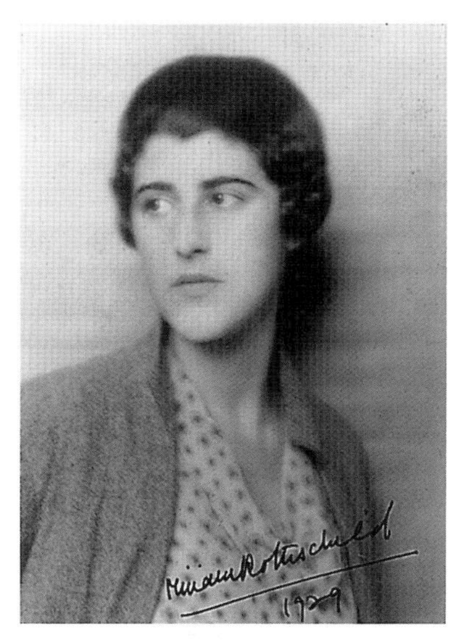

Miriam Rothschild, co-author of *Fleas, Flukes and Cuckoos*, in 1929. (*Hugh Cecil/Natural History Museum*)

To my great consternation, while I was tidying up the index and writing captions for the illustrations, Billy Collins decided that the 'bottom had fallen out of the book market' and he would therefore be unable to publish *Fleas, Flukes and Cuckoos*, which anyhow he now considered an unpopular subject. All I had by way of a contract was his early advertisement of the volume. But the late Julian Huxley was so incensed by this decision and by the aspersion cast upon his assessment of 'a good book' that he threatened to resign (with the whole committee) from the New Naturalists Advisory Board. This helped. After outraged, protracted and acrimonious negotiations the book was eventually published in 1952! I was refused a picture on the jacket, common to the rest of the series; it was considered too costly. However, *Fleas, Flukes and Cuckoos*, without the pretty jacket I hankered after, ran into five editions and became compulsory reading for school biology students .... It proved my only successful book.

One day Billy Collins rang me up: 'I hear you are writing a new book. I would like to publish it.'

'But Billy,' I expostulated, 'this is very flattering, but you don't know the subject let alone the title.'

'Well give me the title,' said Billy.

'*Crooks, Cranks and Collins* – it's quite a lively book.'

Table 8. The New Naturalist Special Volumes and Monographs

Number and title	*Year of publication*	*Author(s)*	*Study area*	*Editor*	*English Sales*
1. The Badger	1948	Ernest Neal	Woods near Rendcomb, Glos, especially Conigre Wood	Huxley, Fisher*	20,590
2. The Redstart	1950	John Buxton	Prisoner-of-war camps in Germany, especially Eichstätt, Bavaria	Fisher	6,000
3. The Wren	1955	Edward A. Armstrong	'Wren Wood', Cambridge	Fisher	3,000
4. The Yellow Wagtail	1950	Stuart Smith	Wet meadows and vegetable plots at Gatley, nr Manchester	Huxley, Fisher	6,000
5. The Greenshank	1951	Desmond Nethersole-Thompson	Strathspey, especially Rothiemurchus	Huxley, Fisher	4,000
6. The Fulmar	1952	James Fisher		author	3,000
7. Fleas, Flukes and Cuckoos	1952	Miriam Rothschild and Theresa Clay		Huxley	8,000
8. Ants	1953	Derek Wragge Morley	Author's parents' garden in Bournemouth	Huxley, Fisher	3,600
9. The Herring Gull's World	1953	Niko Tinbergen	Dutch coast near Leiden	Fisher	12,750
10. Mumps, Measles and Mosaics	1954	Kenneth M. Smith and Roy Markham		Gilmour	2,750
11. The Heron	1954	Frank A. Lowe	Dam Wood, Scarisbrick Estate, Lancs	Huxley, Fisher	3,000
12. Squirrels	1954	Monica Shorten	Oxford area	Fisher	6,000

Table 8 (continued)

Number and title	*Year of publication*	*Author(s)*	*Study area*	*Editor*	*English Sales*
13. The Rabbit	1956	Harry V. Thompson and Alastair N. Worden	Experimental plots at Wye (Kent), Sheepstead (Berks), Breckland and elsewhere	Fisher	5,000
14. Birds of the London Area	1957	The London Natural History Society		Fisher	2,500
15. The Hawfinch	1957	Guy Mountfort		Fisher	2,750
16. The Salmon	1959	J.W. Jones		Huxley	9,250
17. Lords and Ladies	1960	C.T. Prime		Gilmour	2,400
18. Oysters	1960	C.M. Yonge		Huxley	5,500
19. The House Sparrow	1963	J.D. Summers-Smith	Vicinity of author's home in rural N Hants and Stockton-on-Tees, Durham	Fisher	6,500
20. The Wood Pigeon	1965	R.K. Murton	Carlton, Cambs	Fisher	6,000
21. The Trout	1967	W.E. Frost and M.E. Brown	Rivers and lakes in Cumbria and Ireland	Fisher	8,300
22. The Mole	1971	Kenneth Mellanby	Vicinity of Monks Woods, Hunts	author	6,350

* Where Huxley and Fisher are listed together, Huxley had normally read and commented on the MS, while Fisher carried out the formal editorial duties and saw the book through to publication.

9

The Also-rans and the Books-that-never-were

The missing masterpiece

> 'To introduce the microscopic world we would take you, perhaps surprisingly, to the top of the Eiffel Tower. Peer over the edge. Below there appears a different world: minute creatures moving this way and that, some singly, others in two's or three's together, whilst somewhat larger, oblong forms glide among them.
>
> 'Imagine for a moment that we have come, as intelligent beings, from another planet and this is our first close-up view of the earth's surface. And further let us suppose we are of enormous size (for size is only relative) and that this great girder contraption on the top of which we are now poised is our landing apparatus; if the little creatures we first observed were then no larger in relation to our bodies than are the tiny inhabitants of the microscopic world we are about to explore, what would we think of them and their way of life? What indeed are we, in *real* life, to think of those fantastic little animals the microscope will reveal to us?'
>
> *Sir Alister Hardy*. The opening words from Into the Microscopic World, being the first chapter of 'Ponds, Pools and Puddles'(1975)

'Ponds, Pools and Puddles', subtitled 'A natural history of the microscopic world', was to have been the New Naturalist pond life book, covering all the variform small beasts of freshwater from amoebas to water fleas. Alister Hardy had intended to write it shortly after completing his *Open Sea* volumes. He had, in fact, been hoping to write such a book since the 1930s, and had a full set of photographic illustrations ready for it, taken through the microscope by his friend Donald Hutchinson. Julian Huxley, an old friend and Hardy's former tutor (though only nine years older than he), had proposed the title, 'Ponds, Puddles and Protozoa'. Hardy thought that name 'a bit cheap' preferring his own 'Little Waters', as the counterpart of *Great Waters*, the book about his *Discovery* voyages in the 1920s (the book takes its title from Isaiah: 'those that do business in great waters see the wonders of the Lord'). Little Waters had become 'Ponds, Pools and Puddles' by the 1970s, though I suspect that the eventual book would have had a more straightforward title like 'Ponds and Pond Life'.

The trouble was that *The Open Sea* books had consumed Hardy's spare time for years. There was now a backlog of books that Hardy wanted to write. *Great Waters*, the book of the oceans, headed the queue, and after that he hoped to have more time, with retirement approaching, to spend on puddles. At this point we should remind ourselves of the many-sided phenomenon that was Sir Alister Hardy. The word that sticks in one's mind from the excellent biographical essay accompanying his private papers is 'protean'. He was, in terms of our perception, but not his own, a shape-shifter. His Oxford students remember Hardy the brilliant marine biologist who was at heart a wide-ranging Victorian naturalist with an intensely developed natural curiosity. Others knew Hardy the enthusiast. A man of demonic energy, both physical and intellectual, he was an inventor, a watercolourist of considerable talent and an addict of flying machines of all kinds, especially hot-air balloons. Throughout his life he travelled mainly by boat, balloon or bicycle (or, right at the end, by bus). There seems to have been an air of wholesome innocence about this long-limbed, bespectacled, slightly simian figure. One biographer found in his personality 'a tantalising mixture of complexity and simplicity' though he was entirely straightforward in his approach to all that he did. The one point on which everybody who knew him agree is that Hardy was a lovable man, gentle and courteous, and in the widest sense of the phrase, good humoured.

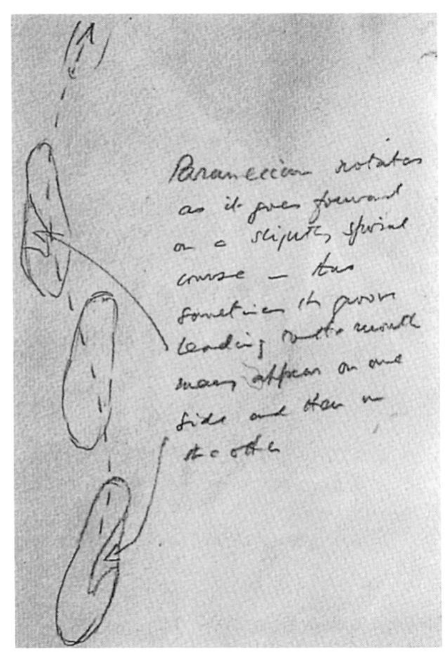

Sketch by Sir Alister Hardy to illustrate the movement of *Paramecium*. (*Private collection*)

There was another Hardy yet, which dominated his published work in the last two decades of his life. Hardy was a nature mystic. Not in any airy-fairy sense but as a student of 'natural theology', which he considered to be a science. As a child he had sometimes felt the presence of an unseen companion 'partly outside myself and curiously partly within myself', especially during his solitary rambles in search of beetles and butterflies (I suspect that this sense of 'otherness' on lone nature rambles is not unusual. I sometimes had it. Did you?). During the First World War, Hardy became convinced of the reality of telepathy. These experiences stayed with him and gave him the idea of shared unconscious experience as an underlying force in animal behaviour. He developed these ideas in the Gifford Lectures in the mid-1960s, which attempt to link natural theology to the process of evolution. Side by side with the lectures, Hardy founded, and then went on to direct, the Religious Experience Unit (now the Alister Hardy Research Centre) dedicated to studying 'encounters with the divine' using scientific methods – which Philip Toynbee memorably compared with trying to catch an angel with a butterfly net. Its work need not distract us further. It is mentioned here only to emphasise that for Sir Alister Hardy retirement was an unusually busy and fulfilling time. Little Waters, the puddles book, was one among many projects, some of which were jumping the queue.

Sir Julian Huxley, of all people, would have known what a masterpiece Little Waters was going to be, the perfect combination of subject and author (for I maintain that only a 'Victorian' naturalist can convey the joys of pond dipping). The small world of pond water lay at the root of Hardy's later exploration of the seas and the air; it was his first love (as it was for so many of us) and he would bring sixty years of accumulated wisdom and experience to the book. In that certainty, the file on Little Waters makes poignant reading. Hardy had suggested the theme as early as 1953, but first he had to finish *The Open Sea*. In 1961, came some 'very good news'. Hardy would be starting on the pond book as soon as he had finished *Great Waters*. But it was not to be, for Hardy plunged instead into his Gifford Lectures and the books based on them, *The Living Stream* and *The Divine Flame*, before getting on with directing the Religious Experience Unit. In 1974, his son Michael joined him in a concerted new stab at the book which now bore the title of Ponds, Pools and Puddles. The Ellises designed a handsome dust jacket for it, and Hardy produced some lovely watercolour sketches (see Plate 14) which suggest that he intended to illustrate the book himself in a similar way to *The Open Sea*. Then he changed his mind and decided to write the book on his own. And all the time, the reader of the file is thinking, for heaven's sake get on with it, man, with the realisation that Hardy had by now celebrated his 80th birthday. Even he would have to slow down soon. First though, he intended to write his autobiography and a book about ballooning. Oh why didn't the Collins team come up to Oxford en masse, and get down on their knees? It was not to be, after all. Sir Alister Hardy died in May 1985 at the age of 89. Of Ponds, Pools and Puddles there is only an introduction, a chapter, the sketch of the dust jacket, and a handful of coloured drawings. They are enough to suggest that the book was conceived along the lines of *The Open Sea: The World of Plankton*, and with the same lightness of touch and Hardean cordiality. There is not, alas, the space to reproduce the fragment here, but perhaps the list of contents below and the watercolour drawings reproduced here will suggest the magnitude of the loss to the series. This, I am fairly sure, is the Missing Masterpiece.

The contents page of 'Ponds, Pools and Puddles' by Sir Alister Hardy MA, DSc, FRS

Editors' Preface
Author's Preface
Chapters
1. Into The Microscopic World
2. The Microscope – and How to Use It
3. Amoeba and its Relatives
4. Sun-animalcules
5. The Flagellates – Plant or Animal, and Mixtures of the Two
6. The Bacterial Background
7. *Multum In Parvo*: The Ciliate Protozoa
8. Desmids and Diatoms
9. The Filamentous Forests
10. Freshwater Sponges
11. Polyps and Polyzoa
12. Wheel Animals and Bristle-backs
13. Flatworms, Round Worms and Worms of Many Kinds
14. 'Water Bears' and Water Mites
15. Water Fleas and Other Small Crustaceans
16. Pond Ecology: Different Waters – Different Beasts
17. Seasonal Succession – Bacteria and Decay
18. Desiccation and Dispersal

Molluscs, marshes and other gaps

In discussing the New Naturalist monographs and the work of the editorial board, I mentioned in passing several titles that never completed their transformation from contract to book – *Fishes, The Fox, Marsh and Freshwater Birds* and several others. These in fact were the tip of a phantom iceberg – perhaps surprisingly, there were almost as many unpublished as published titles, though happily most of the really important ones did get written in the end. Some of the unpublished ones were never much more than a gleam in the eye of one of the editors, or a suggestion that was left on the table. For others, authors were found and books commissioned. Some were never completed and not a few were probably never started. A minority, though, did pass the finishing line but were judged unsuitable for the series – too long, perhaps, or too superficial, or too difficult, or too dull, but at any rate unsuitable.

For one important gap in the series, the story is rather different. This is the book on Molluscs, one of the original titles listed in 1943. Evidently Molluscs was to have been a joint work, with one author tackling land snails and the other freshwater and marine molluscs. When, by 1950, the book showed no sign of progress, the editors despaired of the original authors and sent them an ultimatum. At this point Julian Huxley contacted C.M. Yonge, the world expert on bivalve molluscs, who had recently delivered one of the best books in the series, *The Sea Shore*. Yonge was now collaborating with John Barrett on the Collins *Pocket Guide to the Seashore*, but he agreed with some enthusiasm to take on the mollusc book as his next task, describing it as 'a labour of love'. A year later though, he had had second thoughts about writing a book on the whole of the Mollusca, preferring to confine himself to those he knew best, the marine ones. At this point, then, the New Naturalist Mollusc book became a two-parter, divided between land and sea. But, like his friend Alister Hardy (who had heard about the book and wrote to say 'what fun to hear that you are doing a New Naturalist on Molluscs'), Maurice Yonge was an exceptionally busy man, up to his ears in university administration, public service and world travel. What time he had left for the New Naturalists he devoted to his first love, *Oysters*. Then, for another decade, he was more interested in studying living molluscs than writing yet another popular book about them. Another factor might have been lack of incentive. The terms offered by Collins by the mid-1950s were nothing like as good as in the palmy postwar years, while the colour allowance had shrunk to almost nothing. 'I'd prefer this time to write the book first and consider publication afterwards,' wrote Yonge in 1955, having read the revised terms. Fifteen years on, when he had at last found the time to make a serious start on the book, he received a note from Collins informing him that the 'rise in all costs forces us into quite Aberdonian attitudes .... We are thus going to be in some trouble if the line drawings cost more than £150 .... I'm only too aware of that this kind of fee often represents risibly little return for the work that goes into complex drawings.' This was no way to encourage the author of two best-selling books for Collins, and having received several more penny-pinching notes of this sort, Yonge wrote sharply to Billy (now Sir William) Collins: 'I am dealing with Warnes whose attitude is the precise opposite to yours ... And they haven't made anything like as much money out of me as you have. I must ask them how they do it.' The upshot was that Yonge removed his book from the New Naturalist list since he now believed that he could obtain better terms and more colour outside the series than in it. *Living Marine Molluscs* by C.M. Yonge and T.E. Thompson was eventually published by Collins in 1976. With a touch more tact and beneficence on the part of the publishers, it could have been a very worthwhile addition to the New Naturalist library. As it was, Collins had succeeded in antagonising not only a major author but also Sir Julian Huxley, who was furious at the loss to the series of this title. Perhaps there was something unlucky about molluscs. The contracted-for volume on Land and Freshwater Molluscs never materialised at all, and had to be abandoned in 1968.

There were a few more cases where a disaffected author withdrew his book from the series, but not very many. More frequent was the 'Unsuitable Book', such as the 'art of animal illustration', the zoological counterpart of Wilfred Blunt's well-known book. Here the technical difficulties were great: animal art is a more diffuse subject than its vegetable counterpart. And, while either Wilfred Blunt or William Stearn could have written the plant book on their own, there were no equivalent candidates for the animal book (there might well have been for a book about *bird* art, including James Fisher himself, but the editors wanted something more broad-ranging). In 1947, the task was taken on by Dr F.D. Klingender, and a manuscript was duly delivered five years later under the title 'Animals in Art and General Thought'. James Fisher found it 'remarkable, very detailed', but also 'very esoteric, very Teutonic and without humour'. It was also 180,000 words long, more than twice the length contracted for. While they did not turn it down flat, the Board asked Klingender to have another bash, suggesting tactfully that he threw the 'General Thought' stuff overboard while jollying-up the 'Art'. The second draft was certainly shorter, but it took the story only as far as the Middle Ages! Huxley was in favour of publishing it, though Fisher thought the general reader would find it tough going. Billy Collins was even less enthusiastic, and his opinion was the one that counted most. *Animals in Art and General Thought* was eventually taken to another publisher who specialised in difficult art books.

Despite the Klingender experience, the editors retained the title on their list and continued to seek a suitable author. Robert Gillmor was contacted at one point, but he did not have time for it. Clifford Ellis, too, showed some interest but after reflection decided that the subject called for a larger format and that his interpretation of it would lean towards art history rather than nature. A pity: an Art of Animal Illustration by either artist would have been worth reading.

Among the more prominent missing books were a whole mini-series on bird habitats, of which only two, *Sea-Birds* and (much later) *Woodland Birds* were ever published. In the chapter on photographs, I suggested that one reason for this was the non-availability of colour photographs of sufficiently high quality. For 'Shore Birds', the contracted author, Eric Ennion, proposed to get around that problem by illustrating the book himself. That brought him into conflict with Eric Hosking, who felt strongly that the camera, not the brush, was the handmaid of the New Naturalist library, and he didn't like Ennion's paintings anyway. At any rate, neither 'Shore Birds', 'Freshwater Birds' nor 'Moorland Birds' were ever finished; quite possibly the ornithologists of the day were more at home with birds than with their habitats. The New Naturalist bird books of the 1970s and 80s are more traditional in their treatment, confining themselves to related groups of birds – waders, finches, thrushes, and so on. But that was not how the fathers of the series had intended to tackle the subject.

Another pile of 'Books-That-Never-Were' might come under the broad heading of 'Loose Ideas'. There were usually a lot of these floating about the board room, many of them emanating from the direction of James Fisher. There was a select number of senior naturalists whom the editors might have allowed to write more or less any title they liked for the series, among them Arthur Tansley, Vero Wynne-Edwards, Peter Scott, Harry Godwin and Julian Huxley himself. These were the big fish that were never landed. Other ideas swam in and out of vogue. Perhaps influenced by the popular books he had written for Basil Rathbone, James Fisher suggested broadening the series to include a Junior New Naturalist, a New Naturalist Companion and The New Naturalist on Holiday. Huxley counterpointed in a different key, suggesting the Natural History of Myths and Errors, A Medical Natural History ('including herbs and simples') and The Natural History of Diseases (W. Collins: 'That is a rotten title.' Huxley: 'What about "Life and Death in Britain"?' Collins: 'Excellent!'). In terms of sheer zaniness, Fisher's idea for a set of short 'mini-monographs' takes some beating. For what he had in mind were not the well-known animals and birds of the failing Monograph series, but a more esoteric group altogether: 'The Mammoth, The British Hippopotamus, The Giant Deer and all Five British Rhinos'. This series of extinct Pleistocene wildlife would, he thought, be suited to 10*s.* pocket editions with a print-run of 3,000 to 4,000 copies. To which Michael Walter of Collins was heard to mumble, 'Don't bank on it.'

Some might have wondered why *The Cairngorms* by Nethersole-Thompson and Adam Watson, published by Collins in 1974, was not included in the series, since everything about it smacked of the New Naturalist apart from the title page and dust jacket. The decision was evidently the authors'. They envisaged the book as 'a popular skiers and camper's guide to The Cairngorms', though, in the event, it was mainly about wildlife. Thompson had also fallen out with the Collins editor and had complained about the lack of promotion of *The Greenshank* in Scotland. Another regional book that might have seemed suitable for the series was that on *Upper Teesdale*, edited by Roy Clapham and published by Collins in 1978. Possibly its multiple-authorship militated against it, or perhaps it was judged too specialised. We should note that, by the 1970s, the New Naturalist was not the popular banner it had once been. Some authors may, like Maurice Yonge, have decided that their book would receive better treatment outside the series. Many had experienced irritating delays and misunderstandings with the publisher and, the world of natural history being a relatively small one, these stories probably spread as Chinese whispers. That may be one factor in the exceptionally large number of unfinished titles commissioned in the 1970s. Dust jackets had been designed for 'Bogs and Fens', 'Waysides' and 'Seaweeds', but as the years slipped by it became increasingly clear that there would never be books to wrap them round (see Plate 12). This frustrating non-appearance of titles has continued to the present day, playing havoc with the editors' plans and denying us what in some cases sound like mouthwatering titles (see Chapter 10).

I have included this brief review of the might-have-beens and never-weres of the series as a reminder of the uncertainties faced by any series which depends on busy people writing in their limited spare time – and in the hope that the subject is of interest. The significance of the New Naturalist series lies, of course, in what did appear, not in what did not. Book publishing is like the parable of the sower, with some contracts landing on the stony ground of writer's block and others among the thorns of unsuitability. What was unexpected was that so many New Naturalist manuscripts were of surpassing literary quality and would inspire a whole generation of even newer naturalists. And some of the scattered seed may yet bloom into roses.

Let us now move on to the New Naturalist 'Also-rans': the paperbacks, the journal and the fellow-travellers of the series and the recent hardback reprints in new jackets. None of these are collected to anything like the extent of the main series, but they have their place in the developing story and help to chart the changing expectations of the sellers and buyers of books.

The New Naturalist Journal

'One thing in the world is invincible – an idea whose time has come.' Max Nicholson famously applied these words to conservation, but James Fisher seems to have been imbued with the same confidence when he launched the New Naturalist Journal in 1948. By then, the sales of New Naturalist titles had reached 170,000 (an average of around 20,000 per title). Nature conservation and National Parks were in the news and there seemed to be every good prospect for an illustrated magazine written in the same spirit as the parent books. At any rate James Fisher, always the most optimistic and imaginative of the editors, came to an agreement with Wolfgang Foges whereby he would edit a journal on British natural history designed and produced by Adprint. The journal would be printed by Collins, with the same profusion of colour and monochrome pictures as the contemporary New Naturalist books, though more fully integrated with the text in the fashion of magazines. Fisher enlisted an impressive group of New Naturalist authors and other leading scientists to contribute articles, among them Arthur Tansley, W.H. Pearsall, E.B. Ford, Harry Godwin, Stephen Potter, Frank Fraser Darling, Peter Scott, Ronald Lockley and Brian Vesey-Fitzgerald. Rather than launch the journal on an uncertain magazine market, Adprint decided to bind the first four issues together and sell them in book form at a guinea each. The book duly went on sale at the end of 1948. It was in quarto size and casebound in good quality beige cloth, with a rather dull monochrome dust jacket bearing the 'NN' monograph in giant green letters, and the full title: *THE NEW NATURALIST. A Journal of British Natural History*. After an enthusiastic introduction by Fisher, the quarterly parts of the journal were each devoted to a particular theme: *Woodlands*, *The Western Isles of Scotland*, *Migration* and *The Local Naturalist*. The black and white photographs were first-rate, the articles well written and the overall production attractive. And yet, after this volume, only two more parts, (on '*Birth, Death and the Seasons*' and '*East Anglia*') appeared before the journal ceased publication for good.

Why did it fail? There is no clue in the sixth and final number of summer 1949 that the run would not continue. Unfortunately no contemporary documentation survives. It may be that Adprint or Collins failed to find enough retail outlets for the journal, and that the subscription rate was not high enough for the enterprise to be commercially viable. Or, perhaps, it fell victim to the worsening relations between Collins and Adprint over payments to photographers. Another possibility is that James Fisher's media commitments left him with insufficient time to attend to the journal; he often tended to bite off more than even he could chew. It is likely that the journal was printed in fairly large numbers, for the bound volume is still commonly found in second-hand bookshops. Numbers 5 and 6, published separately in the spring and summer of 1949 are much scarcer, especially No. 5.

So the New Naturalist journal became another of the might-have-beens of the series. There would be nothing to compare with it for many years to come, not at any rate until the appearance of *Animals Magazine* in the 1960s, and not until the launch of *British Wildlife* in 1989 has any journal bridged the divide between scientist and field naturalist half so effectively. Many readers will, I'm sure, already own a copy of at least the first four bound issues of the journal. But because of its special interest in terms of contemporary natural history and its close links with the New Naturalist library, I summarise the contents below.

THE NEW NATURALIST A Journal of British Natural History
Bound volume 1948; 216 pp; 12 colour photographs and 175 illustrations in black and white. Designed and produced by Adprint and published by Collins. 21*s*.

[Number One] **SPRING: *Woodlands***
The New Naturalist. Editorial
British Forests in Prehistoric Times. H. Godwin FRS
British Woodlands. A.G. Tansley FRS
The British Elms. R. Melville
Grey Squirrels in Britain. Monica Shorten
Woodland Butterflies. E.B. Ford FRS
Woodland Tits. Philip E. Brown
Woodland Bird Communities. M.R. Colquhoun
Books and the Amateur Naturalist. Stephen Potter

[Number Two] **SUMMER: *The Western Isles of Scotland***
Editorial
The "Outer" Hebrides. Arthur Geddes
The Climate of the Hebrides. Gordon Manley
The Passing of the Ice Age. J.W. Heslop Harrison FRS
St Kilda. James Fisher
Leach's Petrel. Robert Atkinson
The Natural History of Ailsa Craig. H.G. Vevers
The Atlantic Seal [pictorial]
Science or Skins? F. Fraser Darling

[Number Three] **AUTUMN: *Migration***
Some Problems of Animal Migration. C.B. Williams
Notes on British Immigrant Butterflies. C.B. Williams
Bird Navigation. G.V.T. Matthews
The Migration of Wild Geese. Peter Scott
The Problem of the Corn-crake. K.B. Ashton
Bird Migration Studies in Britain. R.M. Lockley
The Value of Bird-Ringing in the Study of Migration. A. Landsborough-Thompson
The Last Hundred Bird Books. James Fisher

[Number Four] **WINTER: *The Local Naturalist***
On Being a Local Naturalist. Brian Vesey-Fitzgerald
A Directory of Natural History Societies. J.S. Gilmour
The Natural History Societies of the British Isles. H.K. Airy Shaw
School Natural History Societies. David Stainer
Local Journals. W.H. Pearsall
Naturalists on the Air. L.C. Lloyd

Printed in spring and summer 1949, on sale separately at six shillings each.

[Number Five] ***Birth, Death and the Seasons***
The Biology of the Seasons. C.B. Williams
The Breeding Seasons of Animals. A.J. Marshall
Reproduction and Behaviour of the British Amphibia and Reptiles. Malcolm Smith
The Spread of Plants in Britain. Sir Edward Salisbury FRS
Flowering. P.J. Syrett
The Meaning of Bird Song. E.M. Nicholson
The Death of Birds. James Fisher
Insect Hibernation. R.B. Freeman
The Hibernation of Mammals. L. Harrison Matthews
Seasons in the Sea and on the Shore. C.M. Yonge FRS

[Number Six] ***East Anglia***
Flatford Mill. E.A.R. Ennion
The Coast of East Anglia. J.A. Steers
The Draining of the Fenland. H.C. Darby
Aquatic Life in the Norfolk Broads. Robert Gurney
The Shrinking of the Broads. J.N. Jennings and J.M. Lambert
The Broads as a Relict Marsh. E.A. Ellis
The Ecology of Breckland. A.S. Watt FRS
Rushes in East Anglia. Paul Richards
A Naturalist Sportsman in Norfolk. Anthony Buxton
An Ornithological Examination Paper. David Lack

The Fontana paperbacks

Between 1960 and 1975, some 18 New Naturalist titles were printed in paperback by Collins under their Fontana imprint. These books gave a much-needed boost to the series and enabled the author to revise the text more substantially than would otherwise have been possible, so that the Fontana text could then be used as the basis for a new hardback edition. The Fontana titles, in the distinctive green livery used from the mid-Sixties onwards, were essentially a response to the demand for lighter reading from university biology courses, especially, after 1971, by the Open University. They were, in the publisher's slightly patronising words, 'designed mainly for students and the more enterprising sections of the reading public'. Of course, the New Naturalist library had never been *designed* as text-books, but for an altogether more popular market. As T.T. Macan remarked of *Life in Lakes and Rivers*, 'it has caught on at universities in a way I did not foresee.' He had the impression that the university biology was served by so many stolid treatises that teachers were only too glad to recommend something a bit more enjoyable, especially when it was concerned with ecology.

The first Fontana title, published in 1960, was, naturally enough, the best seller of the parent series, *Britain's Structure and Scenery*, which already had a proven market in school geography classes. At that time the Fontana list was short and dominated by rather esoteric titles on history, religion and art. The Fontana 'library', as it was then known, was named after the beautiful eighteenth-century lettering used by Collins from 1936 onwards. That name had, in turn, inspired Eric Gill's elegant fountain emblem, used to decorate the title page of each book published by Collins. The colophon used on the Fontana paperbacks is a stylised version of Gill's fountain. *Britain's Structure and Scenery* was followed in 1961 by *The World of the Soil* and later by other New Naturalist titles, listed below. Presumably their sales were less than spectacular for no more than five titles appeared in Fontana until 1968. It was then that the paperbacks began to take off, helped by the redesigned and much more eye-catching photographic covers, and by the growing demand for such books from ecology courses. The most successful books were those that catered best for this market: *Mountains and Moorlands* (1968), *The Highlands and Islands* (1969), *Pesticides and Pollution* (1970), *Life in Lakes and Rivers* (1972) and three reprinted paperback titles, *Britain's Structure and Scenery, The World of the Soil* and *The Sea Shore.* Each sold over 40,000 copies during the next few years. The best seller was *The Highlands and Islands* at around 80,000 copies. The paperback was published in the same year as Fraser Darling's Reith Lectures, which must have done no harm to its sales.

Table 9. New Naturalist titles in the Fontana library and year of first printing

Britain's Structure and Scenery. 1960
The World of the Soil. 1961
Climate and the British Scene. 1962
The Sea Shore. 1963
The Open Sea: The World of Plankton. 1963
Mountains and Moorlands. 1968
The Highlands and Islands. 1969
The Snowdonia National Park. 1969
Pesticides and Pollution. 1970
The Trout. 1971
The Natural History of Man in Britain. 1971
Life in Lakes and Rivers. 1972
Wild Flowers. 1972
Insect Natural History. 1973
Dartmoor. 1973
The Peak District. 1973
Butterflies. 1975

Other New Naturalist titles in paperback
Fleas, Flukes and Cuckoos. Pelican 1957
The Badger. Pelican 1958

All these books were illustrated by black-and-white plates only. This was the reason why certain popular titles like *The Art of Botanical Illustration* and *Wild Orchids of Britain*, judged to be too reliant on their colour plates, were not printed in paperback. In this, the publishers were probably right, for when *Butterflies* was belatedly reprinted as a Fontana paperback without the colour plates it did not sell. Other books were too long for paperback treatment, or too outdated. Several titles, including *Life in Lakes and Rivers* and *The Sea Shore*, were prescribed by the Open University as set books (the former helped by the advocacy of Margaret Brown, the co-author of *The Trout*), which guaranteed their success. That the sales were overwhelmingly university-oriented is suggested by the relative failure of the National Park books – *Dartmoor, Snowdonia, The Peak District* – on which Billy Collins had set such hopes. These probably missed the tourist market they were intended for in being too 'scientific' and not nearly colourful enough. Indeed, these paperbacks do somehow seem a more difficult read than their parent hardbacks: we miss the colour plates, and the smaller print size does not help either. One curiosity of the Fontana 'library' is the inclusion of David Lack's *Life*

Cover of the Fontana paperback reprint of *Butterflies*.

of the Robin in New Naturalist livery, as though it had been a New Naturalist title (which, of course, it was not). Another was the failure to reprint *The Herring Gull's World*, which, one might have thought, was an obvious choice.

The heyday of the Fontana titles evidently lasted no more than a few years either side of 1970. The last new paperback, *Butterflies*, was published in 1975. In answer to an enquiry a year or two later, Kenneth Mellanby was told that Fontana were no longer interested in printing New Naturalist titles. Mellanby was understandably puzzled by this, knowing that his *Pesticides and Pollution* had sold 45,760 copies in paperback, which most publishers would regard as rather successful. But the sales of these books were declining as other more up to date and better presented books appeared on the market. Most of the Fontana paperbacks were out of print by the early 1980s, and the remaining stock was remaindered in 1985. Curiously enough, they seem to have had no effect whatever on the sales of the New Naturalist hardbacks which evidently cater for a quite separate market of naturalists and bibliophiles.

The Country Naturalist and the Countryside Series

The Country Naturalist series originated in an idea of James Fisher to 'show off' the colour plates taken for the New Naturalist library in a series of cheap paperback titles on various wildlife subjects. The pictures would be accompanied by a 10,000 word text written by the author of the parent New Naturalist title and introduced by James Fisher as editor of the series. The wording of Fisher's standard text, placed opposite the title page as in the New Naturalist books, suggests that he intended the Country Naturalists to be a kind of launch pad for the senior series:

> 'OUR BRITISH ISLANDS, with their wonderfully varied geology and climate, present to the bird's eye an intricate patchwork of woods, fields, moors, mountains, towns and rivers, with a margin of sea coast of great length and complexity. Each of these types of country supports its own peculiar communities of plants and animals.
>
> 'The object of this series of illustrated popular nature books is to give the reader a first introduction to these members of Britain's principal living communities.
>
> 'Each volume in the series is written by a contributor who is not only an expert in the science of its subject, but can analyse and illuminate it for the ordinary reader.'

The books were published between 1952 and 1954. Each contained a mixture of 32 colour and monochrome plates including a cover photograph, most of them taken from New Naturalist titles such as *Butterflies*, *Birds and Men* and *Flowers of the Coast*. The series even had its own little colophon, a sitting red squirrel similar to the individual symbols designed for the senior series by Clifford and Rosemary Ellis. But only five Country Naturalist titles were ever printed, although the advertisement on the back of each title announced that 'many more [are] on their way'. The five were:

Birds of the Field by James Fisher (1952).
Butterflies of the Wood by S. and E.M. Beaufoy (1953).
Birds of Town and Village by R.S.R. Fitter (1953).
Flowers of the Seaside by Ian Hepburn (1954).
Beasts of the Field by L. Harrison Matthews (1954).
Flowers of the Wood by F. M. Day (1953)

Butterflies of the Field by S. and E.M. Beaufoy and *Birds of the Seaside* by James Fisher were advertised, but apparently never printed. Each book sold for 2*s*.6*d*. or 3*s*.6*d*. with a print-run of about 25,000. The cover design contained a colour photograph with 'The Country Naturalist' running down one side with the title of the book at the bottom. Only the first two titles were numbered.

The short life of this series implies either commercial failure or internal dissension. The latter is more likely, since the other New Naturalist Board Members are on record as being less than enthusiastic. Evidently Fisher had pushed ahead with the series in his separate capacity as natural history advisor to Collins. According to the minutes for 20 November 1951, the Board was worried about the potential effect on the sales of the New Naturalist books. Dudley Stamp in particular felt that it 'parasitised' the parent books, and wanted an assurance that they would not appear until at least three years after the relevant New Naturalist volume had been published. After a private meeting with Fisher and Collins, he agreed to a compromise in which the gap was reduced to 'at least a year'. Perhaps to mollify Stamp, the possibility of expanding the series to include geology and climate was raised, and he was invited to contribute a title based on *Structure and Scenery*.

Another problem had arisen over the failure of Collins and Fisher to consult E.B. Ford about the butterfly titles. Since the selling point of these books was the colour plates, Sam Beaufoy had been asked to supply a text to accompany his pictures, and was well able to do so (Eric Hosking had been so impressed with his pictures that he arranged for Collins to publish another collection of them in a book called *Butterfly Lives*, published in 1947). Fisher belatedly wrote a diplomatic letter to Ford, and the Board agreed that 'the position should in future be clarified with the author of the closest New Naturalist book'. Hosking reminded Collins that he would need to pay a reproduction fee to the photographers, and suggested that if the colour plates were all to be filched from the New Naturalist books, then it might be preferable for the sake of originality to furnish the Country Naturalists with a new set of black and whites.

All in all, it would not be surprising if Collins (and Fisher) had had second thoughts about The Country Naturalist series. Though some titles remained in print until the 1960s, no more of them were printed. Quite possibly the rising cost of colour printing in the 1950s was the straw that broke the camel's back.

The Countryside Series, published by Collins between 1974 and 1979, might be deemed the cheap 'n' cheerful second cousins of the New Naturalists. The family

Christopher Perrins, author of *British Tits*, 1993. (*Photo: Christopher Perrins*)

feel was reinforced by the Ellis-designed dust jackets and the fact that three New Naturalist authors, C.T. Prime, William Condry and Christopher Perrins, wrote books for the new series. The Countryside Series was designed principally for the younger naturalist, but the books shared much the same holistic, ecological pitch as the senior series. They were to have been titled 'The Young Naturalist', but that name was dropped after the sales people explained that 'the young don't like to be called young, and the old won't buy books written specifically for the "young"'.

The first title, *Life on the Sea Shore* by John Barrett, was published in 1974, and it was followed by six more titles covering habitats, geology and species biology. Each consisted of around 160 pages, and was reasonably well-illustrated and produced. The series was not, however, the runaway success that Collins had hoped for. Perhaps it fell between two stools, too advanced for the intended market but too elementary for the older student (who already had the New Naturalist library to turn to). The sales people found a scapegoat in the Ellis jackets which, they decided, were too 'upmarket', and so replaced them in 1976 with boring laminated photographic jackets. It made no difference. The series was discontinued in 1979.

For New Naturalist addicts, the interest in this series may lie principally in the Ellis dust jackets. The couple evidently produced six designs in 1973, of which, as far as I know, only three were printed. Two more, 'Ecology' and 'Grasslands' were for books that never materialised. The Countryside Series is not, so far as I know, widely collected, but it may very well become so one day (I have in mind the habits of Observer's Book collectors, which have broadened out into related books and publishing ephemera of all kinds). In that expectation, I list the seven published titles below:

1. *Life on the Sea Shore* by John H. Barrett (1974). Ellis jacket.
2. *Birds* by Christopher Perrins (1974). Ellis jacket.
3. *Woodlands* by William Condry (1974). Ellis jacket.
4. *Plant Life* by C.T. Prime.
5. *Insect Life* by Michael Tweedie (1977).
6. *Rocks* by David Dineley (1977). Ellis jacket designed; was it ever printed?
7. *Fossils* by David Dineley (1979).

The Bloomsbury hardbacks

By 1988, the whole of the New Naturalist library was out of print apart from the most recently published titles. For the first time since 1945 one could visit a reputable bookshop and find not a single New Naturalist title, apart from may be a paperback *Heathlands* or *New Forest* on which the New Naturalist banner was hidden away inside. A few titles, like *Mammals in the British Isles* and *The Natural History of Shetland* were still listed by specialist outlets, but many people not unnaturally assumed that the series was coming to an end. It was to bring some of the classic titles back into print that, in September 1988, Crispin Fisher raised the rather desperate expedient of allowing a remainder company to reprint some of them. The books would be produced in hardback, but with a new jacket design and with the original colour plates in monochrome. When the New Naturalist editors reasonably asked why Collins could not do this themselves, Crispin explained that they could no longer publish such books in-house at a commercially advantageous price. Collins was now geared up to much higher print-runs, and 'structural changes' within the firm had in any case made it impossible. The recent attempt to reprint the series in paperback (see page 227) had not encouraged further in-house experiments. The remainder company, on the other hand, had much smaller profit margins, and would be able to retail the books at around £10 to £15.

The remainder company was Bloomsbury Publishing, an operation of Godfrey Cave Associates Ltd, in which Crispin's brother Edmund was at that time involved. The two thrashed out an agreement that Bloomsbury would print 'a maximum of 10,000' cased copies of each of 24 chosen titles, and, if all went well, would reprint a further 24 titles at a later stage. The Collins part of the bargain was to commission and pay for a distinctive new set of jackets and produce a camera-ready 'lay-down' for printing, which in some cases meant providing file copies of the actual books to copy from. Since the Bloomsbury

Jacket designed by Philip Snow for the Bloomsbury reprint of *London's Natural History*.

hardbacks are reprinted in facsimile, they were not technically new editions but simply reprints of the most recent edition. In choosing the titles for reprinting, Crispin Fisher decided to exclude those recently reprinted in paperback by Collins. The rights of some others had since resumed, while others were badly out of date or were deemed unlikely to sell (the latter consideration ruled out The Monographs *en bloc*). In the end, the 24 titles still reprinted in 1989 and 1990 in batches of six by Bloomsbury Books, were as follows:

Wild Flowers
Mountains and Moorlands
A Natural History of Man in Britain
British Mammals
The Lake District
British Birds of Prey

Woodland Birds
Life in Lakes and Rivers
London's Natural History
Birds and Men
An Angler's Entomology
Insect Natural History

Sea-Birds
Fossils
The Highlands and Islands
Mushrooms and Toadstools
Reptiles and Amphibians
British Plant Life

The Sea Shore
The Peak District
Wild Flowers of Chalk and Limestone
Inheritance and Natural History
The Natural History of Wales
Butterflies

In his standard letter to the authors, asking their permission and offering terms, Crispin explained that reprinting these titles would give the series 'what amounts to a second life. They were seminal contributions to British natural history when they were originally published, and that is how they are still seen, as landmarks in contemporary study.' The new jackets were designed by Philip Snow, and produced in four colours on laminated paper. When the first batch arrived, Crispin enthused over them: 'I can't tell you how excited I am by the first six artworks. I can hardly wait to see them round books.' They were, indeed, nicely executed, in delicate pastel colours reminiscent of magazine illustrations. Gone were the familiar bold, smudgy Ellis jackets, and in their place was an 'ecological' scene of habitats and representative wildlife. All were in much the same style, and, perhaps, the main criticism to be made of them is that they are superficially all rather similar. Nor did they conform well on the bookshelf with the Ellis jackets.

What can be said for the Bloomsbury reprints is that they kept a number of classic titles before the public. In commercial terms they seem to have been a bad misjudgment by both parties. With their poorly printed halftone plates, these books were but a pale imitation of the real thing, and it soon became clear that many New Naturalist fans would be content with nothing less than the original format. They did not sell well and at the time of writing can be picked up in remainder or second-hand shops for a few pounds each – certainly a bargain. After so poor a showing, Bloomsbury Books had no interest in reprinting more titles. Collins came to regret the transaction. The availability of New Naturalist titles in hardback at around £5 each obviously made it so much less likely that the public would be prepared to spend £20 or £30 for a reprint by Collins. Nor did their poor reproduction quality do anything for Collins' reputation. On the other hand, the Bloomsbury reprints made no difference whatever to the second-hand prices of New Naturalists. Most people who collect this series want the originals in Ellis jackets and with their full complement of colour plates; and the Bloomsbury books were no more part of this market than were the Fontana paperbacks.

10

Crisis and Recovery

The New Naturalist series set out to present British natural history in a different way, emphasising the relationships of wildlife and their habitats. Both in terms of style and presentation, the library was a unique innovation: it set new standards and remained at the pinnacle of natural history literature for a generation. By the 1960s, however, its lead was narrowing. While the series maintained its distinctive quality, other natural history books, equally well illustrated, equally informative, were appearing, and advances in colour reproduction were starting to make the older New Naturalists look old-fashioned by comparison. The market, too, was changing. The New Naturalist titles were written for the relatively expert field naturalist, but he (or she) was a declining species, increasingly catered for by the specialist literature. While the faithful continued to buy New Naturalist books, the biggest market for them in the 1960s and early 1970s were schools and universities, especially those titles that had been recommended as set books. By the 1970s, the bird titles were consistently out-selling all the others. Possibly the series became a victim of its own success, encouraging other publishers to commission equally meaty books, sometimes more cheaply and with better illustrations.

The sales had fallen by a degree of magnitude since the heady days of the 1940s. Throughout the 1960s, the first edition of each title averaged between 5,000 and 7,000, compared with upwards of 20,000 each for those published 20 years earlier. And the flow of new titles receded to a mere trickle until, between 1967 and 1971, only one new book, *Nature Conservation in Britain*, appeared. There was clearly some loss of dynamism. It would have been surprising if the five original and now ageing editors had retained all their initial fire and energy; it was greatly to their credit that they stayed together for so long. Stamp died in 1966, and James Fisher in 1970. By then, Billy Collins was semi-retired, and though he continued to attend New Naturalist Board meetings, most of the production side of things was left to his junior editors. The decision to use laminated photographic jackets for *The Broads* (1965) and *The Snowdonia National Park* (1966) was a visible sign of waning confidence, and the monographs simply ceased production.

The series wandered into financial difficulties during the 1970s. Because of diminished sales it was no longer possible to fill the books with colour plates; indeed, some of the 1970s titles had no colour plates at all (in the case of *Pedigree*, no illustrations of any kind). In those inflationary times, the price of the books rose from around £2 each to between £6 and £10 by 1980. Titles which failed to sell quickly enough were allowed to go out of print, and only the most recent or the most popular titles stayed the course. Reprinting older titles was by then becoming too expensive, and so the backlist grew shorter and shorter. By the 1980s, the print-run of new titles had fallen to just a few thousand each except for those where a book club order made it possible to increase the number, as it did with *British Tits* and *British Thrushes*. The sales of *Farming and Wildlife* (1981) and *Mammals in the British Isles* (1982), both well-reviewed books by famous authors, had been particularly disappointing, indeed, bafflingly so for those who had been

Crispin Fisher, Natural History Editor at Collins during the 1980s, photographed in 1987.

involved with the series in the past. 'Professor Mellanby feels there is some mystery about the New Naturalist series,' ran a note from the Fontana office. 'There isn't. It just stopped selling – so we stopped stocking it.'

If the New Naturalist library had been simply about making money, the curtain would have come down on the series some time in the late 1970s.[8] Only two things kept it in play. The first was its high reputation, coupled perhaps with a natural reluctance to end a series that still added lustre to the name of Collins. The second was the 'constitution' of the series with a Board of eminent independent editors as well as the usual in-house staff. The editors were, in a way, electors of the scientific world which still held the series in the highest regard. They were a useful obstacle to purely commercial considerations.

It was at this stage of spiralling prices and diminishing sales that Collins appointed a new natural history editor: Crispin Fisher. It is said that at his interview Crispin warned the panel that, 'If you don't give me this job, the ghost of my father will come back to haunt you!' But he was well-qualified in other respects, too, having spent much of his adult life in book production as a designer and illustrator. As the son of James Fisher, he had, of course, grown up with the New Naturalists. One of his artistic mentors had been Wilfred Blunt, his art teacher at Eton, who taught Crispin to write in an exquisite italic hand and nurtured his graphic skills. Crispin Fisher was by all accounts talented, energetic and likeable. When discussing, as we must, the crisis year of the library, in which the key decisions were Crispin's, we should constantly bear in mind the commercial background and ask ourselves what we would have done in the circumstances. If the 1970s had been a difficult time for the New Naturalist library, the 1980s were worse, and change became inevitable. The latest books looked tired. In design terms, the old division of text and plates was becoming outmoded, and in some books had been replaced by an integrated mix of text and illustration. As a designer, Crispin was at the forefront of the changes in book production. He had strong reasons of sentiment for prolonging the New Naturalist series, but things could plainly not go on as they were. His father would certainly not have allowed the traditional design of these books to stand in the way of sales. He would have wanted a vital, popular series that catered for the broadest possible market compatible with its maintenance of high scientific standards. Crispin was not only the heritor of the Fisher philosophy: he also had the commercial experience to put it into practice.

Crispin decided that the series needed, in his words, 'a kick in the pants'. Or rather, two kicks, both of them designed to reduce costs and increase sales. The first was to reduce the price of the books by switching from hardback to paperback production. The second kick was to adopt 'a new, more contemporary approach towards design and production'. Once they had been assured there would be no concomitant loss of quality, the Editorial Board reluctantly agreed to the changes proposed. But as Crispin went on to explain to them the technique of 'perfect binding', a cheap substitute for traditional book binding, you can sense from the editors' questions the sinking of their hearts. They did at least all agree that the jacket design was sacrosanct: 'they are as contemporary now as they were in the 1940s'.

The first batch of new paperbacks was a set of eight old titles, reprinted as straight facsimile copies of the most recent edition, retaining the traditional Ellis cover-design but with rather muddy halftones in place of the original colour plates. As a token gesture of revision, they were given a new introduction by the author or his reviser. The choice of titles had been left to the editors, though only books that were already out-of-print were eligible. The eight, published in 1984 and 1985, were: *London's Natural History, Mountain Flowers, Dragonflies, Britain's Structure and Scenery, Wild Flowers, Wild Orchids of Britain, Finches* and *The Fulmar*. Only 1,500 of each were printed, and despite some glossy advertising the sales 'were not all they might be'. Crispin therefore decided against printing any more titles in this format. What had surprised everyone was that the fastest seller among the eight was *Dragonflies*. Since the original book had gone out of print there had been a surge in interest in the Odonata, as a sort of birdwatcher's insect, but few dragonfly-watchers had been able to afford the second-hand price of the original.

In November 1984, Crispin outlined his further plans. All future titles would appear in the new format with the photographs integrated with the text. Indeed, the most recent title, *Reptiles and Amphibians in Britain*, had been pasted up in this way as a pilot run, though in the end it was published with the text and plates traditionally separated. In future, the paperbacks would have a photographic jacket which, Crispin thought, would help to attract a wider readership. A much smaller number of hardback books would be printed – he suggested 500 – to satisfy the demand from New Naturalist collectors and libraries, and these would be given an artwork jacket. They would cost twice as much as the paperbacks, and any profit used to subsidise the paperback price. Finally, Crispin proposed a new category of 'star' titles with a larger print-run and a commensurate degree of promotion. These would be confined to topics of wide appeal like 'British Birds' and 'Ecology', and listed separately from the main series. Other subjects, like 'Caves' and 'Estuaries' might fall within the star category, he thought, if they were treated in such a way as to ensure international sales. In the event nothing fell into this category.

There was more than enough impetus here to deliver the desired kick. The doubt lay in where the series might land afterwards. A great deal depended on the reception of the first titles to be issued in the new format, *The Natural History of Orkney* and *British Warblers*. As Crispin wrote shortly before their publication,

> 'The importance of these first two new-style NN titles cannot be over-estimated. If these fail, so will all the others; and that will be the failure of the last available boot up the pants for this series as a whole. This will signal the end of Collins natural history as we know it! Fingers out, chaps!'

The books duly made their appearance in November 1985. Surprisingly few reviewers mentioned the break in tradition, but P.J.B. Slater, writing in *The Biologist*, must have spoken for many when he complained that 'It is a pity the production is not better: the new paperback 'Orkney' is dearer than was the hardback "Shetland" and is not nearly so well put together, with minute print yet needlessly wide margins.' Similarly ill-advised had been the choice of paper, poorly suited for printing halftones and, in the case of *British Warblers*, responsible for the woefully blotchy reproduction of the 'sonagrams' so carefully prepared by the author Eric Simms. It seems that the picture proofs had been satisfactory but while the proofing had been done on coated paper, the pages were printed on inferior paper, presumably to save money. The worst mistake of all had been the ludicrously small hardback print-run of 725 copies each. This seriously underestimated the demand. In consequence, both hardbacks sold out almost immediately (aided by some trade speculation), and Crispin Fisher was reduced to stripping some 500 paperback copies and rebinding them as ersatz hardbacks. The difference between the 'first state' and 'second state' hardbacks was all too obvious, unfortunately, and everyone wanted first state. The print-run had been based on the advance subscriptions obtained from the customary tour of bookshops by Collins' sales representatives. But some shops did not even receive the copies they had ordered in advance, and had to pass this bad news on to their customers. No one seems to have had much idea of the special nature of the New Naturalist market, with its large quota of book-collecting die-hards who wanted hardback copies in traditional livery, and nothing but that, and were willing to pay for them. Crispin did manage to scrape together a few hardbacks of *Orkney* to send to those complainants 'who will commit suicide if they don't get a copy'. But on this issue there was a great deal of ill-feeling, and many collectors and booksellers felt badly let down. Typical of the letters received at Collins at this time was one from a natural history bookshop in Cornwall, remarking that 'This is the second time this has happened to a subscription order, and frankly my confidence in Collins is at a very low ebb.' In the meantime, the much larger number of paperbacks sat stubbornly on the shelves and in the warehouses, despite a large order for *British Warblers* from a book club. The latter title was eventually remaindered, so that while the second-hand price of the hardback quickly soared above the £100 mark, one could pick-up the otherwise identical paperback from a pile in a remainder shop for less than a fiver.

The *Orkney/Warblers* fiasco would have been unfortunate at any time, but in the circumstances of 1985 it presaged disaster for the whole series. Fortunately, production standards did improve, though slowly at first. The next title, *Heathlands*, was a more mainstream New Naturalist subject than its immediate predecessors. With the choice of glossier paper and somewhat improved halftones, it was a distinct advance on *Warblers*, though it retained the irritatingly wide margins and microscopic print. Almost unbelievably, the mistake of under-printing the hardback was repeated, though in this case a proper reprint was ordered ('I have no intention of rebinding any paperbacks in view of the criticism I received from the "public" the last time,' noted Crispin), and first edition hardbacks continued to be available from the Natural History Book Service until the end of the decade. The paperback *Heathlands* was the first to be issued in a photographic cover, and without a series number or New Naturalist colophon. The marketing person explained the reason behind the decision: 'The series once sold amazingly – in Crispin's dad's day – but now struggles along .... Unfortunately the hoi polloi don't give a fig for the New Naturalist concept, so we have decided to give the paperback a brightzappy [sic] cover which will appeal to the millions who visit heathlands each year.'

Table 10. Annual production of new titles (including the monographs) 1945-1994

Year	Titles	Year	Titles	Year	Titles	Year	Titles
1945	2	1962	1	1979	1	1996	1
1946	2	1963	2	1980	2	1997	0
1947	2	1964	0	1981	2	1998	0
1948	2	1965	2	1982	1	1999	2
1949	2	1966	2	1983	1	2000	2
1950	5	1967	2	1984	0	2001	2
1951	7	1968	0	1985	2	2002	3
1952	6	1969	1	1986	2	2003	1
1953	6	1970	0	1987	0	2004	2
1954	7	1971	3	1988	1		
1955	2	1972	1	1989	0		
1956	4	1973	3	1990	1		
1957	3	1974	2	1991	0		
1958	3	1975	0	1992	3		
1959	3	1976	1	1993	2		
1960	4	1977	2	1994	1		
1961	1	1978	1	1995	1		

The new policy settled down over the next few issues so that the printings of softback and hardback bore some relationship to the market – that is fewer of the former and rather more of the latter. The main task now was to maintain a regular flow of new titles – an old problem as one can see from Table 9, though never more so than during the past decade. The publishers managed to achieve their aim of two titles per year in 1985 and 1986, but since then new titles have appeared so erratically that even the faithful began to wonder whether the series had finally run out of steam. *Freshwater Fishes*, advertised and eagerly awaited in 1989, did not surface until 1992, a delay that probably affected sales. Failing to maintain publishing schedules is not always the publisher's fault, and the tendency for deadlines to slip runs through the whole history of the series. What has improved beyond question since 1985 is the illustrations, with more of them in better reproduction, and, above all, more colour. Arguably, the titles, too, have served the series better. *Freshwater Fishes* and *Caves* had long been gaps in the series, and were listed as desirable titles 50 years ago. *The New Forest* and *The Hebrides* are firmly within the New Naturalist tradition in their synthesis of natural history, human history and land-use. *Ferns* and *Wild and Garden Plants* were welcomed as the first botanical titles for many years, the latter with a hopeful eye on the nation's 10 million gardeners. *Ladybirds* recalls the very first book of the series in its blend of traditional natural history and contemporary science. If *The Soil* renews doubts about the advisability of multiple authors and the replacement of old titles, that is not to diminish the scientific quality of a very good book which has sold relatively well. From 1992 onwards, Collins got rid of the broad, wasteful margins introduced seven years earlier. And, as if to mark a turning of the tide, the New Naturalist banner crept back into the paperbacks in 1991. There seems more confidence in the series now, and the sense of despair, almost panic, of ten years ago has receded. I hope the present book will do nothing to harm that trend.

The New Naturalist Board, now consisting of five editors as it did half a century ago, continues to meet about twice a year, usually in Cambridge where, as it happens, four out of the five editors live. As in the past, the editors divide their time between reviewing progress on commissioned titles and holding what

Colin Tubbs, author of *The New Forest*, at Denny Wood, New Forest, 1986. (*Photo: Colin Tubbs*)

Christopher Page, author of *Ferns*, in his element. (*Photo: Chris Page*)

Derek Ratcliffe calls 'headscratching sessions' over new ones. The Library is, after 82 main series titles and 22 monographs, still an open-ended one, and the future possibilities are almost limitless. The greater problem nowadays is in finding experts who write well, or good writers who are also experts. A common complaint of recent manuscripts and synopses is that either they are too dry and technical, or (less often) that they lack the necessary weight expected of this series. The most recent Board meeting at the time of writing went through a list of seven titles in progress, ten more for which authors have been found, and no fewer than 33 subjects which Board Members had suggested themselves, or which someone had suggested to them. They include topics which have eluded the New Naturalist Board since the 1940s: mosses and liverworts, wildfowl, slugs and snails, bogs and fens. Others promise to bring old subjects up to date: pollination, the ecology of weeds and fossils. Among yet others under active consideration are several that might not have been high on the list in the past but which are certainly of interest to naturalists today, such as bats, churchyards, garden natural history and meadows. By the nature of things, not all of these subjects will turn into books; indeed, probably only a minority ever will, and which ones they will be is known only to the god of literature. The point is that the series seems full of life still, if good ideas and good authors are anything to go by. The New Naturalists series will surely chart the changeable waters of natural history for at least the near future as it did in the past. Perhaps it will become for a new generation of naturalists what it was for those of us who remember the 1950s and 1960s, an idea whose time has come – again. But whatever the future may be, a series which has been a natural history standard-setter for half a century, influencing so many to follow in the footsteps of the old masters, deserves our congratulations and our thanks.

11

Nature Conservation and the New Natural History

'Preserve or Conserve – I don't care which: it's all jam to me.'

James Fisher

Like many words which have entered the saloon bars of the land only recently, conservation – 'the wise use of resources' – can mean whatever the sayer wishes it to mean, which in this case will depend on whether you happen to be a naturalist, an arable farmer, a politician or someone making jam. Essentially though, nature conservation is the same thing as the older term, nature preservation. The naturalists and members of the British Ecological Society who championed the cause of preservation in the 1940s, knew that simply putting up a fence around a patch of protected land rarely achieved the desired results. Pearsall, Tansley and others had demonstrated in the wild what every gardener already knew – that nature is dynamic, that vegetation not only grows but progresses, and that pasture will turn into scrub unless something eats or crops the grass. Hence some form of continued *use* was necessary. These pioneers also wished to stress that the aim of nature conservation was not simply to protect, but to protect for a purpose, which to most of them meant scientific study. They began to talk therefore not of preservation but of conservation, which has connotations of activity (later called 'management'), prudence and moral worth.

Organised nature conservation has had but a short history in Britain, though volumes could be written about it. Norman Moore, born in 1923, has played a leading part in virtually every episode in the story, from the foundation-stone laying in the 1940s, through the pioneer years of the Nature Conservancy in the 1950s and the pesticide scare of the 1960s, to the ways and means of reconciling wildlife and agriculture that have dominated in recent years. It is not my intention here to tell that story, even in outline, but to stress the links between nature conservation and the new natural history, as represented by the New Naturalist series. The two grew up together, and both were a product of wartime aspirations. Many New Naturalist authors were closely in-

Norman Moore at Calthorpe Broad, in 1953, 'examining probable *Lestes dryas*'. (*Photo: English Nature*)

volved in the evolution of nature conservation in Britain, and some of them, like Norman Moore, became full-time members of the Nature Conservancy and its field stations. The Nature Conservancy itself was not only the creation of men like Julian Huxley, Max Nicholson and Arthur Tansley, but also the embodiment of their ideas about ecology, planning and the responsible use of land. It represented, to elevate the theme a little, the apotheosis of the new natural history. And a remarkably early one.

Curiously enough, nature conservation did not take the preservation of endangered species as its starting point. As one can deduce from their books, most of the 'professional' New Naturalists were more interested in the processes of nature than in rarities. Hence, their advocacy of nature reserves was not so much as a means of preserving Kentish plovers and monkey orchids, but to provide undisturbed havens for scientific study, with a secondary educational role. We should remember that there was not, in the 1940s, the same urgent sense of habitat loss that there would be half a century later. 'Digging for victory' and military installations had, indeed, destroyed many ancient habitats, but they were regarded by nearly everybody as necessary expedients. No one then realised the implications of continuing indefinitely the wartime measures to boost food production, aided and abetted by a powerful agro-chemicals industry.

In its earliest forms, the crusade was not about safeguarding wildlife so much as creating greater access to the places where the wildlife lived. This issue had been simmering since the nineteenth century. The circumstances of wartime gave the advocates of National (i.e. People's) Parks and better footpaths a platform as part of the promised land fit for heroes which, this time, would follow the war. People like Dick Crossman, Julian Huxley and John Dower had succeeded in persuading the coalition government to create a Ministry of Reconstruction in the interests of maintaining morale at home and support from the egalitarian-minded United States of America. This ministry was more receptive to new ideas than is usual in government. Planning for the peace began in the darkest days of the war in an atmosphere of idealism – and also the same sort of bulldog spirit that the New Naturalist Board possessed as it talked about books and butterflies while the V-bombs whiz-banged into London. One of its members, Dudley Stamp, expressed the prevailing ethos thus:

> 'The better life must include not only food, clothing and homes – the general satisfaction of material needs – but also the satisfaction of the less obvious demands of the spirit. To use much misused words: cultural needs. Daily hunger is countered by daily food, but the need to deal systematically with the demands of the spirit, the vital need for recreation, had long remained less obvious.
>
> 'Planning for the future was in the air. As, one after the other, our cities were bombed, plans were put in hand for their rebuilding. They were to have a green ring of rural land, productively used but not urbanised. There were to be large tracts set aside for quiet enjoyment. But enjoyment of what? Clearly the natural or semi-natural vegetation of mountain, moorland and coast, and with it the wildlife – the animals.'
>
> *Nature Conservation in Britain*. Author's Preface

The pioneers and prophets of nature conservation turned this popular movement to their advantage. Arthur Tansley and Julian Huxley, in particular, argued very persuasively that post-war planning must include reservations for wildlife among all the parks, green belts and beauty spots devoted to human needs.

Largely by coincidence, British naturalists had already put in a great deal of the necessary ground work through the work of the Nature Reserves Investigation Committee between 1942 and 1945, whose members included the ornithologist A.W. Boyd and the Director of Kew, E.J. Salisbury. This is one of the most productive examples on record of amateur and professional co-operation. With the genius for organisation that British naturalists had already demonstrated in the mass bird surveys of the 1920s and 1930s, the Committee tapped the vast reservoir of local knowledge in Britain to produce embryo lists of the best wildlife and geological areas and detailed recommendations for preserving them. Many New Naturalist authors took charge of their own patch. Norman Moore remembers producing a list of sites in East Sussex, including the ditch where he had recently discovered the scarce emerald damselfly. The geologist Frederick North, acting on behalf of the Committee in Wales, produced proposals for a National Park in Snowdonia, in which nature conservation was an important aim. He followed this up with a list of candidate nature reserves in the Principality with the help of other naturalists and scientists, including Bruce Campbell and Ronald Lockley. By the time the Nature Reserves Committee had concluded its investigations in 1945, it had produced a firm factual basis for nature reserve selection in England and Wales that has stood the test of time admirably.

The Nature Reserves Investigation Committee had no formal status, though its findings did help ultimately to shape the direction of nature conservation in Britain. In 1945, however, it was the future National Parks that commanded attention. The case for the Parks had been made by the dying John Dower in his well-known Report, and accepted by Government. But in the usual way, rather than taking immediate action, the government set up another committee to report on the report. This was the National Parks Committee under the Chairmanship of Sir Arthur Hobhouse. By taking clever advantage of the Whitehall system, Julian Huxley got the main Committee to set up two sub-committees, one on footpaths, on which Dudley Stamp was a member, and the other on wildlife: the famous Wildlife Conservation Special Committee, which Huxley chaired himself. This tactic put nature conservation and nature reserves firmly on the postwar agenda, riding on the back of the National Parks. By the same token, it helped to ensure that nature conservation would be seen as something *separate* from the Parks (a development John Dower would have deplored), as a science-led activity rather than one that catered for public amenity. Huxley and his deputy Arthur Tansley ensured that the Wildlife Committee would be manned by people who knew what they were talking about, another novel, some might say unique, innovation in the history of wildlife legislation. Among the team of 11 we note the presence of E.B. Ford, John Gilmour, Max Nicholson and Alfred Steers, with Richard Fitter acting as secretary. Whether by coincidence or not, Huxley was having New Naturalist dealings with Ford and Fitter at just this time, while Gilmour was, of course, a fellow editor. The team got down to work, and its report, published in July 1947, is the famous White Paper on the *Conservation of Nature in England and Wales*, better known by its Whitehall-ese code name 'Cmd 7122'. Its central recommendation was the formation of a Biological Service to set up and look after National Nature Reserves, and to undertake research in support of its duties. There was not much in 'Cmd 7122' that had not been outlined already in one report or another, but this one commanded attention, not only because of its White Paper status but because of the elegance of its language

Max Nicholson, Director of the Nature Conservancy in c. 1965. (*Photo: Bruce Forman/Nature Conservancy*)

and its strong advocatory message. At that time, the government was receptive to such arguments, having among its ranks keen country walkers like Hugh Dalton and Herbert Morrison.

After another wait while Huxley's Scottish counterpart produced a similar but (despite Fraser Darling's participation) less emphatic White Paper, the Biological Service was set up by Royal Charter in 1949. It was called The Nature Conservancy. The Charter laid down that the Conservancy should have permanent staff and be governed by a board of 15 members. Those who served on it under the Chairmanship of Sir Arthur Tansley included E.B. Ford, W.H. Pearsall, J.A. Steers and, later on, Dudley Stamp. Max Nicholson, also a board member, was appointed Director-General in 1952, after the resignation through ill-health of Cyril Diver. He held the post for the following 13 years until his retirement in 1966. It was Nicholson's task to build up the Conservancy's staff virtually from scratch at a time when suitably qualified people were thin on the ground. In 1952, the Geronimos of the Conservancy board might have seemed easier to find than the Indians needed to run the nature reserves and field stations. Gradually, Nicholson built up a close-knit team, among the early members of which were E.B. Worthington (Deputy Director), Norman Moore (Regional Officer for SW England), M.V. Brian (Head of Furzebrook Research Station) and Deryk Frazer (science advisor), later joined by Kenneth Mellanby as the first Head of Monks Wood Experimental Station and Morton Boyd as Regional Officer for Western Scotland.

Max Nicholson's remarkable career as ornithologist, institution-builder and pamphleteer is too well-documented to repeat here in any but the briefest words. A man of supreme self-confidence, formidable intellect and gifts of leadership unusual in a naturalist, he made the Nature Conservancy very much his own show, maintaining its course through a shoal of unfriendly government commissions and predatory departments of state. He is one of two men, both ornithologists (the other was B.W. Tucker), whom the botanist and historian D.E. Allen chose to exemplify the combining strands of amateur and professional natural history in the twentieth century. Though no one would know it, Nicholson's roots lie in the amateur domain – he didn't like the Oxford zoology course and graduated instead in history – but he made a name for himself in the 1920s and 1930s through a series of bird books that were remarkable for their insights into behaviour and relations between birds and man. He was the first to organise a mass-count of a single species, the grey heron, using the media and involving birdwatchers from all over Britain. He helped to found the British Trust for Ornithology in 1932, on which he served as secretary, and took the lead in strengthening the important association between organised amateur ornithology and the University of Oxford, which culminated in the founding of the Edward Grey Institute of Field Ornithology. During the war, he became a Whitehall warrior, and his experiences of government and politicians made him a shrewd and vital player in the organisational game that led to the founding of the Nature Conservancy. He was (and is), in D.E. Allen's words, 'that always rare being – the practical visionary'.

As Director, Nicholson was responsible to the Conservancy's board, though it is easy to be given the impression it was the other way round. But the board members were no non-entities and included some of the best-known names from the world of ecology. One of the most active was W.H. Pearsall, who took charge of scientific policy. Nicholson remembers him as a conscientious and dedicated member, with great influence in academic circles and among the coming generation of ecologists. Perhaps, his most lasting contribution was in education. Recognising that there was an urgent need for suitably trained officers, Nicholson and Pearsall organised a postgraduate diploma course in conservation at the latter's own department at University College London. Launched in 1960, 'the conservation course' has run successfully ever since, and developed into a Master of Science degree. Pearsall's scientific work on lakes, soils and succession might have been tailor-made for the management of nature reserves, though, starved of funds for equipment and full-time wardens, these places seldom received the attention they needed. The exceptions were reserves expressly designated as outdoor laboratories, like Moor House, a vast estate in the North Pennines, purchased (at less than £1 per acre) at Pearsall's personal instigation.

Another influential member of the Nature Conservancy was J.A. Steers, the apostle of coastline studies and, later, of coastline conservation. Steers was single-minded in his dedication to Britain's wild shores. During the 1930s he had travelled the world in the quest of coral reefs and tropical shores, but his *genius loci* had been the Norfolk coast, especially Scolt Head Island (which, not altogether coincidentally, was also made a National Nature Reserve). His contribution to postwar planning had been nothing less than the first comprehensive geographical survey of the whole coastline of Great Britain, acclaimed as a model of patient and meticulous compilation. He wrote a schools textbook and a New Naturalist book *The Sea Coast* largely on the strength of it. His report provided the basic information for coastal planning, and proved its worth during the devastating floods of 1953. Steers' unique knowledge of the coast had made him a useful member of the Huxley Committee, and he went on to serve the Nature Conservancy in the same capacity. Max Nicholson recalls that the National Trust's Operation Neptune project was conceived in his

J.A. Steers (1899-1987), author of *The Sea Coast*. (*Photo: St Catherine's College, Cambridge*)

The Nature Conservancy's England Committee enjoying a picnic lunch at Morden Bog, Dorset, April 1960. Dudley Stamp nearest camera, Deryk Frazer second from left, Norman Moore at the back. (*Photo: Max Nicholson/English Nature*)

own office, and largely at Steers' forceful advocacy. Steers was much involved with the Trust and the National Parks Commission as well as the Nature Conservancy; so far as he was concerned, their aims were (or should have been) much the same: the preservation of natural shorelines, unencumbered by coastal defences and bungalows.

Geographers like Steers must often have felt rather isolated among the ecologists and landowners on these committees, and the role of geographers and geologists in nature conservation is often under-rated. They provided an important academic link with planners, land-use commissions and the small, relatively tightly knit world of professional earth scientists. Among the most active was Dudley Stamp who, late in life, returned to nature conservation, chairing the Conservancy's England Committee and wearing a similar conservation hat on the Royal Commission on Common Land. Like Steers, he was a natural compiler and cartophile (interestingly, they both collected stamps), whether it concerned land utilisation, the footpath network, English and Welsh commons or geological SSSIs. Nicholson remembers him as 'a big benign character – though he could be sharp at times'. National service was second nature to him: he combined a strong sense of duty with an equally strong sense of his worth. Stamp drove himself hard, too hard. He died in harness, while attending an international conference in Mexico City in 1966. It was a paradox, but nonetheless fitting, that it was he, the geographical 'outsider', who wrote the nature conservation book in this series, though he did not live to see its publication.

In 1960, the Nature Conservancy established the most celebrated of its field stations in an enclave surrounded by the National Nature Reserve of Monks Wood. Under the directorship of Kenneth Mellanby, the Monks Wood Experi-

Monks Wood Experimental Station in 1970, visitors' accommodation to the left. (*Peter Wakely/English Nature*)

mental Station had two main functions. The first, under Norman Moore, was to research the effects of toxic chemicals on wildlife. The second was to conduct the kind of research that was needed if the Conservancy was to look after its nature reserves properly, and advise other land-users to do likewise. This was field study elevated almost for the first time to professional scientific research with a specific goal – habitat management. At the same time, the combined efforts of the Monks Wood team produced a considerable amount of new fundamental ecological knowledge for publication in scientific journals. For at least a decade and a half, Monks Wood was a power-house of outdoor ecological research and survey of a kind that has never been equalled. It was an out-going institution, holding seminars on land management and the ecological effects of outdoor recreation, as well as open days and courses of lectures for students. Had the New Naturalist series started 25 years later, in 1970, it might well have been dominated by contributors from Monks Wood. We might then have had a *Butterflies* by Jeremy Thomas, *Wild Flowers of Chalk and Limestone* by Terry and Derek Wells, *The World of Spiders* by Eric Duffey and *Nature Conservation in Britain* by John Sheail and Derek Ratcliffe. Never before, and possibly never again, did so many talented New Naturalists work together under the same roof. As it was, several titles published in the 1960s and 1970s, like *Man and Birds* and *Hedges* are essentially 'Monks Wood' books, and others of similar ilk have followed in more recent times, notably *Heathlands* by Nigel Webb of Furzebrook Research Station and *The Soil* by Brian Davies, the recently retired Deputy Director of Monks Wood. If the Nature Conservancy was the embodiment of the new natural history, Monks Wood in its Mellanby years represented its spirit. Alas, good things never seem destined to last long. The scientific stations were severed from the Nature Conservancy in 1973, and in the drift of events the former have become more concerned with applied research funded by contractors and the latter with the executive and administrative work of nature conservation. Kenneth Mellanby fought hard against 'the split', but other counsels prevailed. 'The split' was a disaster for the new natural history. Under the new 'customer pays' principle, there was no more room for knowledge for its own sake. The new Nature Conservancy Council (NCC) did commission numerous scientific projects from its former scientific colleagues, but these were mostly to do with basic survey work. The people who would have benefited most from the new natural history were seldom in a position to pay for it. It had been a brief golden age.

In his book, *The Bird of Time* (1987), Norman Moore saw nature conservation developing in three stages. The first was 'the pioneer stage, the stage of prophecy' when 'a few individuals observed what was happening in the world and advocated enlightened agricultural, forestry and fishing practices and the establishment of national parks and nature reserves in order to stem the growing destruction wrought by industrial man'. One can trace different aspects of these concerns in the works of Frank Fraser Darling, Sir John Russell and Max Nicholson, and the wise use of resources is at least a minor theme of many of the earlier New Naturalist titles. Moore's second stage, the formation of the National Parks and National Nature Reserves went some way to fulfilling the aims of the postwar naturalists, but, as Moore points out, conservation remained for a long time 'a service for a minority run by specialists'. Since the late 1960s, the environmental bandwagon has begun to roll with ever-increasing speed, though in an uncertain direction. It would be a foolhardy industrialist, land-user or civil servant today who did not claim to be a conservationist. Indeed, the cry for our own times must surely be the recent Transport Minister who boldly claimed that 'roads are good for the environment'. Moore hoped that conservation would provide a potent political idea that would tend to unite mankind. But, he admitted, 'that time has not yet come'.

What is so disappointing about the trend of events during the past 15 or 20 years is that environmental concerns have moved away from their roots in field-based natural history. A large number of young people are evidently against fox hunting, but how many of them know much about foxes, or the relationship between a predator and its prey? Millions watch Bellamy and Attenborough on the telly, but how many of them can read a map or know how to use a microscope? Field study seems in some danger of degenerating into what for most people is a spectator sport. Birdwatching apart, it is becoming a passive activity in which one is led around and shown things by the 'expert'. If one comes across someone pond-dipping or intent among the grass stalks it is much more likely that they are involved in some sort of academic project or ecological evaluation than simply looking at things for the joy of it. We have, it seems to me, lost some of the sense of wonder at nature that the Victorian and Edwardian naturalists had, and which the postwar generation shared in modified form. Conservation and the environment have become for many a moral crusade, but not one based on a better understanding of nature and Britain's wild places.

I think contemporary scientists bear some of the blame for this. The older generation of New Naturalist authors were dominated by full-time academic scholars who felt a duty to communicate their work to a large public, and were able to do so in simple, direct language. Among them were five scientific knights (Darling, Russell, Salisbury, Hardy and Stamp), a dozen Fellows of the Royal Society and a Nobel prize-winner. Though few of them were well-known public figures, these were people whose expertise was known and respected in scientific communities throughout the world. They had helped to widen intellectual horizons and shape the society they lived in. Of more recent authors, only Winifred Pennington and Ian Newton have been made Fellows of the Royal Society. The younger generation have tended to be full-time members of staff of various institutions, while the gifted amateur naturalist has almost dropped out of the list altogether. The only recent holder of the latter ancient flame is Eric Simms, who has made a career from sound recording, broadcasting and writing, in the same spirit as earlier ornithologists of talent, like James Fisher and Ronald Lockley. The younger generation of authors are no less gifted than the older, but in general they are from a different *milieu* and have had fewer opportunities to broaden their horizons and make their mark. Of the greater scientific pundits of recent times, only two have written for the series: Sam Berry and Ian Newton. Where are all the others? Is it that they no longer feel able to communicate with the hoi polloi, or feel that doing so is no part of their social duty, or that they have little to say in any case that would be understood by the untrained? Perhaps, the complex and expensive machinery needed for experimental science today has meant a retreat from field-based study to more theoretical notions. Ecological science, it seems, has 'lost caste'. If so, it seems reasonable to blame the ecologists.

So far as nature conservation in practice is concerned, field work has become not so much the mainstay as the antidote. One of the ironies for today's 'conservationist' is that no sooner had nature conservation become a full-time career employing a relatively large number of people than it lost its original links with field study and was saddled with indoor forms that have nothing to do with nature: budgets, computer programming, paperwork, agendas and, most recently, the unintelligible dalek-croaking of the management culture in operation. If conservation is ever to unite mankind, and mankind and nature, as Norman Moore hoped, we seem to be going about things in a most unpropitious manner. In certain ways, and in certain moods, the divide seems more chasm-like than ever, the scientist at his data, the conservationist caught up in transatlantic managerial gobbledegook, and the average British family hopping into the car to visit a facilitised patch of country that has most of the characteristics of the now vandalised town park.

If human stupidity has for the moment triumphed over the goals of an earlier generation, at least the opportunities to become involved in useful field work have never been greater. The New Naturalist books show how much can be done with a good pair of eyes. When E.B. Ford was working out how evolution (or, if you like, God) works through the passing of genes from one generation to the next, his basic equipment consisted of a butterfly net and a notebook. The lives of the majority of British invertebrates are still little-known, and in his recent New Naturalist book *Ladybirds*, Michael Majerus shows how the amateur can still contribute to unravelling their mysteries. There are messages in the older books of the series that are as relevant today as when they were written. The modern emphasis on the ways and means of conservation so often ignores the question of what we are conserving nature *for*. The postwar pioneers saw national planning and nature reserves as means to an end, not an end in themselves. They looked forward to the British tradition of natural history flowering and flourishing with better-illustrated books, better opportunities for roaming the countryside, and a golden age of field study, informed and inspired by new forms of natural history. They would have applauded the, on the whole, successful building of laws and institutions on the foundations they laid in the 1940s. They would have been puzzled, I expect, by the decline of the amateur field naturalist (though James

Bronze bust of James Fisher by Elizabeth Frink. (*Photo: Clemency Fisher*)

Fisher might have been happy enough with the size of the current membership of the RSPB). They might have hoped to see more natural history journals in which a large proportion of papers were written by amateurs. They might have been disappointed at what they would find at the average High Street bookshop. But they would, I am sure, have been amazed to learn that a certain natural history series published by Collins, was still going strong, having passed the hundred-book mark some time ago. So let us raise our field glasses to the authors and editors of the New Naturalist series, and to the series itself. From *Butterflies* to *Ladybirds* the library has kept bright the flame of British natural history for half a hundred years. Long may that flame endure.

12

Ten Years On

Well, here we are again! The New Naturalist library is now sixty years old, and lies within a short distance of its 'golden goal' of 100 titles – twice as many as the founders of the famous series had once envisaged. To celebrate its 60th birthday, the publishers are reprinting *The New Naturalists*, and have commissioned a new chapter to bring the book up to date. Much has happened since 1995, most importantly no fewer than 14 published titles, and many more in the pipeline. In a different sense, nothing has changed. Though the founders of the series are no longer with us, the spirit in which they launched the series in spring 1945 is very much alive – to communicate 'the results of modern scientific research' in British natural history to the reader in 'the inquiring spirit of the old naturalists'. The synthesis of new science and old curiosity was to prove a winner to the nature-starved audience of 1945. It is still a good idea today, for some good ideas are ageless. The books themselves are obviously no longer at the forefront of modern design technology – indeed, I suspect their old-fashioned look is deliberate. The New Naturalists may soon be given a face-lift, with full-colour printing. But the old face we have grown to love should still be recognisable.

I have taken the opportunity here to review all the new titles, and to add their esteemed authors to the gallery in Appendix 1. I have more to say on how the books are commissioned, written and published (from the standpoint of a bit more experience of the subject than I had in 1995!). In that context, I have also written a short account of a particular book, *British Birds of Prey* by Leslie Brown, which, though not a recent title, probably has the best-documented 'biblio-history' in the series. The 60th anniversary has also been a welcome opportunity to evaluate Robert Gillmor's crucial contribution to the series in the form of those luscious, if light-sensitive, jackets for the hardbacks.

I begin with a prospectus of the current scene with the help of Myles Archibald, editor of the series since 1989.

The New Naturalists in 2005

Perhaps the first thing to mention is that 'Collins' is back! The old name first appeared on the title page of *Seashore* and now lodges beneath the oval on the dust-jacket of *Northumberland* and subsequent books for the first time since 1990. Collins, we learn, is now 'a registered trademark' within the publishing giant whose name I think we need not repeat, nor mention again. *Vivat Collinsiana*!

Fourteen new books have appeared since the first edition of this book was written. What are they about, and where do they take us in the series? As ever, the only order among the titles is the order in which they were written. The subjects are an interesting mix. Eight of the fourteen are new books on old titles, in some cases even sharing the same name: *A Natural History of Pollination*, *Amphibians and Reptiles*, *The Broads*, *Moths*, *Nature Conservation*, *Lakeland*, *Fungi* and *Seashore*. Of the rest, three deal with a particular region or country (*Ireland*, *Loch Lomondside* and *Northumberland*): two with 'plants' (*Plant Disease*, *Lichens*), and only one (*British Bats*) with animals. Putting them all together, five titles deal broadly with places, four with 'plants', three with animals, one with habitats and one with the human domain, if nature conservation can be so categorised. There is a remarkable serendipity in having three books on mycology (the study of fungi and lichens) all at once. There is also the coincidence of two books on 'lakelands', *Lakeland* and *The Broads* (and Ireland, too, is a country full of lakes). Does this amount to a change – a different emphasis, perhaps, or an altered course? Well, the comparable figures for the 14 titles preceding *The New Naturalists* (1995) are two on plants, six on animals, three on places and three on habitats. Zoology, then, has slipped a bit, and it may be significant that there are no recent titles specifically about birds (though the growing number of 'place books' are well-stocked with our feathered friends). The balance is better now, despite three books on fungi. There is little, if any, calculation behind the apparent change of emphasis. What creates a New Naturalist waiting list is only partly the desirability of a given subject; it is more a matter of being able to match subjects with authors willing and able to write for the series.

The pace of publication has quickened slightly. The 'first fourteen', from Mammals to Ladybirds, spanned 14 years. The second, from Pollination to Fungi, took only nine, and in three of those years there were no books at all. The average over the past six years of two books per year is therefore most encouraging. The series has not seen a sustained period of growth like this since the 1970s.

One significant difference between the early days of the series and now is that virtually all the recent authors are from a professional background in science or nature conservation (the exceptions, Eric Simms and Philip Chapman are from the natural-history media). There have been no more books by vicars (Armstrong), bankers (Lousley), schoolmasters (Blunt, Hepburn), classics dons (Raven), retired physicians (Smith) or even museum men (Ramsbottom, Summerhayes). The amateur naturalist seems to have dropped out. The new men are mainly from the universities, from National Park or nature-conservation bodies (Mitchell, Ratcliffe, Lunn), or from Botanic Gardens (Spooner, Roberts). Admittedly the line between the natural-history amateur and science professional was never very sharp; if one thing unites the authors of the series, it might be 'professionalism in the spirit of the amateur'. But I suspect that most would agree there has been a rise in rigour. Most of the new titles are not easy reading. They are admirable for looking things up in, and are manageable enough in small doses, but how often are they read through for pleasure? Are they still significant as literature, as opposed to technical proficiency? And the poor reader is scarcely given a break by the printing, which in recent titles has been 52 lines per page – it used to be 39, and printed by black letterpress rather than grey filmset. There is, of course, a pressing reason for using a smaller font: authors are writing longer books.

Still, the series is in better shape today than for a long time. Thirty-five years ago, it seemed to have spluttered to a halt. Twenty-five years ago it came close to being wound up. Fifteen years ago, many believed it had run out of steam. Yet it has survived, and even recovered lost ground. Not only has production quickened but the hardback print run has doubled from 1,500 in 1995 to 3,000 today – and they still sell out within a couple of years. If any single individual can take the credit for this modest increase, it is Myles Archibald, editor of the series since 1989. He works on the first-floor wing of the Collins office on Fulham Palace Road, which opened in 1992. His office is lined with natural history and reference books published by the company, including the famous field guides and pocket 'Gem' guides. Prominent among them is a long shelf of New Naturalists in their gleaming Gillmor jackets. They face his desk, and he is obviously proud of them. Myles believes the future is reasonably bright. His aim, over the past fourteen years, has been to 'rebuild the reputation of the series'. "We're now, I think, producing the right books on the right subjects. The series is being managed properly, with a better understanding between us and the Editorial Board. And we've developed a very broad range of contacts, which helps us find the right authors." However, it seems harder than ever to find sufficiently knowledgeable people who also write well. As science becomes ever technical and intellectually demanding, the need for first-rate communicators and popularises is greater than ever. There are such people, of course, but they are not necessarily excited by the prospect of a microscopic advance for a New Naturalist. To write for this series today, you really have got to want to.

The ways in which books are sold is changing rapidly. Time was when the New Naturalist reader or collector bought his books in a shop, after a leisurely browse and a natter. Unfortunately the economic facts of life do not currently favour the stocking of serious books on natural history, apart from field guides and TV spin-offs. The profit margins of the High Street chains are too wide to stock low-selling titles, and ordering is done with an eye on quick sales. Our traditional standby, the independent bookshop, is usually short of space, and there, too, the bookseller is forced to pick and choose carefully. You are more likely to find New Naturalists in a secondhand bookshop, or, for those unluckily remaindered paperbacks, a seconds shop. Only in university and specialist bookshops are they still displayed prominently. Fortunately, says Myles, "the target audience of the New Naturalists is fantastically good at finding them outside retailers". Many books, particularly the hardbacks, are sold by direct mailing, using the order form on the flier that normally arrives when publication is imminent. The Collins New Naturalist mailing list has over 2,000 names. Inserts in magazines like BBC Wildlife generate more custom. Unfortunately, the company sometimes forgets to issue a flier, the packaging is, in my experience, by no means knock-proof, and the reader has to pay the full price. An increasing number of collectors are turning to online services like Amazon, who somehow manage to knock off up to a third of the recommended retail price. In fact, selling the New Naturalist hardbacks to its core audience is relatively easy. It is reaching a wider readership that has proven much more difficult. Sales of the paperbacks have been consistently disappointing. It was hoped that they would find a lively market on university field courses, as the Fontana paperbacks did so successfully in the 1960s and early 1970s. Perhaps they are too expensive; perhaps their failure is due to the decline in field-taught natural science and the rise of molecular biology and applied science. Perhaps such books need energetic promotion in today's market. As it is, the fate of all too many New Naturalist paperbacks has been remaindering after two years or so.

What about the recent price-hike to £40? Myles denies that its purpose is to subsidise the paperback price. It was necessary to raise the price to cover production costs: the quality buckram-cloth bindings, the large number of illustrations, the dishy jackets. He thinks it is still good value. He points out that the price has gone up roughly in line with inflation: twenty years ago, when postage stamps cost about 12p, and a pint of beer was just over a pound, the books cost just under £20. The price went up to £30 in 1988, and £35 in 1996. In 1945, the books cost 16 shillings (which was considered expensive), when postage stamps cost a penny-halfpenny, and beer a shilling. He has a point. Whether he is right in assuming New Naturalist readers are entirely happy about paying £40 for the books I do not know. But the knowledge that the book will soon be worth twice whatever you are paying for it must be some compensation.

The good news, as they say, is that the series is on the brink of the greatest advance since 1945. The New Naturalists are to be printed in full colour! This has been a dream, for many of us, for a long time. Some of the recent titles, such as Pollination and Lichens would have looked lovely in sharp, accurate colour. The increasingly old-fashioned segregation of colour to a few central plates has long been a weak point, especially when the designers insist on arranging pictures on broad acres of white, like a stamp album (an arrangement in which the authors have little or no say, and can only gnash their teeth). Moreover, most half-tones today are made from colour transparencies, and the results are all too often disappointing (did I, on observing what the printers had done to *Nature Conservation*, did I in my helpless anguish rend my garment? Did I cast my cloak over my face? I can't remember, but I'm sure it was one or the other). Full colour printing would require printing the books overseas, probably in the Far East. At the Editorial Board meeting in April 2003, Myles suggested 'it would be quite a statement' to produce the 100th volume in full colour and set the style for the future. It is possible, however, that all-colour printing may be even closer; in spring 2004 it was being considered for the very next volume, *Northumberland.*

What has sustained the series all this while? I had always assumed that prestige, the glory it had brought to the House of Collins, was the fuel that fed the engine. I was therefore a little taken aback by Myles' reply. Certainly, while Billy Collins was alive, this might have been so. The company was a family business, and he could do as he liked. After his death in 1977, Collins Ltd. entered a period of financial crisis. "We had to make 'economic decisions'," said Myles. The New Naturalists were no longer selling well. Crispin Fisher, with all the loyalty to his father's legacy that his appointment as natural history editor implied, did his best to rescue the series. Though most of his ideas for promoting it were unsuccessful, he helped it to turn an awkward corner. Very slowly, the series picked up again, helped by cheaper printing costs and the swallowing up of Collins within the vast Murdoch publishing empire. Within that empire, a series like the New Naturalists is a valued but very minor entity. It keeps going because Myles, as associate publisher, wants to keep it going. And it's as simple as that. So long as the series isn't losing company money, there is no pressure to stop. He has even been able to resist 'internationalising' the series (an old chestnut) in pursuit of foreign sales: "As far as I'm concerned, it's about British natural history, and always will be" (British in a natural-history context, means of course the British Isles, and includes Ireland).

Even after 60 years and nearly 100 books (plus 22 monographs), the New Naturalist survey of Britain's natural history is not yet truly comprehensive. Some topics, like mosses and liverworts, bogs and fens, slugs and snails, rivers and natural woodlands, seem forever tantalisingly out of reach, (though *Mosses*, at least, is now written and, moreover, looks very promising). Myles talks dreamily about a New Naturalist on British crows; my own dream is one on the small life of ponds, puddles and birdbaths. Forthcoming books include the natural history of gardens, British game birds, British wildfowl, plant galls, the Isles of Scilly, seaweeds, and a new weather book including the effects of climate change. New titles for old include fungi (due for publication in spring 2005), dragonflies and Dartmoor.

Predicting which of them will finally complete the long, slow race, let alone in which order, is as impossible as ever. But the current queue of books in various stages of completion makes the publishers aim of two new titles a year seem quite achievable. And that, as they say, is good.

As for any rumours of stopping when we reach No. 100, Myles tells me they are, on the contrary, thinking of stopping when they reach 1,000.

William Stearn (1911–2001), botanist, bibliophile and historian, co-author of the new edition of *The Art of Botanical Illustration* (Antiques Collectors Club, 1994).

Writing a New Naturalist

On 24th April 1995, the House of Collins held a party in the lobby of its vast office to celebrate the half-centenary of the New Naturalist series – effectively fifty years of natural-history publishing. There were speeches, and flowers, and an exhibition of printing and artwork, which included some of the original jacket designs by Clifford and Rosemary Ellis. Champagne flowed and conversation buzzed. You realise how small the natural history world still is when you notice that nearly everyone recognises everyone else. I counted at least a dozen New Naturalist authors among the assembled litterati: beaming, red-faced Willie Stearn with his shock of white hair; Richard Fitter, whose book, *London's Natural History*, was rolling off the presses on that very day, fifty years ago, when Hitler was in his bunker and the 14th Army was entering Mandalay; and dear Bill Condry, who roared with laughter when I creepily told him I was looking forward to his next book. David Attenborough made a gracious speech. He even asked me to sign his copy of *The New Naturalists*, though any hopes of getting him to film my life and times faded when he was led away by some minder. Being the centre of attention for a change was very pleasant, and the evening passed all too quickly.

Most people at the party had either written books, or had commissioned and published them. They all looked happy enough, but in some ways times are hard for natural history writers (and publishers). There are not enough serious reading-naturalists for the writing-naturalist to make much money, unless you happen to be an established television star (in which case it doesn't matter what you write about, or how badly you write it), or are writing a field guide with potential international sales, or are writing about birds for a mass market. People who write New Naturalist books are certainly not doing it for the money. Yet the books take a long time to write and need a lot of thought and preparation. How is it, then, that there has nearly always been a sufficiently large pool of authors to sustain the series for sixty years?

The recent authors of the series are older than the earlier generation. By my reckoning, since 1993 the average age of the author on the publication of his book was 55½, compared with about 45 for the series as a whole. The youngest recent writer was Michael Majerus (then 40), followed by David Lack (then 43) and me (44). The oldest was Noel Robertson, who tragically died a few days after receiving the page proofs of his book, aged 76. All the authors since 1973 have been men. There are remarkably few female authors in the series (precisely 4 in the mainstream series, and all co-authors; and 5 women wrote or co-wrote monographs). The books are therefore written from the perspective of mid to late career, or even retirement. They embody considerable personal experience, as well as a review of the literature on a subject. They seem to be by people who enjoy teaching, in whatever form, and yearn to popularise a subject that means so much to them. Some titles, I suspect, are written, consciously or not, with an immediate audience in mind – one's students, or people on field courses or wildlife outings, or conservation volunteers. They are writing the kind of book *they* would have wanted when they were students. Enthusiasm is a strong characteristic of nearly all the books – as is much of the correspondence behind them. Dedication is also a necessary attribute, since only the truly dedicated will ever complete a New Naturalist in today's market (the incentives, fifty years ago, were considerably higher). I doubt whether ambition plays much part, unless it is literary ambition; in terms of career development, the time spent writing a New Naturalist could doubtless be put to better use. Besides, most of the authors have already climbed the greasy pole. Pride, however, must be a serious motive. Most people still regard it as an honour to be invited to write for such a long-running and prestigious series. There may also be an element of paying one's dues. Oliver Gilbert and Derek Ratcliffe, I know, were glad to be able to pay public homage to their mentors in natural history, and I suppose the book you are reading could be regarded as one long obeisance to the living and departed great. On the whole, it is perhaps a mixture of enthusiasm, duty and flattery that keeps the engine moving. It works almost as well today as it did in 1945.

The Editorial Board play a key part in author-hunting – in finding suitable authors and then persuading them to write a book. The five of them, together with the associate publisher, Myles Archibald, and the managing editor, meet roughly twice a year, usually in Cambridge, where (not by coincidence) three of them happen to live. In return for a small royalty, the editor traditionally acts as a fairy godmother to the book, encouraging the author, commenting kindly and positively (but, if necessary, firmly) on his manuscript, and being as helpful as they can. Sarah Corbett normally supervises the entomological titles, Derek Ratcliffe the ones about uplands or nature conservation; Max Walters takes on some of the botanical ones, and Richard West and David Streeter share the animal books and the more broadly ecological ones. But it can vary. For instance, as a bryologist, Derek Ratcliffe assumed responsibility for the forthcoming volume on mosses and liverworts.

The bulk of each meeting is spent reviewing the progress of each forthcoming title – and that can be a lot of titles: 34 of them at a recent Board meeting. For every book that is ready for publication there is normally a string of titles in every state of preparation from a bare synopsis to a nearly-complete manuscript. The editors also take the opportunity to propose new titles and discuss them around the table. The minutes record the action decided on: 'to chase' one title, to approach a preferred author, 'to remind' another of an overdue manuscript. Directions must be given to avoid too much overlap between titles. For example, Peter Hayward's book on the *Seashore* is deliberately light on seaweeds, because a book on that subject is in preparation. Similarly the *Fungi* volume had to steer around plant diseases and lichens. What the Board cannot very well do is orchestrate the rate of production. The rate of growth of the series is governed by a wish to publish two new titles per year, which depends in turn on whether authors finish on schedule (which is not unknown). While some titles will drop off the list, a large proportion of the titles commissioned do, eventually, pass the finishing post. It is a long, slow race, with all sorts of obstacles and pit-stops, but the lead contenders come home in the end. Twenty-first century writers tend to finish their books. Eventually.

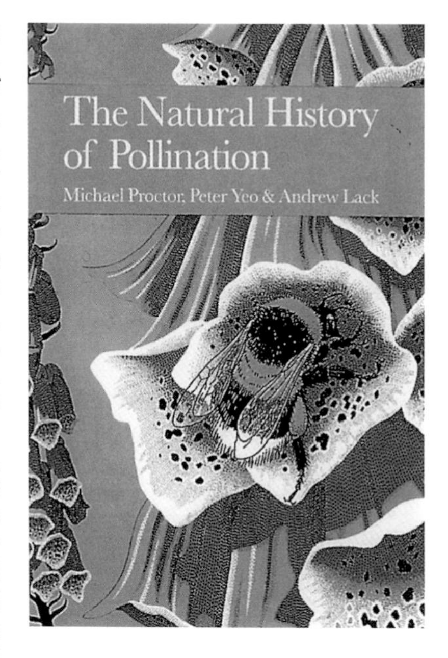

The three recent botanical volumes, *The Natural History of Pollination*, *Plant Disease* and *Lichens*, offer insights into how and why such books are written for the series. All are semi-specialised, and leaning towards the university or field course end of the spectrum. All three needed, and I think, received, good illustrations to make them more palatable to a mainstream natural-history audience. But each has its individual traits. *Pollination* was originally intended to be a revised edition of the classic Proctor & Yeo *Pollination of Flowers*, published in 1973 as No. 54 in the series. But, as the authors quickly realised, the growth in the subject since then really requires a new book, or at least a new text grafted onto a well-pruned stem. One potential problem for a specifically British natural history is that the scientific literature on pollination is worldwide, and more of it was done in America or even Australia than at home. However much of this work offers new insights into how British flowers and insects exploit one another. Pollination is in fact the key to the myriad shapes, colours and scents of flowers, as well as why some insects have decided to get social and live in a hive. Hence, the examples need not all be British, so long as they illuminate general principles. To help bring the book up to date and provide broader perspectives, the original co-authors, Michael Proctor and Peter Yeo, drafted in a third member, Andrew Lack, who had studied the pollination of knapweeds for his PhD thesis under Yeo.

Given the nature of the subject – the book's micro printed references run to 36 pages – as well as the nature of co-authorship, this was not going to be one of those books one dashes out during an unusually wet summer. In a letter to me, Andrew Lack remembered how "Peter and Michael first asked me to suggest changes and what to cut out, which I did, and moved things round a lot". Then, he went on, "I had to add 'my' chapters, and they to rewrite theirs, although by then we were consulting each other about everything anyway. It took a long time, but towards the end, when Michael in particular was in full steam, he took to ringing me up at 10.30 in the evening, and discussing a few arcane points for an hour … I think at least once I picked up the phone and said 'Hello, Michael' before he said anything! The book seems to have gone down well, with quite a bit of the credit down to Michael's photographs, and people have generally been very nice about it – rather flatteringly, I heard it described as the 'Bible of pollination'".

There is, alas, no picture of the three 'pollination people' together. "We tended to get together only for book discussions, and often it was just the two of us."

Written as a review, in scientific, if reasonably clear and well-explained, language, *Pollination* was never likely to be read straight-through – and especially not in Collins' dense, eye-straining print. It is primarily a book for clever A-level or university students, or naturalists with a scientific bent. I can attest that it is, however, a wonderful book for looking things up. When I wanted to find out why flowers smell as they do for a television programme (in which I was masquerading as 'the expert') I didn't need to look any further. It is one of those books that brings new pleasures to the shortest ramble in the countryside, or even the garden.

Plant Disease is another work of co-authorship that gets you looking more closely at plants than you did before. It came about through an idea of Max Walters. The 'grand old man' of the series had written *Wild and Garden Plants* as an attempt to bridge 'the artificial apartheid' dividing 'wildlife' from whatever grows (or lives) in a garden. One of the things that wild and garden plants share is their diseases. Max persuaded a distinguished Cambridge plant pathologist, David Ingram, to contribute a book on the perpetual warfare between green plants and the 'treacherous, mutable' army of cankers, blights, wilts and rotters attacking them. Ingram brought in his old friend and mentor, Noel Robertson as co-author. Like Proctor, Yeo and Lack in *Pollination*, they wrote their own chapters 'because we

both wished to show the other what we could do', and then worked on each other's 'to lead to a more homogeneous style'. The joins do not seem intrusive, nor even obvious. The problem with plant pathology, even more than in pollination biology, is the technical vocabulary used for structures and processes that go unnoticed except by specialists. But the book's chapter heads are certainly inviting: 'curls, scabs, spots and rusts' – 'the dark and secretive smuts and bunts' – 'bringing down the trees'. The series needs such books, and if this is inevitably one of the more challenging titles, it is arguably as readable as it can be without becoming oversimplified. In the past, amateur naturalists have certainly engaged with the microfungi of plants, and indeed have discovered most of those that live on wild flowers. Ted Ellis, author of *The Broads*, was a great field micro-mycologist; and his late son, Martin, together with his wife, Pamela, wrote the standard work of identification, *Microfungi on Land Plants*.

Oliver Gilbert, pioneer of 'Adventure Lichenology', author of *Lichens* (2000).

By comparison, *Lichens* is fun to read, even though the species are unfamiliar to most of us, and known only by Latin names. One reason may be that there is a single author, but the main reason is that the author is Oliver Gilbert. A gifted writer, and a populariser *par excellence* of his arcane subject, Oliver invented the new science – or is it a sport? – of 'Adventure Lichenology'. He and a small, fraternal band of fellow-enthusiasts delight in exploring the remotest parts of Britain – mountain-tops, small islands, woods hanging on cliffs – as well as some less conventional habitats, such as the drip-zone under motorway crash-barriers, the remains of a crashed aeroplane on St Kilda, or even the 'dog-zone' of certain trees! Oliver Gilbert has brought to this book a refreshing approach based on habitats and scenery rather than taxonomy. He has even managed to convey what one reviewer described as 'a white-knuckle ride' of exploration. The achievement of this band of ardent lichenologists in terms of mapping, recording and discovering new species is one of the brighter corners of contemporary natural history. The layout of *Lichens* is also impressive: Oliver told me he deliberately arranged the pictures so that you were never confronted with consecutive pages of dense text (how good it would have looked in full colour!). The book would have been better still if the editors and publishers had been persuaded to include the short 'interludes' he wrote after each chapter. These gave a flavour of what lichen-hunting is like: the companionship and fun of working in remote places; struggling to scribble notes with frozen fingers; accidentally knocking the best specimen into the breakfast pan; being dropped by helicopter onto the summit of Ben Nevis, appearing as a sun-scorched rock above an ocean of cloud. I have a jolly good mind to include one of Oliver's 'interludes' here, so you can see what you missed (And I have – see Note No. 9).

Perhaps the interludes would have been retained had there been a face-to-face meeting. When writing *The New Naturalists*, I was necessarily in more or less frequent communication with the Collins office in London, especially my editor Isobel Smales, and this contact not only helped to ensure that the book was going in the right direction but conveyed an assurance that the editors were almost as caught up with the book as I was. This was strangely comforting. By the time I was writing *Nature Conservation*, this cosiness had ceased; there were no more bistro lunches in Fulham, and my occasional queries were answered with, it seemed to me, hesitation and reluctance. Of course, the real umbilical cord is supposed to be between author and scientific editor, but valuable and positive though that is, it brightens the experience when there is a sense of shared participation in the project by the publisher: an organic growth, like a human baby, rather than some production process, with the books rolling off the line like tins of paint.

I think Oliver Gilbert felt this too. He was disappointed, he told me, to have had no direct contact with the publishers nor the Board: "Everything was done by post or phone. I was even asked to deliver the finished manuscript by post". Here, however, Oliver dug in his heels. "Determined to meet my editor – who had changed twice during the two years – I said I was bringing the MS down myself, thinking we might have an agreeable lunch together.... In fact the two of us had a working lunch of sandwiches and bottled water."

We have come a long way, alas, from those early meetings when Julian Huxley or James Fisher would wine and dine their authors at The Traveller's Club, or when Fraser Darling, hearing a knock at his remote croft house in Strontian, opened the door to William Collins himself.

The 'place-books' (plus some animals)

The most recent New Naturalist titles include no fewer than five books about places, whether a whole country (Ireland), a county (Lakeland i.e. modern Cumbria and Northumberland), or a National Park (The Broads, Loch Lomondside). The inclusion of four National Parks, for large parts of Cumbria and Northumberland are so-designated, suggests a continuation of the policy made very early in the series to include special volumes about each one. In practice, it seems to be coincidence – the regional titles of the series are simply about places which seem to have reasonably broad readership appeal. Undoubtedly, the recent declaration of Loch Lomond and The Trossachs as Scotland's first National Park strengthened the case for a volume on Loch Lomondside. But titles waiting in the wings include Galloway and the Scottish Borders, the Isles of Scilly, and, more distantly, Breckland, the Wye Valley, and the Gower, none of which are National Parks. The proportion of 'place-books' in the series has slightly increased from a quarter (3 out of 12) in 1985-95 to a third (5 out of 15) in 1995-2005. But over the lifetime of the series, place-books have shown a significant increase from about a fifth to a third, though there are long sequences – volumes 14 to 25, 28 to 43, 54 to 63 – with none at all.

Ireland (1999), David Cabot's *tour-de-force*, is one of the most remarkable, certainly one of the longest, titles of recent years. It is the first comprehensive review of Ireland's natural history for more than fifty years, and probably only David Cabot could have written it. He manages to confine so large an area within a single book by emphasising Ireland's peculiarities and differences – its waterscapes, its limestone pavements, its callows, slobs and turloughs, and miscellany of special plants and animals, such as St Daboec's heath and the Killarney slug. In the spirit of the

grand surveys of the nineteenth century, he begins in the clouds of Ireland's mountainous rim and moves gradually towards the sea through lakelands, bogs, valleys and farms. Though densely 'fact-packed', and scientifically precise, the book is always readable. David Cabot is an ornithologist, with twenty years experience as head of conservation science in Ireland's state planning department, and his obvious love of the outdoor life, as well as his partiality for antique natural-history books and manuscripts, is right for the series. The origins of the book go back to the moment when Cabot stepped foot in Ireland from the Rosslare Ferry over forty years ago, and, spotting a hooded crow 'frisking some rubbish', began to wonder why British crows are mostly black and Irish crows are grey; Robert Gillmor was right to put the 'hoodie' on the jacket. *Ireland* was one of those titles that required a long, slow marination. Cabot had been pencilled in as a possible author as early as the 1960s on the strength of a testimonial from Professor David Webb of Trinity College, Dublin. But a busy career intervened, and the ink did not really start to flow until the mid-1990s. A vast amount of reading must have gone into the book, despite Cabot's extensive first-hand experience: the book includes a bibliography of nearly 30 pages (and an index of 28). It arrived at a crucial time for the series, helping to lift it out of the doldrums with sales brisk enough to justify an early reprint. David Cabot is now working on *British Wildfowl*, another title with a long history of failed commissions and a succession of over-busy authors.

The Broads (2001) is another interesting and original book. One of the problems for the author was that there were already two large books on the subject; one of them in the series itself. Ted Ellis's well-known book, published in 1965, described the landscape history and wildlife of the Broads, but had little to say about the downturn in water quality and the growing tension between recreation and conservation. Its planned successor by Martin George described that conflict much more closely, as well as the broader social history of Broadland. But the resulting book was too long for a New Naturalist, and it was published elsewhere in 1992. Brian Moss, author of the new book, therefore needed to steer an original course through water that was already fairly turbid. As he puts it, the early place-books of the series 'could write of man the curious rural inhabitant. I must attend to man the powerful, competitive species in the environment of man the creative and destructive modifier ... I have tried to present what a reader might realistically see'. In other words, this natural history is by no means an escape from real life into the other-worlds of birds, plants and insect life, as its predecessor was. This *Broads* presents human life at the centre, fair and square, and how wildlife adapts to it or goes under. The result is a refreshingly honest book, a kind of social-cum-

Brian Moss, freshwater ecologist, author of *The Broads* (2001).

ecological history of how The Broads came to be as they are, not without plenty of wrong-turnings and the kind of commercial shortsightedness Moss characterises as 'struthious', that is, ostrich-like. While at the University of East Anglia in Norwich, Moss came up with the bright idea of 'bio-manipulation' or ecological engineering to clear up the plankton blooms that were suffocating the Broads. Algal grazing could be stepped up by thinning out the top predators, the fish, or by encouraging water fleas and other zooplankton with fish-free bunds or 'shelter-woods' of branches. Such measures, together with suction-dredging the phosphate-rich mud, worked quite well on a small scale, especially where the water could be isolated, but most of the lakes were still murky and weedless when Moss left East Anglia in 1989. The Broads are now promoted rather glibly as 'a place for the cure of souls', to which Moss drily comments, "perhaps".

Curiously enough, *The Broads* does not come across as a 'downbeat' book. In Moss's account, the modern custodians of the area, 'the officer, visitor, minister, sage' of his part 4, are the successors of a thousand years of human water-life, from 'hunter, pirate, marshman, monk' in part 1 to the more recreational 'gunner, yachtsman, wherryman, naturalist' of part 3 (the tacit implication that the 'naturalists' are now the 'officers' wasn't lost on this reader). That the future need not be all that bad is suggested by the last two startling, almost surreal, colour plates showing the future of Broadland – the epoch of belief' and 'the spring of hope', complete with timber-barges, water buffalos and wide open spaces. They refer to the imaginative 'postscript' of the book, which Moss had to fight hard to retain, in which the author pretends to be writing fifty years from now, and reflecting on the experiences of the past century. In this scenario, the historic continuity of the Broads as 'a working landscape' has been resumed after 'the doldrums years between 1960 and 2025', when successive tribes of officials struggled against the 'inevitable and inexorable'. You do not have to accept all the prognosis, but this is a well-considered and digested personal journey, written in an agreeably clear and lively style.

That *The Broads* and *Lakeland* should be published within eighteen months of one another is another of the little serendipities that befall the series from time to time. The Broads and the Lake District are two of England's three natural lakelands (the third, less well-known, is the West Midlands meres, with Cheshire at their heart – a landscape described, as it happens, in *A Country Parish*). Like Brian Moss, Derek Ratcliffe had to tiptoe round another New Naturalist book, this time *The Lake District* (1973) by W.H. Pearsall and Winifred Pennington. *Lakeland* takes a different perspective in two ways. It broadens the subject-matter to include the whole of Cumbria, and so embracing the coast and its estuaries, lowland farms and the bordering moors and fells of the North Pennines, as well as its mountainous heartland. And, like *The Broads*, the book is a personal interpretation by one who has tramped over practically every inch of it, and spent many a long day inspecting the nests of peregrines and ravens, or photographing flowers and ferns. Some subjects, fully covered by the earlier book, such as climate and geology, could be dispensed with. Like *Ireland*, *Lakeland* is habitat-based, and, also like *Ireland*, is topped-and-tailed by a full chapter on Lakeland naturalists and another on 'conservation problems and approaches'. Ratcliffe grew up near Carlisle – 'Lakeland was my youthful stamping ground' – and he was able to return there frequently after retirement in 1989. Hence, there is a strong 'then-and-now' element, which is always interesting, if, inevitably, sometimes rather depressing. There is also the sense of a long and agreeable ramble over hill and dale, in the company of a quiet-spoken but exceedingly knowledgeable and observant master (with, one might add, a photographic memory). By keeping an eye on its production, Derek Ratcliffe ensured relatively high standards for this book, with sharp half-tones and a larger font size.

John Mitchel, author of *Loch Lomondside*, at the sign of The Chequered Skipper.

As I write (in April 2004), *Northumberland* by Angus Lunn lies in the future. It and Cumbria are the back-to-back counties of northern England; soon, if all goes well, they will be joined by Derek Ratcliffe's current book on Galloway and the Borders. Apart from the Isles, Scotland has been rather neglected by the New Naturalist series. The grand survey of *The Highlands and Islands* by Darling and Boyd has served as the series' Scottish book for almost as long as the series itself. At last, in 2001, it was joined by *Loch Lomondside* by John Mitchell, who for many years served as the resident nature warden (nature's ambassador is a better title) in that enchanting area. Mitchell and his editor, Derek Ratcliffe, had to fight hard for this book. The other editors thought it too specialised, despite Loch Lomond's proximity to Glasgow and its two million inhabitants. John Mitchell is a clear and interesting writer, with eclectic interests and a fascination for delving into what might be called 'unconsidered trifles' (as the historian Simon Schama has noted, it's the little details that can illuminate whole landscapes). When I lived in Scotland, I sought out some of Mitchell's papers on such things as Victorian fern-collecting in the Moffats, or the unlikely escaped animals that had established themselves on the Bonny Banks, which included, if I remember correctly, the kangaroo. An all-round naturalist, he knows his Scottish docks as intimately as his shelducks and herons, and also enjoys researching the area's human past: its cornstone workings and charcoal woods, and the ancient communities of the lake isles.

Equally important, he gets on well with its present occupants – farmers, keepers, foresters, fishermen and 'officers' of every kind. By happy chance, *Loch Lomondside* coincided with the opening of Scotland's first National Park. The 'long struggle' that got there in the end is described in the final part of the book on 'Conservation: past, present and future'.

So much for the place-books. We are left with the four new books on animals and invertebrates, all with nice short titles: *Bats*, *Seashore*, *Moths* and *Amphibians and Reptiles* (short, that is, on the wrapper and binding. On the title page, *Seashore* becomes *A Natural History of the Seashore*. Which one is the official title is unclear). The only wholly new subject is bats. When Leo Harrison Matthews wrote *British Mammals*, very little indeed was known about the natural history of living bats – and a third of our native mammals happen to be bats. The series needed a bat book. The editors had pencilled it in as a possible topic as early as 1959, but few zoologists then specialised in bats. Ten years on, Eric Hosking had found Robert Stebbings, who, he urged, would be able to write them a splendid book, illustrated with the matchless photographs of S.C. Bisserot. However, when enquiries in the trade indicate there was insufficient demand for a bat book, the idea was shelved. By the mid-eighties bats had grown enormously in popularity, thanks mainly to legal protection and enhanced status as 'honorary birds'. Local bat groups were springing up to record their numbers and distribution, and advise householders how to look after their bats, or get rid of them without harming them. Two or three potential authors were contacted; a contract was produced for one of them; but nothing happened. Finally, an author was found in John Altringham, an authority on 'biomechanics', that is, the means by which animals get around, and especially on the flight of bats. He was also familiar with bat natural history and conservation work, from loft inspections to roost-box design. Appropriately, given Eric Hosking's notion of matching writer and photographer, Altringham was able to draw on Frank Greenaway's splendid photographs of bats in flight and at rest (as well as some very nice text-drawings by Tom McOwat). The book, one of the shorter ones of recent years, takes us through the collective biology of bats, as well as one by one, 'in a functional and evolutionary context'. Like other new New Naturalists, the reader needs to concentrate, but if the going proves a bit austere, one can dip instead of swim, and there is a good balance of text and pictures.

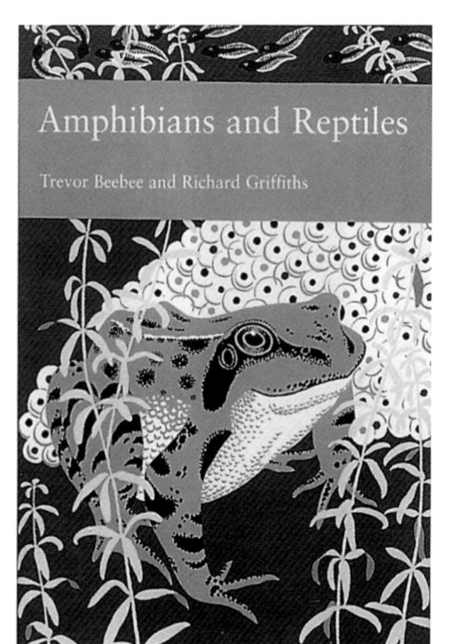

Reptiles and Amphibians, or, in their latest incarnation, Amphibians and Reptiles, have been treated generously by the New Naturalist library. There are now three books and nearly 800 pages, or roughly 75 pages per native species. The first one, by Malcolm Smith, commissioned in 1948, was to have been a monograph, but was promoted to the mainstream series at the last minute. At that time, what was known about 'herptiles' was based on bones, dissection, or, if the animal was lucky, an aquarium. Field study eventually picked up, and now our tiny but much- loved herpetofauna has acquired its own conservation trust and a dedicated band of followers. 'Herptiles 2' or, more properly, *Reptiles and Amphibians in Britain* by Deryk Frazer, published in 1983, was more of a review than Smith's book (Smith had very little to review), and included many new studies of ecology and behaviour, as well as the puzzling species-complex of 'pool frogs', and recently established species. It was an ill-fated book, held up for a year for technical reasons, soon out-of-print and not reprinted. A few years earlier, a Collins editor, Robert McDonald, had noted that 'As a general point, I am not sure that it is wise for us to be thinking of replacing older volumes almost as a matter of course. With the New Naturalists suffering from rising costs and falling sales, we need to be rather careful about what we take on. ... Perhaps this is just a gloomy thought occasioned by Monday morning'.

McDonald's advice was obviously not heeded, and, in different commercial climate of today even he might have had second thoughts. Replacing some of the old, classic titles with modern, up-to-the-minute books could be seen as a mark of confidence in the series. Which titles are in fact replaced relies largely on chance and opportunity. For example, *Moths* came about because there was a Michael Majerus able and willing to write a new book in the same spirit as E.B. Ford, but 'concerned more with the place of moths in the biological world ... dealing with their behaviour, ecology and evolution'. Similarly, Peter Hayward has given us an entirely new *Seashore*, not a replacement of C.M. Yonge's book, published in 1949, but one 'which offers another view from another time'. With characteristic modesty, Hayward hoped his book would complement, rather than supplant Yonge's. It occurred to me that it also supplements Hayward's own Collins Field Guide to the Seashore rather nicely – as a biological background to a fine set of pictures.

But back to 'Herptiles 3'. Perhaps feeling that Herptiles 2 never got a fair crack of the whip, David Streeter asked his Sussex University colleague, Trevor Beebee, whether he would like to write a new book. Beebee's is one of the best-known names in modern herpetology; he has written a string of popular books on ponds and amphibia, as well as the standard monograph on the natterjack toad. He is also a New Naturalist in the fullest James-Fisherian sense of blending the latest, often highly technical, advances in scientific research, with a solid grounding in field study and conservation – which for these animals entails a lot of digging and scrub-bashing, especially in winter. Beebee invited Richard Griffiths and a third colleague, who later dropped out, to share the load – which was substantial. As they explain, or perhaps warn, British 'herps' are not only studied for themselves, but as 'model organisms' for 'research into behaviour, especially sexual selection, thermoregulation and community ecology'. Our tiny herpetofauna, one of the smallest in the world, is also the most intensively studied in the world. The new book reviews the recent scientific literature admirably, without rehashing too much of its predecessors. One wonders, though, whether some of these titles are not becoming a bit too exclusively scientific. The 'herps' in particular have a large and enthusiastic popular following, whose activities vary from maintaining garden frog-ponds to rescuing or 'reintroducing' crested newts and sand lizards. They have, in short, entered the public domain, the broader human consciousness. Is this broadly 'cultural' aspect of a subject a legitimate subject for the series? Is it time to broaden the approach of a new natural history from the purely scientific preoccupations that were so fashionable in 1945? But, if it is, where do we find the authors? We live in a world full of specialists gazing narrowly, but there are evidently fewer talented communicators who possess breadth as well as depth of knowledge, and the capacity to convey it in interesting, readable English. And, where they exist, how do you persuade them to write a book which is neither mass-market nor a cornerstone academic textbook, but something inbetween? The wonder, perhaps, is that the additions to the New Naturalist library are as good as they are.

An author explains....

The book you are reading was easy to write: I wanted to write it even before I was given the opportunity. I wanted to write it so much that I devoted six months of my life to it, at a time when I really should have been earning a proper living. It felt like the working-out of fate, as though I had no choice in the matter. It would be untrue to claim that I spent these months on a rolling surf of unalloyed pleasure, because any big book is an awful slog, and writing them makes for a very boring life. But researching the book, mainly by interview or forensically sifting huge, dusty piles of papers, was undeniably fascinating. It was a privilege to meet or correspond with some of the people who made the New Naturalists what they are. And it was creatively satisfying to investigate the background and make history from it.

It was also as good a time as any to take stock. After five decades, the series seemed becalmed, and I wondered whether I was writing its funeral sermon. Yet, by its very existence, *The New Naturalists* encouraged fresh interest in the books, and I'm told, helped to create a new dynamic. The success of *Ireland*, a few years later, enabled our editors to increase the print-run and get to grips with the series more effectively.

As for writing the book, I have few memories other than the sunshine outside, which looked especially inviting in that warm summer of 1994. Each day, I started as early as I could (which wasn't all that early), and stopped around 4, at which time I often played a game of badminton and afterwards enjoyed a large gin-and-tonic. Are such things of interest? I had two deadlines: one, the official one, was the end of September, to enable the publishers to flourish the book at the celebrations. This was tight, but the other deadline was tighter still because the money from the advance would, I calculated, run out in June or July. By taking on some small commissions for ready money, I eked out my income to the due date, and, on the whole, it was a happy journey.

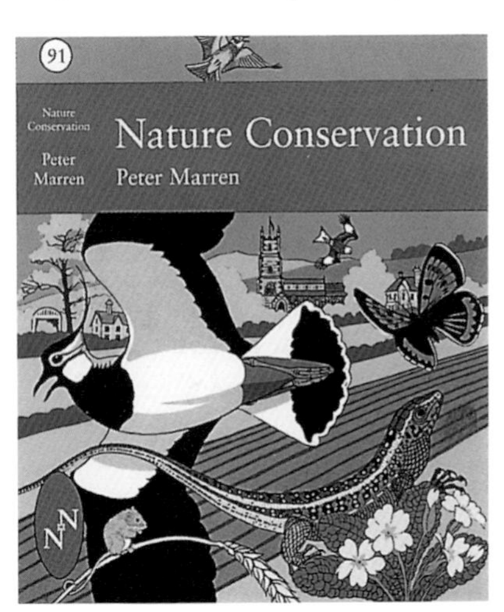

I hesitated longer before accepting the publishers' invitation to tackle *Nature Conservation*, the intended successor to Dudley Stamp's venerable and posthumous book published in 1969. Clearly I was being engaged on the strength of *The New Naturalists*, for, although I had worked for nature conservation bodies over many years in different capacities, I was neither an originator nor a policy maker as Dudley Stamp was. Indeed, my career as a scientific civil servant in the Nature Conservancy Council had met with a crashing lack of success. On the other hand, I seemed to know a lot about it, and had no particular axe to grind: unusually, among the conservation crowd, I was unaligned. I wanted to justify some of those dreary days I had spent notifying SSSIs and writing annual reports. In the end, I accepted because I felt the opportunity would not come again, and that, with my skates on, I might be able to repeat *The New Naturalists* challenge and polish it off in six months. Not that I had any choice about that. When you write for a living, the last thing you want to do with your spare time is more writing. So, once again, it looked like another journey against the clock, taking on enough extra work to see it through without starving. Heroic stuff, I thought.

It was, of course, a much harder book to write than *The New Naturalists*. While the latter was a nice narrow-but-deep subject, like a well, that enabled me to wallow happily in the detail, nature conservation has grown impossibly broad and fluffy – where does it end? The gardener who uses compost, or a more 'environmental' weed killer, thinks he is helping to conserve nature. The Government used to claim that building more roads was good for the environment. Recently, English Nature surprised us all when it told us that even house-building could, in certain circumstances, benefit wildlife. So many people – so much conservation! (it sounds better in Latin). I decided it would avoid a lot of fruitless grasping for definitions if I focussed on the interesting question of whether all this conservation had been good for wildlife. You might think it obvious that it has been, but it is surprising how little hard information there is on how this or that policy has worked out in the field. Even nature reserves are not necessarily success stories; not at any rate if half the wildlife has fled since you set up the fence.

An extra note of acerbity may, I admit, have crept in. Some of the papers I waded through when researching *The New Naturalists* would have been very dull except to an ardent bibliophile, but that was nothing compared with the numbing effect of what passes as the literature of the conservation movement. Saving the natural world seems to rely on producing joyless micro-managing 'strategies', from 'action plans' for the most obscure moss or snail to confident utopian tomes promising to redesign whole landscapes with the aid of 'joined-up thinking'. Yet the sense they give is not the countryside we all know and love but a kind of drawing board around which faceless figures are busy pulling levers and making adjustments. Is this where Dudley Stamp, Max Nicholson and all the other pioneers of scientific nature conservation in Britain thought we would end up?

I didn't think a simple review format would work for *Nature Conservation*. Instead of trying to cover everything, I selected themes, and freely offered my own opinion on what it all amounted to. I tried, as far as I could, to say not only that such a thing had happened, but why it happened. And I also tried to give readers plenty of space to make up their own minds. The book is about the same length as *The New Naturalists* (about 125,000 words) and took about the same time to write – but twice as much effort. If I was to do it again, I would include more from the voluntary perspective. But in terms of conserving species and habitats, we live in a top-down country.

I dedicated the book to Derek Ratcliffe, one of the towering figures of postwar nature conservation, but I don't think he liked the book much. He and some others found it too downbeat: "too many of your apples have maggots in them". I was conscious while writing it that I wasn't going to please everyone, especially those who had dedicated their professional lives to trying to save what is left of the natural world. All the same, I think the book has its merits: it bounces along at a brisk, lively pace (perhaps a bit too fast; there are places where I should have paused), that its analysis and judgements look reasonably sound to me, and that it reaches places other books don't. It was well received, where it was reviewed at all, and is not, I think, a bad book by any means. If you find its conclusions a bit pessimistic, I can only say that, as I see it, nature conservation isn't a success story. In the eternal contest between making money and saving something for the future, supply and money will usually win, witness the dried-up rivers and spreading housing-estates of rural England. To be honest, I think a lot of conservation activity is a waste of time. The movement has become a job-creation industry as much as anything. Which is not, of course, to deny that nature conservation has achieved a great deal, and that the world would be the poorer without it.

The tale of a book

Few books in the series are better-documented than Leslie Brown's *British Birds of Prey*. Not only was the author a prolific (and entertaining) letter-writer, he left us a detailed account of how he came to write the book in *The Birdwatcher's Handbook* for 1974. I was unable to trace a copy (not even in the Bodleian Library in Oxford) when writing *The New Naturalists*, which, for that reason, scarcely did justice to Leslie Brown nor his fine and important book. A few years afterwards, my friend Bob Burrow found a copy of the elusive book in which Brown's article appeared and kindly sent me it as a gift. The following piece is by way of making amends, and also to reveal a little more about the process in which this particular New Naturalist was written.

British Birds of Prey first published in 1976, was the best-selling New Naturalist title since the 1950s. Some 20,000 copies were printed between 1976 and 1982, including 5,500 for the Reader's Union book club. The first edition of 9,000 sold out within a few months, and was reprinted the same year. The book was even given the rare honour of a prepublication serialisation in *Animals* magazine, the leading natural history periodical of the day. Perhaps its success was not surprising. Birds of prey are eternally popular celebrities among the avian kind, both for their fierceness and beauty and as potent symbols of wildness and freedom. Yet few have attempted to review our birds of prey as a whole, as opposed to writing in depth about individual species; since Leslie Brown's book, first published in 1976, only one book has done so (*Birds of Prey in a Changing Environment*, 2003) and that one required the services of five editors and over forty authors; Leslie Brown's book, by contrast, was very much a personal interpretation. That is, in part, what makes it so readable. Birds of prey are indeed a knotty subject in several ways. The scientific literature was enormous, even in the 1970s. Every species had its own expert, and by no means everything about them had been published. For example, there were

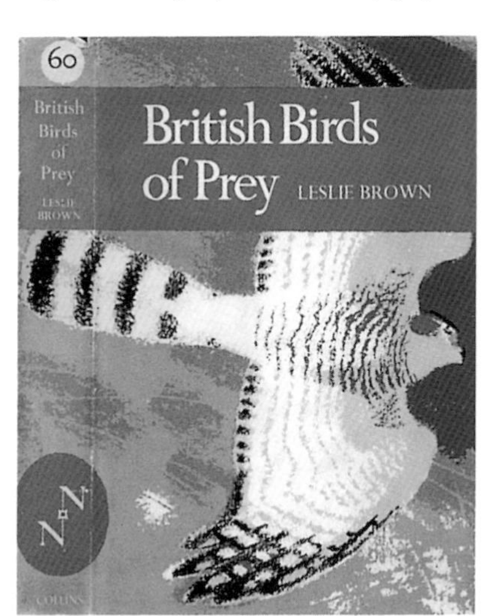

long-term studies on the red kite and hen harrier, based on years of work, which had not been written up. In the case of the honey buzzard, very little indeed had been committed to paper; the experts on this rare bird had committed themselves to a kind of masonic pledge of secrecy (perhaps it all went back to the famous episode of the robbing of the honey buzzard's nest in BB's *Brendon Chase*). Birds of prey were, besides, an unusually sensitive subject. Few birds compete so openly and fearlessly with mankind. Though all are protected by law, birds of prey are, notoriously, still persecuted. Keepers blame hen harriers for a dearth of grouse on the moor; pigeon-fanciers are not always well-disposed towards peregrines; some have even called for a cull of sparrowhawks on the grounds that they are eating too many garden songbirds. Anyone writing about birds of prey in Britain has to do so against a background of scientific controversy and inflamed passions, to which even ornithologists are not immune.

Leslie Brown (1917–1980), eagle-watcher extraordinary, author of the best-selling *British Birds of Prey* (1976).

The New Naturalist editors had had no settled policy on how to tackle the birds of prey. In keeping with the ecological ethos of the series, they originally planned to produce a series of habitat-based bird books: the published *Sea-birds* and *Birds and Men* (which was effectively about farmland and towns) were to be followed by volumes on the birds of woodland (Bruce Campbell, later W.P. Yapp), sea-shore (Eric Ennion), moorland (James Fisher) and marsh and freshwater (R.C. Homes). Individual species were to be covered by the New Naturalist Monographs, which, by 1953, included commissioned volumes on The Peregrine (J. Walpole-Bond and James Ferguson-Lees) and The Golden Eagle (Seton Gordon). In practice, birds of prey slipped through the net completely. The problem was that the sort of people who tramped the hills in search of eagles and peregrines, and their nests, were not always the kind of naturalist-scientists capable of fusing field observation with scientific insight. 'Jock' Walpole-Bond had asked sceptically of James Fisher: 'is he a cragsman?'. And Fisher retorted by bluntly informing Bond that his literary style would make them all a laughing stock.

By 1969, the commercially unsuccessful New Naturalist monograph series was coming to a close; the death of its main protagonist, James Fisher, the following year lost the series its main advocate. In its place, the natural history editor at Collins, Michael Walter, planned to publish a series of bird books in the mainstream series based on groups of related birds: finches, tits, thrushes, waders and so on. He had already contracted one author, Ian Newton, to write about finches. But the new policy would benefit from a really good book on the most popular birds of all – the birds of prey – all the more so since the series had so far barely touched on the subject. Michael Walter already had an author in mind. Leslie Brown had a reputation as an authority on birds of prey throughout the world through his many papers in learned journals and his authorship of the much-praised standard work, *Eagles, hawks and falcons of the world* (1968). He was also an authority on British golden eagles, and had, with Adam Watson, written a definitive paper on the distribution of eagles in relation to their food supply. He had also, as it happened, recently written a book on African birds of prey for Collins. The only problem was that Brown lived in Kenya, where he had been a senior agriculturalist in the colonial government until the independence of that country in 1963. Thereafter he had been an independent advisor on land-use and conservation matters in East Africa, but spent an increasing amount of his time studying and writing about birds. Brown's knowledge of British birds, even golden eagles, was not recent. On the other hand, he was a first-rate communicator, fast, fluent and opinionated as a writer, vivid and unstoppable as a raconteur, and known to be hardy and indefatigable – and short-fused. He had the rare ability to blend science and accurate observation with undiminished enthusiasm. Many of his studies of flamingoes, pelicans and African eagles had been done under arduous conditions in remote country. Though published in learned journals like *Ibis*, they were always readable. In 1967, the BBC made a film about him entitled 'Bird-watcher extraordinary', showing him on the trail of migrant pelicans in Ethiopia, clambering up a waterfall to a red-winged starling's nest and, after an all-day trek in the Scottish hills looking for eagles, leaping naked into a Highland burn.

With the agreement of the Board, Michael Walter wrote to Brown asking him whether he would like to write a New Naturalist. The idea appealed to Brown, who replied promptly that he would be very pleased to do it, though pointing out that there were others, living in Britain, who would feel they could do it also. Brown certainly felt flattered to be asked to contribute to such a prestigious series, but it also appealed to the poacher in him. He was a keen amateur pheasant poacher, and liked the idea of stealing the subject from under the noses of the birds-of-prey establishment. As he put it in his preface to *British Birds of Prey*,

> I have gathered that several British ornithologists might have liked to sink their teeth into this particular cock pheasant [but] have not felt able to spare the time and effort from other more essential tasks to bag it legitimately. So it has fallen to me to climb over the fence with my concealed weapon and snatch the bird from under the noses of British experts. Those who know me well will merely observe that I am a creature of habit, and would not be able to resist it. However, when they have seen me making off with the bird they have all encouraged me in my nefarious task and, if I faltered, have plied me with meat and drink, bed and board, that I might poach the better (Brown 1976).

Following his acceptance, Brown was sent a contract stipulating a completion date by the end of 1971 and was given an advance large enough to cover his travel expenses for a short stay in Britain. The timetable was hopelessly optimistic. It did not take long for Brown to discover that the task was larger than he had anticipated. His visit to Britain in the fall of 1970 allowed him time to consult back-runs of *Bird Study*, *British Birds* and other journals in the library of the Edward Grey Institute in Oxford, and inevitably they showed up the apparently gaping lack of knowledge of some species. There was also, he knew, a great deal of work that had not been published, which would require a second visit to pick up, and delay the book for at least a year.

Brown wrote about his second visit, in spring 1972, in great detail in an essay for *The Bird-Watcher's Book* (1974) called 'A bird of prey odyssey'. An odyssey is, of course, a tale of wandering, a journey with many turns and adventures. Brown planned to tour the country interviewing experts, inspecting monitoring reports and other unpublished data, and also to see as many birds of prey as possible, from buzzards in Devon to hen harriers in Orkney, from Welsh kites to East Anglian marsh harriers. Unfortunately it turned out to be one of the wettest springs in memory. Brown's dormobile odyssey was, he wrote, a 'grinding slog'. He liked to write up what he learned while it was fresh in his memory, and, if possible, show it to the relevant expert before moving on. Hence a full, and probably wet, day in the field might be followed by several hours in the cramped dormobile or in a friend's spare room, banging out a report on his portable typewriter (by all accounts a fairly unreliable machine). The odyssey began in Exeter, where Brown spent a couple of days poring over Peter Dare's highly regarded PhD thesis on the buzzards of Dartmoor. Turning east, Brown inspected the BTO library at Tring to extract data on migration and nest records, and also the RSPB library at Sandy, before taking off to Minsmere to watch, with some help from the warden, Bert Axell, a real bird of prey: a bigamous marsh harrier cock tending his two sitting wives. On a day of rare May sunshine Brown dropped in on his old friend John Buxton, of *Redstart* fame, to talk about harriers and spend a day on a trout stream with a dry fly. He caught the permitted maximum of three trout, including a big one-and-a-half pounder, 'one of my rarest and keenest pleasures'. Thence he headed north to meet Ian Newton (*Finches*), who was studying sparrowhawks in Dumfries, talk to Douglas Weir in Speyside about buzzards, eagles, peregrines and merlins, and inspect, with derision, the RSPB's somewhat chaotic osprey records at Loch Garten. While in Speyside, Brown took the opportunity to revisit an eagle eyrie he saw in 1945 and found it occupied by an eaglet tucking into its dinner: a grouse, a hare and a mallard drake with its bright orange webbed feet. He was gloomy, however, about the future of the golden eagle: 'Gamekeepers and their employers are about as hopeless a proposition in 1972 as they were in 1872', he decided. After a day or two adding, 'a little titbit of knowledge' observing an eagle's attempt to extend its nesting territory in a remote Sutherland glen, Brown caught the boat to Orkney and spent three days roaming the hills with a taciturn Eddie Balfour looking for hen harriers. He also looked for, but failed to find, the phone box inside which he and two fellow birders had spent an excruciatingly cold night in 1937, longing to be arrested for the comparative comfort of a cell.

Brown returned south, with a detour to Wales to find out about kites ('I saw seven, probably about 10 per cent of the whole population'), and another to the New Forest to discuss buzzards and hobbys with Colin Tubbs (*The New Forest*). That day, the first hot day of summer, and in the company of Richard Fitter (*London's Natural History*), he was lucky enough to witness a pair of honey buzzards performing their unique aerial display, repeatedly raising their wings vertically over the back, somehow without losing height. Finally Brown came to port at Chris Mead's hospitable home in Tring, where he wrote much of the book, hammering the keys 'round the clock from seven to seven'. 'I was dead beat at the end of each day [but] felt I must complete all the chapters on individual species before leaving for Kenya, where I could do the rest.... All but a few of the experts to whom I sent revised chapters for review came up with useful criticisms.' 'Those that did not,' he added, characteristically, 'have, of course, blasted me already for misquoting them.'

Brown was critical about bird watching in Britain. Instead of watching birds closely to learn more about their behaviour, most bird watchers wasted their energy dashing from one place to the next to tick off another rare bird, contributing 'little that is useful, but antagonising farmers and others by leaving gates open and trampling crops'. Brown completed the book at home in Kenya and sent the manuscript to Collins towards the end of 1972. Michael Walter found it not only 'a most filthy manuscript' but 'a dangerously long book' full of tables and references. Brown, meanwhile, reflecting on the whole experience, had vowed never again to try to write 'anything like a new naturalist book' on British birds except in the somewhat unlikely circumstance of being based at a British institution and enjoying a good regular salary. 'The result of my Odyssey won't be as good as I, or others, might like it to be, though I think it is fair and most of my expert referees have said so. For good or ill, it's done....'

Unfortunately the book was not done. Three years passed before the 'filthy' manuscript of 1972 was transformed into a book, a frustrating delay which goaded Brown into a fury, especially as he was already, he calculated, £2,500 out of pocket over it. There were, of course, communication problems with an expatriate author who was often away on his travels as an agricultural advisor in east Africa – and who also, at one stage, lost all his valuable notebooks to thieves. But the main reason for the delay seems to have been Michael Walters' insistence that the book was too long, contained too much data, and needed drastic cutting before it could be published as a New Naturalist volume. The last straw for Brown was when, the contracted indexer having turned it down, he was asked summarily to produce the index himself, just as he was setting off for Ethiopia for several weeks. Brown insisted that his preface should reveal the date he had actually written it – 1972. As he pointed out, one British bird of prey, Montagu's harrier, had ceased to breed since then! His editor, Kenneth Mellanby, begged him to reconsider, pointing out that the text had been much revised, and that the discrepancy between Brown's date of 1972 and the published one of April 1976 would imply that the book was out of date and prejudice its inception. Eventually Brown conceded, but the experience had evidently entered his soul. He resolved to give up writing books – 'I simply can't stand the strain of dealing with publishers from this range'. Whether it was Collins or another errant publisher that formed the subject of the 'numbing' four-hour tirade remembered by his friend Jeffery Boswall (Boswall 1982), Brown had certainly added book publishers to his list of *bêtes noires* along with 'pompous landowners, constipated functionaries, and fools generally'.

The odyssey was not quite over even then. *British Birds of Prey* had been eagerly awaited by a large ornithological public and was well received. It combined an in-depth study of the subject in a readable style and with many personal touches, especially when Brown was describing a bird he knew well, like the golden eagle. His carefully considered judgements came over well, and in places it is hard not to see a little of the bird-of-prey in the beady-eyed, narrowly gazing, high-wired author himself! What drew more criticism was the presentation. To confine such a long book (for Brown had eventually had his way as far as cutting went) within 400 pages, the text had been micro-printed, 50 lines to the page. There were no colour plates, and the paucity of text illustrations gave it the hallmarks of an academic book. The book was also published at a time of rampant inflation, so that the not inexpensive initial price of £6 had, by 1982, escalated to a decidedly steep £11.50. Even the much-praised dust wrapper design by Clifford and Rosemary Ellis was reprinted in Spain as an economy measure, on inferior paper and paler colours. The book, of course, did well – so well indeed that large book club editions were also ordered for the forthcoming specialist volumes on British Thrushes and British Tits. The modest first edition of just over 5,000 sold out very quickly. With reprints, book club editions and a small US edition published by Taplinger, total sales came to 20,000. A quarter-century later, although the book is clearly out-of-date in many ways, it is still the standard popular work on British birds of prey.

When *British Birds of Prey* was published, Leslie Brown was on his back in hospital, having suffered a 'bad but not massive' coronary. Expecting 'bad and ungenerous, if not actually libellous' reviews from rival ornithologists, he was pleased that, on the whole, they agreed with him that it was 'a darned good book'. For the second reprint, he added two pages of notes (including a stubborn reference to his having completed the manuscript in 1972). There was also what amounted to an apology to Desmond Nethersole-Thompson who had found himself fortuitously libelled as an egg collector. A lively correspondence between Brown and his publishers continued, sometimes good-humoured, often apoplectic. Brown partly blamed the decline of his health in the late seventies on books and publishers, though that did not prevent him from working on another enormous literary project, a multi-volume *Birds of Africa*. A few months before his death from a heart attack in August 1980, aged 63, he had returned to his beloved Speyside and watched golden eagles again. One reason for the visit was an intended revision of *British Birds of Prey*. As he had remarked in his last letter to Michael Walter, 'I should prefer to revise it myself than see anyone else do so, though I may of course be incapable of so doing for reasons beyond my control. If so, it won't matter so much'.

The New Naturalist series has had a record of good judgement – and perhaps good luck – in matching the right author with the right subject. Leslie Brown and *British Birds of Prey* is a good example. Although there were more obvious candidates for the job in terms of experience and scientific qualifications, few could have rivalled the amateur Brown's lucidity and sense of quest. As Brown himself described his personal odyssey in his *Encounters with Nature*:

> The more one learns the more there is to learn, and one bit of new knowledge just leads on to another… anyone can make discoveries; it needs no PhD or even a degree, just a pair of observant eyes and ears, and a questing mind. Some of the very best naturalists of recent times have been, scientifically speaking, rank amateurs. Although much natural history literature is dated almost as soon as it appears, it should still be published, for it is the springboard from which someone else may dive deeper into a new and unknown sea.

New Naturalist jackets 1995-2005

Nowadays Robert Gillmor's name is spoken by New Naturalist fans with the same reverence that his illustrious predecessors, Clifford and Rosemary Ellis, enjoy. Among words that describe his designs, I've heard 'luscious', 'tactile', even 'sensual'. By the end of 2004 there were a full two dozen of them, spanning nearly twenty years, from *British Warblers* to *Northumberland*. Separately they are admirable enough, but together they are likely to be the most attention-grabbing books on anyone's shelf. Using only three colours, plus black, Gillmor creates designs that are detailed enough to satisfy the naturalist, yet also find graphic forms to express the colours and shapes of nature with sufficient clarity and boldness. Good examples include the swaying reeds on *The Broads*, rendered in rhythmic form, or the bold plumage of the bittern on the same jacket, both created using linocuts. Or the artful stippling to create the texture of old stone on *Lichens*, or the ivy-leaves of *Plant Disease*, an ideal subject for lithography with their bold outlines and flat planes. We all must have our favourites. *Lakeland* and *The Hebrides* are hugely like-

Robert Gillmor in his studio at Cley-next-the-Sea.

able with their birds and cool tones, but for me the palm still goes to the half-abstract design of *Ferns* with its satisfying textures and forms. Oh, and of course I like the two jackets he did for me, too!

Robert (for I can't call him Gillmor), a versatile and successful bird artist and book illustrator, was asked to take over the design of the New Naturalist jackets after the death of Clifford Ellis, who, with his wife Rosemary, had produced consistently imaginative and colourful lithographic designs for the books for over forty years. The then editor, Crispin Fisher, gave him the greatest possible amplitude for interpreting each book. His only suggestion was that he should keep the title band and the oval containing the NN logo. Fortunately, Robert saw no pressing need to change the basic design at all. Less fortunately, his jackets were, with the exception of *Warblers*, restricted to the hardback 'collector's editions'. The paperbacks were given uninspiring designs based on colour photographs, which, it was supposed, would appeal to a wider book-buying public. The differences between Gillmor and Ellis jackets are mainly due to technique. The Ellises based their characteristically smudgy designs on pure colour, and rarely used line. Gillmor uses line more often, and overlaps solid blocks of colour to achieve extra colours and tones. For example, at its simplest, a yellow printed over blue produces green. Though no more 'photographic' than the Ellises, his technique does allow more detailed designs which often feature a range of birds, animals, insects and flowers in more or less realistic colours and settings. As always, they are produced and printed by craft methods. The New Naturalist jackets are a rare example of a computer-free zone in the modern book-publishing world.

Since 1985, Robert has supplied the printer with his artwork in the form of a series of blocks or 'colour separations', in which each printed colour is rendered as black on white. Only when the jacket is printed do the colours emerge and combine to form the picture. For this reason, the artist normally does a detailed colour sketch from which the blocks are made, and it is this sketch, rather than the printed jacket, that has been used on the fliers and advertisements of some forthcoming titles (my own *Nature Conservation* was an example). The first time the artist actually sees the design in printed form is when he is shown a kind of proof, known as a chromalin, produced by inkjet laser. This enables him to decide on any last-minute adjustments to the colours before printing. Technology helps that task, too – a gadget placed over the colour registers a number which enables the printer to adjust the tone to the exact level required. The jackets themselves are printed by the technique known as offset lithography, in which the printing inks are added as the paper passes between whirling rollers. Such printing machines are nowadays becoming museum pieces, and are used mainly for reproducing art-

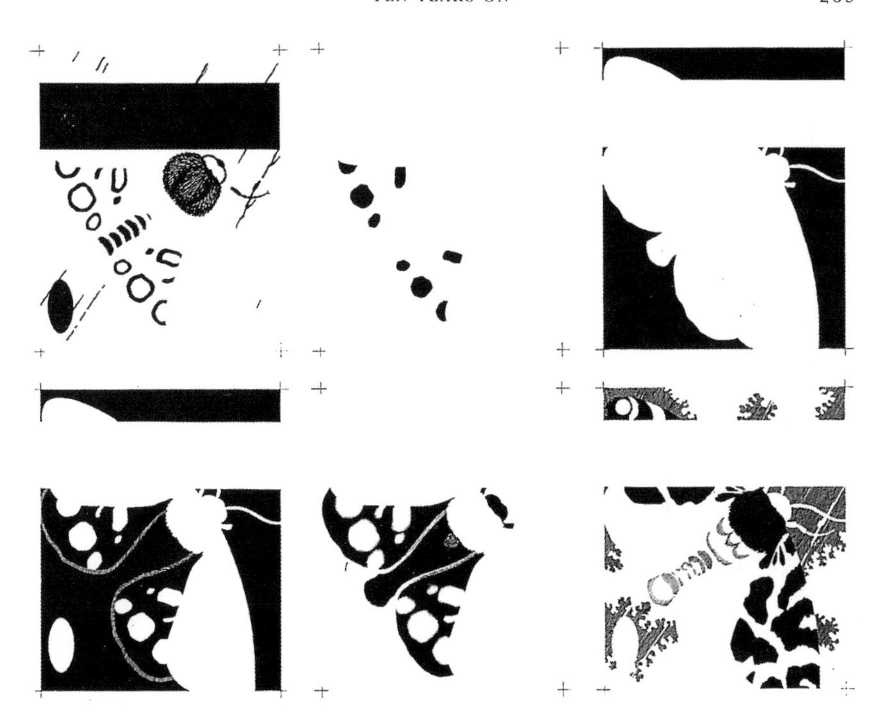

The six pieces of artwork used to make up the four plates for printing the jacket of *Moths*. The 30 per cent black tint was printed over the yellow to make the dull green background. The mixture of the 80 per cent red and yellow created orange. The dark grey centres to the black wing patches are an 80 per cent black tint. The solid black and two black tints are combined in one plate, as is the red with its 80 per cent tint. See plate 12 for the finished jacket.

work as prints. Between 1985 and 1996, the jackets were printed by Radavian Press in Reading, and since then by the Norwich-based Saxon Photolitho Ltd. By good fortune, both were close enough to Robert Gillmor's homes for him to supervise each printing to ensure that the right combination of colour tones are selected. The naturalistic colours of the jackets depend on very precise tones chosen from a wide range in the Pantone matching system. In recent times the process has become even more critical, since Robert has taken to using tints, artful layers of dots visible only under a strong lens, to squeeze a wider range of colours and tones out of his four basic colours.

The Gillmor jackets offer an intriguing mixture of design ideas, from naturalistic scenes, like *The New Forest* and *Lakeland*, to more graphic designs, like *Ferns* and *Plant Disease*. The jacket of *The Natural History of Pollination* could be a homage to Clifford and Rosemary Ellis, who used a similar bee-and-foxglove motif for *Bumblebees*. With *The Broads*, where the brilliant colours of the spine bleed into a monochrome waterway, the jacket even contrives to tell a story. How do such ideas surface? Work on the jackets normally begins after the text is nearly ready, at least in draft form, but several months in advance of the publication date. The artist is normally sent a synopsis of the book by the publishers, together with the introduction and a sample chapter. This gives him a general idea of the book

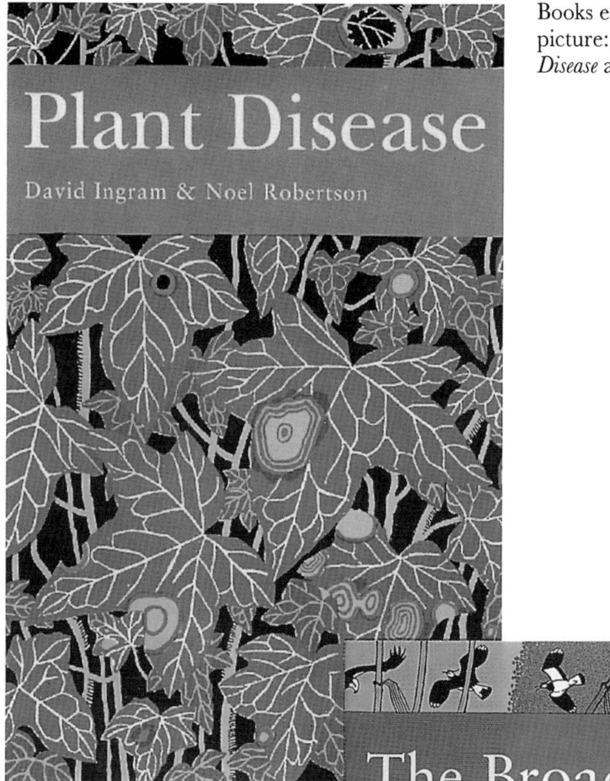

Books encapsulated in a picture: Jackets of *Plant Disease* and *The Broads*.

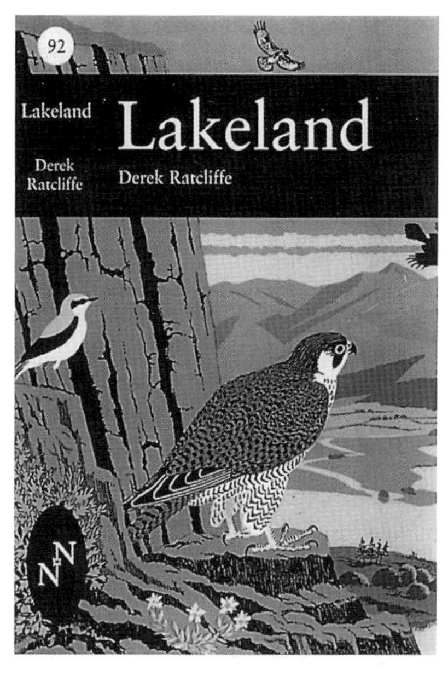

from which a suitable design may emerge by doodling numerous sketches. There is generally some discussion between artist and author. A chat with the author often results in agreed motifs which the artist can work on, and it also helps to ensure the author is happy with the chosen design. For example, for *Lakeland*, Derek Ratcliffe sent Robert a list of suitable subjects, including various species of birds, insects and plants, along with a colour transparency of a view from Falcon Crag, above Derwentwater. The prominence of a peregrine falcon in the finished jacket is a reference to Falcon Crag, but may also be a compliment to Ratcliffe as the author of a well-known book about that remarkable bird. For my own book, *Nature Conservation*, the original design had a plainer background with a superimposed outline of the British Isles. I wasn't keen on this, and, to my delight, Robert substituted a beautiful landscape setting for the finished book, which, from various clues in the architecture, wildlife and distant hills, we agreed must be somewhere in west Somerset! In several cases, the jacket is, in fact, based on a real place. The church at Wiveton, Robert's next door village on the North Norfolk coast, forms the background to *British Bats*, while it is the ancient stones and tombs of Binham Priory that adorn *Lichens*. By the by, at the request of the author, Robert added a little lichen-mimicking moth to one of the stones on the latter design. Did you spot it? (I didn't.) The *Loch Lomondside* jacket shows the early evening light of midsummer playing on Ben Lomond. *Pollination* incorporates a bee that is recognisably *Bombus hortorum*, the garden bumblebee. For *Plant Disease*, the authors' reference to target-spot disease of ivy inspired Robert to examine some ivy on a wall a few yards from his front door, and almost immediately found it, in all its spotty glory. Other designs took longer to work out. Robert abandoned his first versions of *Amphibians and Reptiles*, and *Freshwater Fish*, featuring several amphibians and game-fish respectively. *Moths*, on the other hand, needed no modification. The bold colours of the garden tiger-moth are so perfect for lithography that it virtually chose itself.

Northumberland was one of the more protracted commissions, partly because new methods of production had been imposed on the artist (see below), but also because there was less in the book about birds than had been expected. Robert had originally produced a design in which birds were prominent, including a fine ring ousel on the spine and an eider duck (as the iconic bird of St Cuthbert of Lindisfarne). To bring the jacket closer to Angus Lunn's text, some of the birds were shooed off the design. Robert added an outline of the county in white, which one of the editors hilariously mistook for bolts of lightning – it would have been an extremely dramatic thunderstorm, but may be the weather can be like that, up there in Northumberland.

For *The Seashore*, Robert Gillmor was asked to produce a design that said

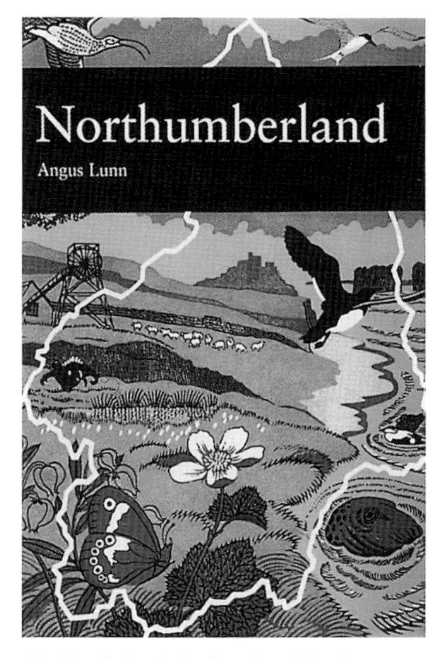

Northumberland: the hardback jacket was reproduced from a linocut and screen-printed in a single process.

Seashore: the hardback jacket was based on a linocut and printed by offset lithography.

'seashore' to the reader, clearly and unambiguously, but without seeming clichéd. He began by making pencil sketches of starfish, crabs, limpets and other shore life, combining the different elements into a coherent design. Barnacles, in particular, proved a good subject for a linocut, with their bold, sculptured shells and flickering gills. But perhaps the most telling feature was the addition of concentric white lines to create ripples, as when the observer, peering down into the miniature world of the rock-pool, disturbs the surface with his finger. The jacket forms the greatest possible contrast with the original *Sea Shore* of 1949, for which the Ellises produced an austere design of a sun-bleached crab's claw lying half-buried in the sand.

Over the years, the artist's technique has become more complex and demanding, no doubt to bring an element of challenge to the commission, but also reflecting the development of Robert's wider work. One innovation has been the use of tints, first employed on *Ireland*, to create subtle variations in tone as well as extra colours. For *Moths*, a hand lens reveals that the apparently even olive-green background is in fact a tint obtained by overprinting yellow with fine black dots. Tints require a degree of exactness that borders on the virtuosic. For the recent jackets, he has produced up to nine separate pieces of artwork, from which the printer makes the four printing plates.

Another change has been the use of linocuts to produce bold, expressive line artwork, full of character. Linocuts are made by cutting into a sheet of linoleum with an awl-like tool to leave a raised design, from which a print is obtained. They first appeared on *The Broads* where the luscious colour on the left contrasting with the monotony on the right owes much of its effect to the technique. The black ink plates for *Nature Conservation* and *The Seashore* were both linocuts.

In future, linocuts may become the

The black plate used for *Seashore*.

The Original linocut for *The Broads* jacket.

basis for the whole design, as in *Northumberland.* In 2004, the publishers asked Robert Gillmor to produce ready made full-colour artwork for the printer to screen and print as a single process, rather than making the plates from a number of separate drawings. This may become necessary if, as is currently planned, the books are colour-printed. Colour-printing would almost certainly have to be done overseas, and there are commercial advantages in printing the jackets and the books together. Robert, an experienced printmaker, intends to make more use of linocuts which can also be made as camera-ready artwork. He hopes to transfer the same rich effect of the original hand-printed work to the printed jacket. It does, however, require immense technical skill (and time) to produce a piece of lino intricately carved in relief, resembling, to the unskilled observer, a detailed diagram of Hampton Court maze. The slightest mistake is hard to rectify. Another problem, though hopefully not one for the foreseeable future, is that craft skills are being replaced by computer techniques, and fewer artists are skilled in traditional techniques, such as linocutting. In my view, the jackets, on which so much of the New Naturalists' attractiveness to collectors is invested, are not a wise place to make economies, nor radical changes.

Despite the great care that goes into their production, some jacket designs have tiny technical defects. Robert pointed out to me one of his own mistakes on the *Nature Conservation* jacket. Under the body of the large copper butterfly, the hedgerows should be green, not blue; he had missed out a small section of the yellow block. Since the book happens to be written by me, I must have looked at that jacket a thousand times without spotting the mistake (did you?). Some improvements would be possible, in retrospect. Robert is unhappy about the colour tone of the cliffs on *Lakeland*, and thinks a darker colour on the main rock face would have created a better contrast with the foreground crag. The aberrant coloured book-number on *Lichens* could have been avoided by substituting a grey sky tone in the open 'windows' above the title band.

Far more annoying to collectors is that the design occasionally stops slightly short of the spine, so that there is an ugly white margin to the left, especially if the jacket has slipped on the book. Though Robert takes pains to prevent this from happening, the exact size of the book is hard to gauge before it is printed. The Ellises got around the problem by smearing colour onto the back, but this is impossible with Robert Gillmor's more detailed and intricate work. Much more serious is the tendency of the inks to fade or change colour over time, and more quickly if placed in direct sunlight. By some perverse trick of fortune, the most colour-sensitive titles tend to be the most valuable ones, and a faded spine can reduce the book's value by scores, if not hundreds, of pounds! I spoke to the current printers about this, and was told that the reason for using non-colourfast inks is the printing method, which requires very exact colour tones. It means that the printers select from a very wide range of ink colours, most of which are not made to be colourfast. The big advantage of using the half-tone screen process is that it requires only the four basic printing colours, black, magenta, 'cyan' and yellow, so that the jackets can be printed with colourfast inks, as the early Ellis jackets evidently were. But the collector is left with the conundrum of what to do with numbers 73 to 94. The more valuable they become, the more reason to conceal them. Has anyone designed a transparent wrapper that filters out ultraviolet light? Should one make several laser copies right away, and consign the original to a bank vault? Or is the investment value of a book immaterial, and, like a car, you accept that condition depreciates over time. The answer probably depends on why you bought the book in the first place.

Selected Bibliography

A note on written sources

I decided against using references or footnotes in this book for two main reasons. The first is that books in the New Naturalist series more scholarly than this one, have not done so. But mainly it is because so much of this book is based on unpublished files and papers not easily available to the reader, who would not, therefore, be in a position to verify what I have said. The main sources for unpublished or elusive material is as follows:

1. The New Naturalist archive held at the printing department of HarperCollins in Glasgow, and the current files in its natural history section in London. The files cover nearly every book in the series, but they have been severely 'gutted' and much of interest has probably been destroyed, especially before 1960. These files now deal mainly with the book's printing history, but contain extraneous correspondence between authors and editors.

2. The minutes of the New Naturalist Board. A full set of Board minutes exist from 1951 to date in the HarperCollins archive. A full run from about 1955 to 1970 also exists among the private papers of James Fisher, now in the care of the Natural History Museum. Attempts have been made to trace any earlier Board minutes but none have been found, though very likely some record was kept at the time.

3. Private papers of Sir Alister Hardy and E.B. Ford (Bodleian Library, Oxford); Sir Maurice Yonge and James Fisher (Natural History Museum); and Clifford Ellis (private collection). Between them these collections yielded much of interest, particularly on the early years of the series. I have also been shown private correspondence relating to several other authors.

Published memoirs and obituaries exist in varying degrees of detail for the majority of deceased New Naturalist authors. Those found at any reference library will be the relevant volumes of the *Dictionary of National Biography*, *Who was Who* and *The Times Obituaries*. The most detailed and interesting are the Biographical Memoirs of deceased Fellows of the Royal Society, available as off-prints from the Society. The authors and editors listed are: H.J. Fleure, A.D. Imms, Sir Alister Hardy, Sir Julian Huxley, Roy Markham, L. Harrison Matthews, W.H. Pearsall, Sir John Russell, Niko Tinbergen, W.B. Turrill, Sir Edward Salisbury, K.M. Smith, C.B. Williams, S.W. Wooldridge and Sir Maurice Yonge. A Memoir of E.B. Ford is in preparation. Obituaries of most of the deceased ornithological authors are in the relevant volumes of *British Birds* and *Ibis* (notably Armstrong, A.W. Boyd, Buxton, Brown, Campbell, Fisher, Hosking, Murton, Nethersole-Thompson, Stuart Smith and Tinbergen). Obituaries in *Watsonia* include Gilmour, Lousley, Prime and Summerhayes. I list the most detailed memoirs in the selected bibliography that follows.

Allen, D.E. 1976. *The Naturalist in Britain. A Social History*. Allen Lane, London.
Allen, D.E. 1986. *The Botanists*. St Paul's Bibliographies, Winchester.
Armstrong, E.A. 1940. *Birds of the Grey Wind*. Oxford University Press.
Baker, J.R. 1976. Julian Sorrel Huxley 1887-1975. *Biographical Memoirs of Fellows of the Royal Society 22*: 207-238.
Bath Academy of Art. 1989. *Corsham. A Celebration. The Bath Academy of Art 1946-1972*. Michael Parkin Gallery, London, 1989. (Contains an account of the career of Clifford and Rosemary Ellis).
Beaufoy, S. 1947. *Butterfly Lives*. Collins, London.
Blunt, W. 2nd ed. 1971. *The Complete Naturalist – A Life of Linnaeus*. Collins, London.
Blunt, W. 1983. *Married to a Single Life*. Collins, London. (Autobiography).
Blunt, W. 1986. *Slow on the Feather*. Collins, London. (2nd volume of autobiography).
Blunt, W. & Stearn, W.T. 1994. *The Art of Botanical Illustration*. New edition revised and enlarged. Antique Collectors' Club in association with The Royal Botanic Gardens, Kew. (contains introduction by Stearn about Wilfred Blunt).
Boyd, A.W. 1946. *The Country Diary of a Cheshire Man*. Collins, London.
Boyd, J. Morton. ed. 1979. *The Natural Environment of the Outer Hebrides*. Royal Society of Edinburgh.
Boyd, J. Morton & Bowes, D.R. 1983. *The Natural Environment of the Inner Hebrides*. Royal Society of Edinburgh.
Boyd, J. Morton. 1986. Fraser Darling's Islands. Edinburgh University Press. (a part biography of Darling).
Boyd, J. Morton. 1992. *Fraser Darling in Africa: A Rhino in the Whistling Thorn*. Edinburgh University Press.
Brown, L. 1974. A bird of prey odyssey. In: *The Bird-Watcher's Book*, ed. J. Gooders. David & Charles, Newton Abbot.
Campbell, B. 1979. *Birdwatcher at Large*. Dent, London.
Chisholm, A. 1972. Kenneth Mellanby. The extreme English Moderate. *In: Philosophies of the Earth. Conversations with ecologists*. Chiswick & Jackson, London. 76-87.
Clapham, A.R. 1971. William Harold Pearsall 1891-1964. *Biographical Memoirs of Fellows of the Royal Society 17*: 511-540.
Clapham, A.R. 1980. Edward James Salisbury 1886-1978. *Biographical Memoirs of Fellows of the Royal Society 26*: 503-541.
Collins, W.A.R. 1946. The New Naturalist series. A major publishing project. *The Bookseller*, 18 April 1946: 588-592.
Corbet, P. 1962. *A Biology of Dragonflies*. Witherby, London.
Crowcroft, P. 1991. *Elton's ecologists. A history of the Bureau of Animal Population*. Chicago University Press.
Darling, F. Fraser. 1937. *A Herd of Red Deer*. Oxford University Press.
Darling, F. Fraser. 1940. *Island Years*, George Bell, London.
Darling, F. Fraser. 1955. *West Highland Survey: an essay in human ecology*. Oxford University Press.
Darling, F. Fraser. 1970. *Wilderness and Plenty*. BBC Reith Lectures 1969.
Department of Health for Scotland. 1947. *National Parks and the Conservation of Nature in Scotland*. Report by the Scottish National Parks Committee and Scottish Wild Life Conservation Committee, Cmnd. 7235. HMSO, Edinburgh.
Dewsbury, D.A. ed. 1985. *Leaders in the study of animal behaviour*. Bucknell University Press, Lewisburg. [Contains autobiographical chapter by Niko Tinbergen].
Dony, J.G. 1977. J. Edward Lousley (1907-1976). *Watsonia 11*: 282-286.
Elliott, J.M. & Humpesch, U.H. 1985. Dr T.T. Macan 1910-1985: a short biography. *Archiv. f. Hydrobiologie*, Bd. 104: 1-12.
Elmes, G.W. & Stradling, D.J. 1991. Michael Vaughan Brian (1919-1990). *Insect Soc. 38*: 331-332.
Elsden, S.R. 1982. Roy Markham 1916-1979. *Biographical Memoirs of Fellows of the Royal Society 28*: 319-345.
Fisher, J. 1940. *Watching Birds*. Penguin Books, Harmondsworth. New ed. 1974, revised by Jim Flegg and illustrated by Crispin Fisher. T. & A.D. Poyser, Berkhamsted.
Fisher, J. 1966. *The Shell Bird Book*. Ebury Press and Michael Joseph, London.
Freethy, R. 1983. Collins New Naturalist. In the Beginning. *Country-side*, autumn 1983, 183-189.
Garnett, A. 1970. Herbert John Fleure 1877-1969. *Biographical Memoirs of Fellows of the Royal Society 16*: 253-278.
Gilmour, J. 1944. *British Botanists*. Britain in Pictures. Collins, London.
Goldsmith, E. 1978. What makes Kenny [Kenneth Mellanby] run? *New Ecologist* May-June 1978: 77-81.

Gooders, J. ed. 1974. *The Birdwatcher's Book.* David & Charles, Newton Abbot. (contains an account of the writing of NN *British Birds of Prey* by Leslie Brown).

Goodier, R. ed. 1974. *The Natural Environment of Shetland.* NCC, Edinburgh.

Goodier, R. ed. 1975. *The Natural Environment of Orkney.* NCC, Edinburgh.

Hamilton, W.D. 1994. On first looking into a British treasure. W.D. Hamilton celebrates the Collins New Naturalist series. *Times Litt. Supp.* 12 August: 13-14.

Hardy, A.C. 1967. *Great Waters. A voyage of natural history to study whales, plankton and the waters of the Southern Ocean in the old Royal Research Ship 'Discovery'.* Collins, London.

Hardy, A.C. 1965. *The Living Stream: a restatement of evolution theory and its relation to the spirit of man.* Collins, London.

Hardy, A.C. 1966. *The Divine Flame: an essay towards a natural history of religion.* Collins, London.

Harrison, R. 1987. Leonard Harrison Matthews 1901-1986. *Biographical Memoirs of Fellows of the Royal Society 33*: 413-442.

Hayter-Hames, J. 1991. *Madame Dragonfly. The life and times of Cynthia Longfield.* Pentland Press, Durham.

Hinde, R.A. 1990. Nikolaas Tinbergen 1907-1988. *Biographical Memoirs of Fellows of the Royal Society* **36**: 549-565.

Hooper, J. 2002. *Of Moths and Men. Intrigue, Tragedy and the Peppered Moth.* Fourth Estate, London.

Hosking, E. 1970. *An Eye for a Bird.* (autobiography) Hutchinson, London.

Huxley, J.S. ed. 1940. *The New Systematics.* Clarendon Press, Oxford.

Huxley, J.S. 1942. *Evolution: the modern synthesis.* Allen & Unwin, London.

Huxley, J.S. 1970. *Memories.* Allen & Unwin, London. [Disappointingly little on the New Naturalists].

Huxley, J.S. 1973. *Memories II.* Allen & Unwin, London.

Imms, A.D. 1925. *A General Textbook of Entomology.* Methuen, London.

Kassanis, B. 1982. Kenneth Manley Smith 1892-1981. *Biographical Memoirs of Fellows of the Royal Society 28*: 451-477.

Keir, D. 1952. *The House of Collins. The story of a Scottish family of publishers from 1789 to the present day.* Collins, London.

Lipscomb, J. & David, R.W. eds. 1981. *John Raven by his friends.* Privately printed.

Locket, G.H. & Millidge, A.F. 1979. William Syer Bristowe 1901-1979. *Bull. British arachnological Society 4*(8): 361-365.

Marren, P. 1999. *The Observer's Book of Observer's Books.* Peregrine Books, Leeds.

Marren, P. 2003. *The Observer's Book of Wayside and Woodland.* Peregrine Books, Leeds.

Marshall, N.B. 1986. Alister Clavering Hardy 1896-1985. *Biographical Memoirs of Fellows of the Royal Society 32*: 223-273.

Matthews, L. Harrison. 1951. *Wandering Albatross. Adventures among the albatrosses and petrels in the Southern Ocean.* MacGibbon & Kee, London.

Matthews, L. Harrison. 1977. *Penguin: adventures among the birds, beasts and whalers of the far South.* Peter Owen, London.

Mellanby, K. 1976. *Talpa the Mole.* Collins, London.

Ministry of Town and Country Planning. 1947. *Conservation of Nature in England and Wales.* Report of the Wild Life Conservation Special Committee (England and Wales), Cmnd. 7122. HMSO, London.

Moore, N.W. 1987. *The Bird of Time. The science and politics of nature conservation.* Cambridge University Press.

Morton, B. 1992. Charles Maurice Yonge 1899-1986. *Biographical Memoirs of Fellows of the Royal Society 38*: 379-412.

Mountfort, G. 1991. *Memories of Three Lives.* Merlin Books, Braunton.

Nature Conservation in Britain. 1943. *Report by the Nature Reserves Investigation Committee.* Society for the Promotion of Nature Reserves, London.

Nethersole-Thompson, D. 1971. *Highland Birds.* Highlands and Islands Development Board, Inverness.

Nethersole-Thompson, D. & Watson, A. 1974. *The Cairngorms. Their Natural History and Scenery.* Collins, London.

Nethersole-Thompson, D. & Nethersole-Thompson, M. 1986. Waders: their breeding, haunts and watchers. T. & A.D. Poyser, Calton. (contains biographical preface by D. N-T.'s sons).

Nethersole-Thompson, D. 1992. *In Search of Breeding Birds. Early essays by Desmond Nethersole-Thompson reprinted from the Oologist's Record.* Peregrine Books.

Nicholson, E.M. 1957. *Britain's Nature Reserves.* Country Life, London.

Nicholson, E.M. 1970. *The Environmental Revolution. A Guide to the New Masters of the World.* Hodder & Stoughton, London.

Nicholson, E.M. 1993. Ecology and

conservation: our Pilgrim's Progress. *In*: F.B. Goldsmith & A. Warren eds. *Conservation in Progress*. Wiley & Sons, London, 3-14.

Oldham, T. 1989. The New Naturalist Series. *Books, Maps and Prints*, May 1989: 37-40.

Osborne, R.H., Barnes, F.A. & Doornkamp, J.C. eds. 1970. *Geographical essays in honour of K.C. Edwards*. Dept. of Geography, University of Nottingham. (includes a biographical introduction).

Pearsall, W.H. 1917-1918. The aquatic and marsh vegetation of Esthwaite Water. *J. Ecol* *5*: 180-202; *6*: 53.

Peterson, R.T. & Fisher, J. 1956. *Wild America. The record of a 30,000 mile journey around the North American continent by an American naturalist and his British colleague*. Collins, London.

Raven, J. 1971. *The Botanist's Garden*. Collins, London.

Raven, S. 1982. *Shadows on the Grass*. Blond & Briggs, London (for portrait of John Raven).

Rothschild, M. 1991. *Butterfly Cooing Like a Dove*. ('an autobiographical anthology') Doubleday, London.

Russell, E.J. 1956. *The Land Called Me*. (autobiography). Allen & Unwin, London.

Russell, F.S. & Yonge, C.E. 1928. *The Seas. Our knowledge of life in the sea and how it is gained*. Warne, London.

Salisbury, E.J. 1935. The Living Garden. G. Bell & Sons, London.

Salisbury, E.J. 1952. *Downs and Dunes. Their plant life and its environment*. G. Bell & Sons, London.

Scott, P. & Fisher, J. 1953. *A Thousand Geese*. Collins, London.

Sheail, J. 1976. *Nature in Trust. The history of nature conservation in Britain*. Blackie, Glasgow.

Sheail, J. 1985. *Pesticides and Nature Conservation. The British Experience 1950-1975*. Clarendon Press, Oxford.

Sheail, J. 1987. *Seventy-five years in ecology. The British Ecological Society*. Blackwell Scientific Publications, Oxford.

Shorten, M. & Barkalow, F. 1973. *The World of the Grey Squirrel*. Yale University Press.

Simms, E. 1976. *Birds of the Air. The autobiography of a naturalist and broadcaster*. Hutchinson, London.

Smith, M. 1947. *A Physician at the Court of Siam*. Country Life, London.

Stearn, W.T. 1981. *The Natural History Museum at South Kensington*. British Museum (Natural History), London.

Steers, J.A. 1946. *The Coastline of England and Wales*. Cambridge University Press.

Stone, E. 1988. Ted Ellis. *The People's Naturalist*. Jarrold, Norwich.

Summers-Smith, J.D. 1988. *The Sparrows: A Study of the Genus* Passer. T. & A.D. Poyser, Calton.

Summers-Smith, J.D. 1992. *In Search of Sparrows*. T. & A.D. Poyser, London.

Sweeting, G.S. ed. 1958. *The Geologist's Association 1858-1958: A history of the first hundred years*. Colchester.

Tabor, R. 1992. Urban Trailblazer. Richard Fitter in conversation with Roger Tabor. *Country-side*, 27-31.

Taylor, J.H. 1964. Sidney William Wooldridge 1900-1963. *Biographical Memoirs of Fellows of the Royal Society 10*: 371-388.

Tenison, W.P.C., Bellairs, A. d'A. & others. 1959. Obituary. Dr Malcolm Arthur Smith. *British Journal of Herpetology 2*: 135-148; 186-187.

Thornton, H.G. 1966. Edward John Russell 1872-1965. *Biographical Memoirs of Fellows of the Royal Society 12*: 457-477.

Tinbergen, N. 1958. *Curious Naturalists*. Country Life, London.

Walters, S.M. 1989. Obituary of John Scott Lennox Gilmour. *Plant Syst. Evol. 167*: 93-95.

Walters, S.M. 1992. W.T. Stearn: the complete naturalist. *Bot. Journal Linnaean Soc. 109*: 437-442.

Wigglesworth, V.B. 1949. Augustus Daniel Imms 1880-1949. *Biographical Memoirs of Fellows of the Royal Society 6*: 463-470.

Wigglesworth, V.B. 1982. Carrington Bonser Williams 1889-1981. *Biographical Memoirs of Fellows of the Royal Society 28*: 667-684.

Williams, C.B. 1930. *The Migration of Butterflies*. Oliver & Boyd, Edinburgh.

Williams, C.B. 1964. *Patterns in the Balance of Nature*. Academic Press, London.

Wilson, D. 1988. *Rothschild. A study of wealth and power*. Andre Deutsch, London.

Wilson, D.P. 1935. *Life of the Shore and Shallow Sea*. Ivor Nicholson & Watson.

Worthington, E.B. 1983. *The Ecological Century. A personal appraisal*. Clarendon Press, Oxford.

Yonge, C.M. & Thompson, T. 1930. *A Year on the Great Barrier Reef*. Putnam & Co., London.

Yonge, C.M. 1976. *Living Marine Molluscs*. Collins, London.

Appendix 1

An Updated New Naturalist Biography

These short notes list nearly all the authors of the New Naturalist library (apart from two or three that I failed to track down), and also include their publisher, editors and dust jacket designers. My sources have varied from entries in *Who's Who* and obituaries in *The Times* to *curricula vitae* and personal contact. The responsibility for their accuracy is, of course, mine alone. Within the limits of space, I have tried to emphasise the part of the writer's life that has a bearing on the corresponding New Naturalist title. The reader will also find phrases intended to connote the 'jizz' of the person concerned, as in a field guide. Those who are the subject of this method will, I hope, find facility in the method even when they deplore its execution. I have updated this Who's Who of the New Naturalist authors to include all the books up to number 94, *The Sea Shore*, and also to include details of ongoing careers as well as record the sad loss of several authors since this book was written ten years ago.

Altringham, John D. [Derek] b. 1954. Zoologist, university professor, author of *British Bats* (2003). Graduated from York (BSc 1978) and St Andrews (PhD 1981). Researcher in biomechanics and animal locomotion in the USA and UK, before joining staff of School of Biology at Leeds University, where he was successively lecturer, Reader, and now Professor of Biomechanics. Scientific Medal of the Zoological Society (1994) for work on fish swimming. An 'extracurricular 'passion' for bats led to a book (*Bats: biology and behaviour*, 1996, OUP), a second research theme in bat ecology, and increasing involvement in bat conservation. Counterbalances the frustrating aspects of university life by studying bats, and going caving with his 'dazzling team' of PhD students. Lives on the edge of the Yorkshire Dales National Park with his wife Kate and two children, and looks forward to the day he can move into the heart of the Dales.

Armstrong, Edward A. [Allworthy]. 1900-1978. Priest, theologian and naturalist, author of *The Wren (1955) and The Folklore of Birds* (1958). One of the latest in England's long line of distinguished parson-naturalists, and an authority on bird song and bird behaviour. An Ulsterman by birth, trained in philosophy and psychology in Belfast before taking Holy Orders. Apart from two years as Chaplain of St Andrew's in Hong Kong, spent all his ministry in England, in Yorkshire and from 1943-67 as Vicar of St Mark's in Cambridge. There he studied wrens in nearby gardens and woods, showing how intimate were the connections between song, courtship and breeding behaviour to the bird's environment and food-supply. Later applied these ideas to birds in the tundra of Iceland and Lapland, on offshore islands in West Scotland and Ireland, and in East Africa. His meticulous observation won him a lasting reputation as a great bird behaviourist and a posthumous place in the *Dictionary of National Biography*. Books included a classic *Birds of the Grey Wind* (1942), about Ulster, studies of Shakespeare and St Francis, suggesting that neither worked from personal observation, and original works on bird song and bird psychology. Bald, meticulous, with a perpetual sense of wonder, he believed that understanding the interconnectedness of living things helped gain better insight into the mind of the Creator.

Beebee, Trevor b. 1947. Herpetologist, author of *Amphibians and Reptiles* (2000 with Richard Griffiths). Devotee of amphibia since childhood. A Lancastrian displaced to southern England, he read biology at University of East Anglia and obtained PhD in biochemistry at University College London. Since 1976, lecturer in biochemistry at University of Sussex, and now Professor of molecular ecology there. Since 1970s active in research and conservation of Britain's rarer reptiles and amphibians, especially Natterjacks, from scrub-bashing to population genetics. Especially interested in the effects of habitat fragmentation on the genetic health of isolated animal populations. Founding trustee of the Herpetological Conservation Trust. Author of six books and 130 scientific papers. Also interested in how plant and animal distributions took shape in Europe after the last Ice Age. Admits to 'a minor obsession' with the larger kinds of water-beetles and bugs, 'chasing them with unwieldy pond nets'.

Berry, R.J. [Robert James, called 'Sam'] b. 1934. Geneticist, naturalist and Christian, author of *Inheritance and Natural History* (1977), *The Natural History of Shetland* (1980, with J.L. Johnston) and *The Natural History of Orkney* (1985). Educated Shrewsbury School and Caius College, Cambridge, gaining PhD in 'slipped disc and water on the brain in mice' (1976). Lecturer, then Reader, then Professor of Genetics Royal Free Hospital School of Medicine 1967-78; Professor Genetics University College, London 1978-2000. Member human fertilisation and embryology authority 1990-96, NERC 1981-87, and president at different times of Linnaean Society, British Ecological Society, Mammal Society and Christians in Science. FRSE (1981). Much involved in ethics and, as a member of the General Synod of Church of England 1970-90, reconciling science and Christian religion. Chose islands for research for their genetically isolated populations, but likes life in remote places. Has also studied rats on radioactive sands in India and limpets on the Antarctic shore. Books include *Orkney Nature* (2000), *Science, Life and Christian Belief* (1998) and *God's Book of Works* (2002). Intelligent and clear thinking, wry, gossipy. Believes in 'the first and last Adam'. Recreation: 'remembering hill-walking and then dreaming'.

Blunt, Wilfred [Jasper Walter]. 1901-1987. Artist, art teacher, writer and museum curator, author of *The Art of Botanical Illustration* (1950). Polymath and aesthete, educated at Marlborough, Oxford and Royal College of Art. Initially, an experimental painter in Paris, became gradually more conservative specialising in calligraphy and botanical illustration. Taught art at Haileybury 1923-32, then Eton 1932-59; curator of Watts Gallery, Compton 1959-85. FLS (1969). From 1940s author of biography, travel, Islamic art and botany, including a notable life of Linnaeus (1971), a 'botanical novel' *Of Flowers and a Village* (1963) and two candid volumes of autobiography, *Married to a Single Life* (1983) and *Slow on the Feather* (1986). Cultured, genial, gossipy. Lifelong fascination with flowers, but not one to dirty his hands weeding: once posed by waist-high ornamental urn commenting: 'This is the sort of gardening I like.'

Boyd, A.W. [Arnold Whitworth]. 1885-1959. Countryman and naturalist, author of *A Country Parish* (1951). One of the last of the old-style amateur naturalists; agent for the family yarn business of James Boyd & Son. A Cheshire man to his fingertips, the Gilbert White of Great Budworth. Great field skills; among the most active pioneer bird-ringers, original papers on tree sparrow, greenfinch and swallow; put sewage farms and the Staffordshire lakes on the birdwatcher's map. Left vast archive of notebooks and nature diaries, all in his characteristic neat minuscule hand. Editor of *British Birds* 1944-58, and a leading light of the BTO, BOU and RSPB in 1940s and 1950s. His long-running column in *Manchester Guardian* resulted in *The Country Diary of A Cheshire Man* (1946). Master of country lore; observed ancient local customs, always raising his hat to magpies as he rode along in his high car. Served in Lancashire Fusiliers at Gallipoli, losing an eye and gaining a MC. The most modest of men, kind, affable and without enemies. Uncle and mentor of James Fisher.

Boyd, John Morton. 1925-1998. Ecologist and conservationist, author of *The Highlands and Islands* (1964, with F. Fraser Darling) and *The Hebrides* (1990, with his son, Ian Lamont Boyd). A pioneer of nature conservation north of the border, scientist, traveller, romantic and ardent lover of nature. Son of an Ayrshire master builder married to a textile designer, educated locally. Original plans to become an engineer interrupted by wartime service in RAF, after which contact with nature on camping trips led him to enrol at Glasgow University reading geology and zoology under his mentor, Maurice Yonge, going on to study earthworm ecology in the Hebrides for a PhD. Also studied seals, field mice, sea-birds and Soay sheep, and explored the islands of Ronay, St Kilda and Tiree (where he later owned a holiday home). Appointed the Nature Conservancy's first regional officer for the West Highlands and Islands 1957, becoming its Scottish director 1971-85. Somehow combined busy life as a scientific civil servant with travels to Jordan, Uganda and the Serengeti, Aldabra, Christmas Island and other ocean islands, which occupy half his memoirs. Much involved in reintroduction of sea-eagles, establishment of National Nature Reserves, impacts of North Sea oil and gas and other developments, as well as with the National Trust for Scotland, including its lecture cruises. On retirement became consultant on forestry and hydro-electric schemes. A man of many parts, writer, watercolourist, poet (he was among the few scientists with an honorary doctorate in literature), a compelling lecturer, a Church of Scotland Elder, and something akin to a nature mystic, responding to 'the voice in the mountains'. 'He led from his heart.' FRSE (1966), CBE (1987). 'Morton in full flow was a magnificent sight, his body seeming to shake with enthusiasm, his beard vibrating with the passion of it all … an enthusiasm, it has to be said, which was often unstoppable.' His posthumous memoirs are *The Song of the Sandpiper* (1999).

Brian, M.V. [Michael Vaughan, called 'M.V.B.']. 1919-1990. Entomologist and ecologist, author of *Ants* (1977). Lifelong student of social insects as lecturer in zoology at Glasgow University 1946-53 and officer-in-charge of Furzebrook Research Station 1953-82. 'A true intellectual', developed use of statistics and conceptual ecology as well as advancing his own special subject: the ecology and behaviour of British ants. OBE (1982). Painter and pianist; keen gardener and beekeeper. Bearded, guarded expression; 'a modest man with a retiring disposition'. Loved gossip, but kept close about his own (several-times married) life.

Bristowe, W.S. [William Syer, called Bill]. 1901-1979. Industrial administrator and doyen of spider studies, author of *The World of Spiders* (1958). Unrivalled in studies of spider behaviour which began at age of five: 'our first spider ecologist'. Science graduate at Cambridge (though then more interested in sport, especially athletics). Joined Brunner Mond (later incorporated into ICI), retiring in 1962, after 14 years as head of central staff department. Well-travelled in pursuit of spiders, especially to islands, mountains and the Far East. Over 100 publications on spiders, over 56 years, including the two-volume *Comity of Spiders* (1939, 1941). Gained doctorate in science from Cambridge for work on spiders. Also wrote series of well-researched articles on neglected early entomologists such as Joseph Dandridge and Eleanor Glanville. Possessor of most of the virtues of a great observer: good memory, patience, accuracy and a logical mind; clear and entertaining writer; the only thing he couldn't do was draw. Ploughed a lone furrow as a naturalist. Non-arachnological interests, indulged mainly in retirement, included genealogy, antiquities, gipsy caravans and Sherlock Holmes.

Brown, Leslie [Hilton]. 1917-1980. Ornithologist, agriculturalist and writer, author of *British Birds of Prey* (1976). Of Scottish descent, he was born in India and spent most of his life in Africa, in Nigeria and Kenya, which was his adopted home after 1946. Educated at Oundle School, graduated in zoology at St Andrews University and went on to study tropical agriculture at Cambridge and Trinidad. Received PhD from St Andrews in 1973. Joined Colonial Agricultural Service 1940; from 1956, Deputy Director for Agriculture, then Chief Agriculturalist in Kenya, retiring in 1963 (OBE 1963). Thence an agricultural advisor for World Bank and other aid organisations, and also studied birds, especially pelicans, flamingoes and eagles (Crowned Eagles nested in his garden at Karen house, Kenya). Indefatigable, despite intermittent ill-health, wrote a dozen books and 30 scientific papers for *Ibis* and other journals in 1970s alone. His masterworks were the two-volume *Eagles, hawks and falcons of the world* (1968), and the first volume of *The Birds of Africa* (1982). A great raconteur, clever, courageous, a born communicator, 'not a glad sufferer of fools, pompous landowners, constipated functionaries, imperceptive editors or slow publishers'.

Brown, Margaret E. [Mrs M.E. Varley, called Peggy]. b. 1918. Fish biologist and university lecturer, author of *The Trout* (1967 with Winifred Frost). Educated Malvern Girls School and Girton College, Cambridge, gaining PhD (1945) on the nutrition and growth of trout, which began as a contribution to wartime food production. Lecturer and demonstrator in zoology and Fellow of Girton College 1946-50; visiting scientist at fish research laboratory in Uganda 1950-51; lecturer at Kings College, London, 1951-55, thence lecturer and demonstrator in zoology and supernumerary Fellow of St Hilda's College, Oxford, and tutor to Open University. Married Professor George Varley 1955. Research primarily experimental, but also studied wild fish as biologist to the Salmon and Trout Association (though she never became an angler). Served on Councils of Freshwater Biological Association, Institute of Biology, Linnaean Society etc. and an active contributor to the International Biological Programme in 1960s and 1970s. Visiting Buckland professor, writing *British Freshwater Fishes* based on her lectures.

Butler, Colin G. [Gasking]. b. 1913. Entomologist, author of *The World of the Honeybee* (1954) and *Bumblebees* (1960 with John B. Free). Authority on bees of world repute. Graduate of Queens College, Cambridge, life career at Rothamsted where Head of Bee Dept 1943-72 and Head of Entomology 1972-76. OBE 1970, FRS 1970. Discoverer of 'the queen substance', the mystery at the heart of bee behaviour. Active member of Royal Photographic Society, Royal Entomological

Society, beekeeper's associations and international social insect study groups. First-rank nature photographer and pioneer of close-up work on living insects.

Buxton [Edward] **John** [Mawby]. 1912-1989. Scholar, university teacher and ornithologist, author of *The Redstart* (1950). Read Greats at New College, Oxford (Fellow 1949), but academic career interrupted by war. Joined commandos, taken prisoner in Norway and spent next five years in prison camps watching birds, especially redstarts. On return to Oxford, taught history and literature, specialising in sixteenth and seventeenth centuries. Active promoter of bird observatories and ringing; introduced mist-net to Britain. A cultured man, impressive looking; quiet mannered but another non-sufferer of fools: 'a snort of contempt was characteristic'. Owned the best handwriting of the New Naturalist authors.

Cabot, David [Boyd Redmond] b. 1938. Irish naturalist and writer, author of *Ireland* (1999). Educated at Dartington Hall School, Devon and Trinity College, Dublin where he studied natural sciences, going on to University College, Galway, where obtained a PhD in ecology of bird parasites. For twenty years worked as ecologist, later Head of Conservation Research in the State Institute for Physical Planning (An Foras Forbartha) in Dublin. There responsible for Ireland's National Heritage Inventory and developing nature conservation policy, later also as special environmental advisor to two Irish Prime Ministers. More recently helped develop environmental policies in countries soon to join the European Union, as well as independent consultant at home in Ireland. All-round field naturalist and ecologist, specialises in seabirds and wildfowl, enjoying the arduous life in Arctic and other wild spots. Film maker, author of 11 books, one-time university lecturer, and presenter on radio and TV. Collects antiquarian books. Probably the best Irish naturalist since Praeger. Believes 'life on the edge keeps you sharp'.

Campbell, Bruce. 1912-1993. Naturalist, writer and broadcaster, author of *Snowdonia* (1949 with F.J. North and Richenda Scott). English-speaking Highlander, one of the first field naturalists who was also a trained scientist – he graduated in forestry but gained doctorate in comparative bird studies. Learned his birds in Ardnamurchan, but mostly lived in Wales or Oxfordshire. Sometime schoolmaster, first full-time secretary for BTO 1948-59, active on BOU, British Ecological Society and conservation bodies. Expert bird-nester, pioneered nest-box study of pied flycatcher at Nagshead, Gloucestershire. Writer, broadcaster, BBC producer, natural history adviser to *The Countryman*. Books included an excellent *Dictionary of Birds* (1985 with Elizabeth Lack) and an autobiography, *Birdwatcher at Large* (1979). OBE (1976). Knew everything about wildlife, genial, inveterate list-maker, socialist.

Campbell, R. [Ronald] **Niall.** b. 1924. Zoologist and conservation officer, author of *Freshwater Fishes* (1992 with Peter S. Maitland). Lifelong Highland naturalist, younger brother of above. Graduate in zoology at Edinburgh, infantry officer and instructor 1942-7; fish and fisheries researcher 1951-63; Nature Conservancy regional officer for NW Scotland 1966-84; now consultant for fish and wildlife biology. Expert on trout, charr and sticklebacks; angler and countryman, founder member of Fisheries Society of British Isles and Institute of Fisheries Management. Proud Campbell, wry, funny, laid-back.

Chapman, Philip [called Phil]. b. 1950. Ecologist, photographer and BBC producer, author of *Caves and Cave Life* (1993). Born in West Africa, graduated at Bristol University. Eleven caving expeditions, mainly to tropics, and the publication of field studies established him as an authority on cave wildlife. Conservation worker with IUCN and WWF. Sometime school teacher, education officer and university researcher, now working for BBC.

Clay, Theresa [Rachel]. b. 1911. Entomologist, author of *Fleas, Flukes and Cuckoos* (1952 with Miriam Rothschild). Educated at St Pauls School, graduated in zoology at Edinburgh University. World authority on bird lice. Entomologist on scientific expeditions to Africa, Middle East and Arctic with Colonel Meinertzhagen and others 1935-38 and 1946-49. Unofficial worker at Natural History Museum from 1938, becoming full-time senior staff member from 1949, cataloguing Museum's bird-lice collections. In 1955 promoted to Principal, in charge of Apterygote insects, lice and termites. Deputy Keeper of Entomology 1970-75. On retirement, married Mr R.G. Seabright.

Collins, W.A.R. [William Alexander Roy, called 'Billy']. 1900-1976. Publisher of the New Naturalist library 1943-76. Born into the family business, gradually taking over management in 1930s. More interested in book publishing than the bibles/stationery side of business. Built up contemporary fiction list and led firm towards religious publishing, world literature, paperbacks and natural history. Bought and restored Hatchards bookshop in Piccadilly. Scottish businessman: commercial, non-intellectual, hard working, nose for detail; drove a hard bargain and knew what he wanted. Patriarchal ruler, 'abrupt, impatient, but his was a kind heart'. CBE 1966, knighted 1970. Keen naturalist, sportsman and gardener. Said to have regarded the New Naturalist library as his most significant achievement.

Condry, William M. [Martin], called 'Bill'. 1918-1997. Naturalist and writer, author of *Snowdonia National Park* (1966) and *Natural History of Wales* (1981). Born in Birmingham, son of a socialist diamond-setter, but spent most of his life in Wales. He read French and English at Birmingham and London Universities, and taught at a prep school before taking up a career of nature writing and conservation. Helped establish bird observatory on Bardsey in 1952, and was warden of RSPB's Ynys-hir nature reserve 1969-81. Always a clear and graceful writer, he was a columnist for *Country Life*, nature correspondent in *The Guardian*, and in demand as a lecturer and broadcaster. His books include biographies of Thoreau, a strong influence, and the Welsh botanist, Mary Richards, as well as on travels in Africa and in his adopted home in the Welsh hills. His autobiography, *Wildlife, My Life* (1995) is a gentle celebration of a lifelong love of nature.

Corbet, Philip S. [Steven]. b. 1929. Entomologist and university professor, author of *Dragonflies* (1960, with Cynthia Longfield and N.W. Moore). Born in Malaya, son of a well-known lepidopterist, graduating at Reading University and gaining PhD in 1953 (St Johns, Cambridge) on the seasonal ecology of dragonflies. A true Commonwealth scientist, having held research posts and senior academic appointments in four continents. Research on fisheries and mosquitoes in Uganda 1954-62; entomologist at research institutes in Canada from 1962, becoming Professor of Biology, Waterloo University (Ontario) 1971-74. Set up and directed first centre for environmental studies, Canterbury, New Zealand 1973-78; Professor of Zoology, Dundee University 1980-90. Major research interests centred on mosquito and dragonfly biology. Author of *A Biology of Dragonflies* (1962), now being revised and updated. Has served on numerous advisory bodies in Canada, New Zealand and Scotland concerned with public health, pest management and nature conservation. Still working on dragonflies and mosquitoes as research fellow in Edinburgh. Bearded, meticulous, thoughtful, a quiet, pleasant manner concealing incisive mind and great energy.

Darling, Frank Fraser. 1903-1979. Animal ecologist, conservation prophet and environmental sage, author of *Natural History in the Highlands and Islands* (1947). Legendary career: trained in agriculture and animal genetics, but made his name in 1930s studying deer, seals and seabirds in remote places and writing books about them. Pacifist, isolated on Summer Isles during war, then promoter and director of West Highland Survey 1944-50. 1950s travelled widely among US and African National Parks, neglecting his home duties at Edinburgh University. Appointed vice-president US-based Conservation Foundation 1959-72. Reputation at home restored with 1969 Reith Lectures *'Wilderness and Plenty'*, honorary degrees and a knighthood (1970). A much-admired but enigmatic man, melancholic, reflective, perhaps more influential as a philosopher than scientist. Lived primitive life on islands, but liked his comforts – books, claret, pets and collections of jade, carpets, bronzes and glassware from around the world. See: *Fraser Darling in Africa: A Rhino in the Whistling Thorn* (1992) by Morton Boyd, *Fraser Darling's Islands* (1986), also by Boyd.

Davis, Brian [Noel Kittredge]. b. 1934. Ecologist, author of *The Soil* (1992 with Norman Walker, David Ball and Alastair Fitter). Educated at Sherborne and Aberdeen University reading zoology. Postgraduate research at Nottingham on soil fauna of reclaimed ironstone workings (PhD 1961). First member of well-known Toxic Chemicals and Wildlife team under Norman Moore at Monks Wood, later specialising in ecology and biogeography of plant-eating insects, habitat creation and restoration, and environmental effects of spray drift. Deputy Head of the field station 1984-94. Editor of *Biological Conservation*. Good field naturalist in the old Monks Wood tradition, as well as professional ecologist. Enjoys hiking in mountain scenery and playing the cello.

Edlin, H.E. [Herbert Leeson, called 'Bill']. 1916-1976. Forester and writer, author of *Trees, Woods and Man* (1956). Mancunian, graduated as forester; early career as rubber planter in Malaya and district officer in New Forest. Best known as prolific populariser of forestry. FC Publications Officer 1945-76; from 1944 author of stream of books on trees and forestry, including the excellent *British Woodland Trees* (1944), *Woodland Crafts in Britain* (1948) and *Collins Guide to Tree Planting and Cultivation* (1970). MBE (1970). Active member of all four British forestry societies. Temperate, quiet, friendly, deaf. 'His conversation was punctuated by humorous anecdotes, but these were never reflected in the serious business of writing.'

Edwards, K.C. [Kenneth Charles]. 1904-1982. Geographer, university teacher and promoter of regional studies, author of *The Peak District* (1962). Entire career spent at Nottingham University where appointed by Swinnerton and successively lecturer, reader and Professor of Geography 1925-69. Promoter of field studies, introducing 'easter camps' in 1930s, and active member of ramblers' youth hostel and footpath associations. Took part in Dudley Stamp's land utilisation survey, and much involved in regional and industrial planning. Many academic ties abroad, especially Luxembourg (subject of his doctorate 1947). CBE (1969). Ruled his department with paternal kindness; called 'K.C.' Dependable and encouraging, 'the man who kept the geographer's feet solidly on the real earth'.

Ellis, Clifford [Wilson]. 1907-1985. Art educationalist, designer (with his wife Rosemary) of the New Naturalist dust jackets. Qualified and later taught at The Polytechnic, Regent Street School of Art. Moved to Bath (and later Corsham Court) where Principal of Bath Academy of Art 1938-72. From 1930s, he and Rosemary designed series of lithographic posters, mosaics, book dust jackets and art wallpaper, always under the joint cipher 'C & RE'. An influential promoter of art, which he interpreted very broadly; taught students a sense of vocation 'like a gardener tending his plants'. Exceptional in his range of sympathies which included the history of art and natural history. Gifted with an acute visual memory, used words with great precision. Ruled Bath Academy with paternal benevolence. Voracious reader, long distance walker, habitué of galleries, museums and zoos. 'He had style.'

Ellis, [Dorothy] **Rosemary.** 1910-1998. Artist, teacher and co-designer, with her husband Clifford, of the New Naturalist dust jackets. Daughter of cabinet maker, Frank Graham Collinson, married Clifford Ellis in 1931 while art student in London and shared in all his work as artist and teacher. Taught art at Longleat during war, afterwards joining Clifford on the staff at Bath Academy of Art at Corsham Court, where specialised in art education. A quick, critical mind; a natural sense of colour, tone and composition is evident in her individual sketches, pottery and poster work. Later became skilled photographer. Retired with Clifford in 1972.

Ellis, E.A. [Edward Augustine, called 'Ted']. 1909-1986. Naturalist, writer and museum curator, author of *The Broads* (1965). Self-taught natural history all-rounder, Keeper of Natural History at Norwich Castle Museum 1928-54. Daily diarist for Eastern Daily Press for 40 years; from 1964, monthly diarist in *The Guardian* and regular broadcaster on radio (BBC Midland's 'Nature Postbag') and, latterly, on local television. From 1945, lived by Wheatfen Broad, which he preserved as a nature reserve, becoming a popular venue for school parties and excursions. Sometime president of both Norfolk and Suffolk naturalists trusts. Conferred DSc *Honoris causa* by University of East Anglia, 1970. A local hero: 'the people's naturalist'. Cottage, archive and surrounding fen now preserved as field centre by Ted Ellis Trust. Like many great all-rounders had a speciality – micro-fungi – discovering a number of new species. Charming and lovable man; bird-like features; smoked pipe; wore hat of coypu fur; memoir: *The People's Naturalist – Ted Ellis* (1988).

Fisher, Crispin. 1941-1989. Editor, graphic designer and artist, editor of New Naturalist library 1982-1989. Son of James Fisher, studied art at Eton under Wilfred Blunt and later at Leicester College of Art. Book designer and illustrator with Hodder and Stoughton, later Andre Deutsche and Phaidon. Joined Collins as head of natural history 1982, responsible for the Attenborough TV books and introduced the New Generation field guides. Illustrated his father's classic books *The Migration of Birds* (1966) and *Watching Birds* (1975). Charming and conscientious, a chip off the old block.

Fisher, James [Maxwell McConnell]. 1912-1970. Ornithologist and writer, among the foremost advocates of field study and conservation through publishing, wireless broadcasts and television. Editor of New Naturalist series 1943-70; author of *The Fulmar* (1952) and *Sea-birds* (1954 with Ronald Lockley). One of the great naturalists; as a popular influence second only to Peter Scott. Made his name with

Watching Birds (1940) and early work with BTO on nest records and sea-bird census. Tall and broad-shouldered, climber and island-hopper; member of dozens of conservation bodies (he could be said to have 'collected' them); leading light of RSPB and IUCN; member of National Parks Commission, which led to being vice-chairman of the newly-formed Countryside Commission 1967-70. Bibliophile and assiduous note-taker, learned in bird art and literature, amassed vast archive material on population changes of fulmar, gannet and collared dove; also expert on fossil birds. Affable and charming, 'a master of one-upmanship' (Stephen Potter), 'a natural communicator'. A rare combination of brilliance, dynamism and affability; life crowded with 'what-ifs', including its tragic end in a road accident. Time we had a biography, published by Collins.

Fitter, Richard [Sidney Richmond]. b. 1913. Writer and editor of natural history books, author of *London's Natural History* (1945) and contributor to *Birds of the London Area* (1957). Founder, in Britain, of the modern illustrated field guide with 'a genius for compression'. Graduate of London School of Economics, writer of Political and Economic Planning reports, later joining *The Countryman* as sub-editor before going freelance as full-time writer. The most influential urban naturalist since W.H. Hudson. Active with his wife Maisie on many conservation bodies, most notably the Flora and Fauna Preservation Society (Secretary 1964-81). Author of over two dozen books on all aspects of British, European and world natural history, best known for his Collins pocket and field guides. Friendly and well-liked all-round naturalist, lifetime devoted to observing 'wild and human life'.

Fleure, H.J. [Herbert John]. 1877-1969. Geographer, anthropologist and university teacher, author of *A Natural History of Man in Britain* (1951). From a long-lived Channel Island family (his father was born in 1803), but loyal to his adopted Wales, where Professor of Geography and Anthropology at University College 1917-30; later Professor of Geography at Manchester University 1930-44. One of the great geographers by dint of his uniquely broad approach; memorable teacher, first geographer elected FRS (1936). Student of man's physical and social evolution; writer (especially after retirement) and editor of *The Corridors of Time* 1927-56. Numberless international honours. Wretched health in childhood, frail but indestructible as adult: 'a well-revered figure', one-eyed, with 'sharply pointed white beard and strong head of creamy white hair'. Methodist, fond of music and family life. Worked till he dropped, 'a quiet personality concealing an unending store of energy'.

Ford, E.B. [Edmund Brisco but to intimates he was 'Henry Ford'] 1901-1988. Geneticist and entomologist, author of *Butterflies* (1945) and *Moths* (1955). Born in Cumberland but career based at Oxford, by turns lecturer, reader 1938-63, director of genetics laboratory and Professor of Genetics (1963-69). Fellow of All Souls, FRS (1946). Founder of environmental genetics through studies in natural variation of butterflies and moths. His *magnum opus*, *Ecological Genetics* (1964), was fruit of 30-years' research. In speech, as in writing, sharp, meticulous, clear-headed, parochial, prejudiced. Disliked newspapers, fish and chips, women students, impertinent students. Liked medieval churches, heraldry, Oxford, butterflies and white Cinzano. First-rate mind, selectively generous, gimlet gaze, dark horse, always uncompromisingly himself: 'you never forgot Henry Ford'. A hero of this book.

Frazer, [John, Francis] **Deryk.** b. 1916. Zoologist and conservationist, author of *Reptiles and Amphibians* (1983). Born London, the son of a prominent professor of anatomy. Educated at Lancing College and Merton College, Oxford where he read mathematics before switching to physiology. He also qualified as a medical doctor. Wartime service as sole GP onboard HMS Helford, where he played rugby in four countries. Combined lecturing in physiology at Charing Cross Hospital with animal study, specializing in growth rates and reproductive cycles. Also completing a PhD on toad movements. Moved to rural Kent in 1954, where helped to found the Kent Field Club, which he chaired for many years, and the Kent Naturalists Trust (now Kent Wildlife Trust). In early 1960s switched career to become a conservation officer in the Nature Conservancy, later serving as its head of International Conservation, retiring in 1978. Spent much of the next ten years surveying newt ponds in Kent as well as rescuing amphibians threatened by development. Sometime President of Mammal Society and British Herpetological Society. Married, four daughters. Also wrote *The Sexual Cycle of Vertebrates* and a volume on amphibians for the Wykeham Scientific Series.

Free, John B. [Brand]. b. 1927. Entomologist, author of *Bumblebees* (1959 with Colin Butler). Lifelong promoter of bees and beekeeping, doctorate under Colin Butler on bumblebees; permanent research worker at Rothamsted 1951-87, working on social life of bees and crop pollination. Hon. Professorial Fellowship, University of Wales (1984). Well-travelled, advising on beekeeping in the tropics; academic ties in Canada. Author of five books about bees and pollination.

Frost, Winifred E. [Evelyne, called 'W.E.F.']. 1902-1979. Fish biologist, author of *The Trout* (1967 with Margaret Brown). Cheshire-born biologist, graduating in zoology at Liverpool University, thence to Irish Fisheries Dept, Dublin (1929-39) where she produced a series of papers on trout and the river Liffey, and was taught fly-fishing by Rowland Southern. Fisheries scientist at Freshwater Biological Association, Windermere 1939-69, working on food and ecology of charr, pike, eels, trout and minnows. Awarded doctorate by Liverpool University 1944. Studied eels in Kenya 1948/49. Leading member of Salmon and Trout Association; President of Windermere District Angling Association. Well known in fish biology circles world-wide. Unmarried, 'a great joker', noted for droll and sometimes 'outrageous' sayings, both intentional and unintentional ('I want to marry an Irish peer with good alkaline water').

Gilbert, Oliver [Lathe] b. 1936. Lichenologist, author of *Lichens* (1999). Educated at Exeter University, Imperial College, London and Newcastle University, where he obtained his now-classic PhD study of air pollution and lichens. Lifelong interest in applied ecology with special reference to lichens, began as deputy warden at Malham Tarn Field Centre, later as lecturer, and then Reader in landscape, in the Department of Landscape Architecture, Sheffield University. A true naturalist with a good eye allied to literacy gifts. 'Taught his landscaper students to respect the environments they were manipulating.' British Lichen Society editor and past-President. His books include *Habitat Creation and Repair* (1998, with Penny Anderson) and *The Ecology of Urban Habitats* (1989), as well as a lichen flora of Northumberland. Unlike most leading lichenologists, his approach is habitat-based, specialising in neglected mountains, islands, freshwater and urban places. Cheerful disposition, fond of hill walking, rock climbing and reading poetry.

Gillmor, Robert [Allen Fitzwilliam] b. 1936. Wildlife artist, designer of the hardback dust-jackets from 1985 to date. Based at Reading until 1998 when he moved to north Norfolk, he is well-known for his book and magazine illustrations and linocut prints. Educated at Leighton Park, Reading, and at The Fine Arts Dept, Reading University, he returned to the former to teach, 1959-65. Freelance artist since 1965. Founder member and later President of Society of Wildlife Artists. Past Council member of three national bird societies, active member of local ornithological club and Reading Guild of Artists. Art editor of the nine-volume *Birds of the Western Palearctic.* Held joint exhibitions with his wife, the landscape artist Susan Norman in 1992 and 1999. A versatile artist, he works in a variety of mediums, with pen and ink, watercolour, linocuts etc where his well-observed and expressive work is easily recognised as 'a Gillmor'. Likes to paint finished watercolours from life, one eye on the bird, the other on his sketch pad.

Gilmour, John [Scott Lennox]. 1906-1986. Plant taxonomist and horticulturalist, author of *Wild Flowers* (1954 with S.M. Walters) and editor of the New Naturalist series 1943-79. Cambridge graduate and later Fellow of Clare College, made his reputation in scientific horticulture and systematics. Assistant Director at Kew 1931-46, at Wisley 1946-51 and Director of University Botanic Gardens, Cambridge 1951-73. Though involved internationally in the naming of plants, remembered as a pioneer conservationist and association with a group of talented Cambridge botanists in 1950s; he was 'socially cohesive': affable, humanist, courteous; late in life prone to melancholy.

Goldring, Frederick [called 'Fred']. 1897-1997. Guest-house owner and photographer, author of *The Weald* (1953 with S.W. Wooldridge). Lived in the Weald from age of three; ran Timberscombe Guest House 1926-59, near Midhurst, much frequented by London University field excursions. Talented amateur photographer, sometime recorder of Wealden churches and historic buildings.

Griffiths, Richard b. 1957. Herpetologist, author of *Amphibians and Reptiles* (2000, with Trevor Beebee). Born in London, his first encounters with amphibians were as a boy fishing in ponds in North London and Hertfordshire. Like Trevor Beebee, his lifelong interest in amphibia was first inspired by Malcolm Smith's classic New Naturalist title. Read zoology at Westfield College, University of London, and obtained PhD on amphibian behaviour at Birkbeck College, under John Cloudsley-Thompson. Research fellowships followed, and since 1999 he has been Senior Lecturer in Biodiversity Conservation at the University of Kent, specialising in conservation biology. Involved with amphibian and reptile conservation projects around the world and at home, where his main interests are Crested Newts and Natterjacks. Has penned over 500 herpetological publications, and is currently editor of the *Herpetological Journal.*

Hale, W.G. [William Gregson, called 'Bill']. b. 1935. Zoologist and university lecturer, author of *Waders* (1980). Lancashire ornithologist, based at Liverpool University from 1963, later dean of science faculty and Director of Natural Sciences. Holds personal professorship in animal biology (1979). Active in university life and biological education. Specialises in wader populations and their food. In 1973 began long-term study of redshank on the Ribble estuary and instrumental in securing that site as a nature reserve. Described graduate life as 'beer, ornithology, cricket and ideas'.

Hardy, Alister [Clavering, variously called 'Prof', 'Ali' and 'Uncle Mac']. 1896-

1985. Marine biologist and religious thinker, author of *The Open Sea: The World of Plankton* (1956) and *The Open Sea: Fish and Fisheries* (1959). Oxford graduate; spent much of 1920s afloat studying herring food in North Sea. Zoologist on *Discovery* expedition to the southern seas 1925-27. Successively Professor of Zoology at Hull (1928-42), Aberdeen (1942-5) and Linacre Professor at Oxford (1942-61). FRS (1940), knighted 1957. His business was always living marine animals; invented marine plankton tow-net; introduced fish tanks, aviary and a complete termite's nest to Oxford. After retirement, wrote series of books on marine biology and his ideas on natural religion. Founder of Religious Experience Unit researching 'encounters with the divine', the subject of his Gifford lectures 1963-66 and award of the Templeton Prize in 1985. Gifted writer and watercolourist, 'at heart a wide-ranging Victorian naturalist'. Tall and rangy, wild look behind spectacles, demonic energy, yet warm and courteous with permanent sense of fun: 'a lovable man'. Lectures notorious for well-worn jokes, including a stuffed 'mermaid'. 'Couldn't administrate himself out of a paper bag' but created a wonderful atmosphere for academic study out of sheer enthusiasm.

Harris, J.R. [J. Richard, called 'Dick']. b. 1910. Irish entomologist, fishing consultant and tackle merchant, author of *An Angler's Entomology* (1952). Keen angler and tier of flies from boyhood, sometime merchant seaman, journalist and freshwater biologist. Demonstrator in limnology, Trinity College, Dublin. A director of Garnetts & Keegans Ltd, firm of gunsmiths and fishing tackle. 'Perhaps the greatest living Irish angler-entomologist.' A large, affable man 'with a sharing attitude towards his whiskey and a colourful manner of expressing his trenchant views on fishing, fishermen, journalists, rugby, life and other matters'.

Harvey, L.A. [Leslie Arthur]. 1903-1986. Zoologist and university teacher, author of *Dartmoor* (1953 with D. St Leger Gordon). Zoological all-rounder and field biologist with lifelong association with Exeter, where lecturer and later Professor of Zoology 1930-69 at the then University College of South West. Graduated in zoology at Imperial College, London; postgraduate research on egg development in marine invertebrates. Always bore a heavy teaching load; began field courses under canvas with wife tackling the botany. From 1940s, involvement with a variety of conservation bodies, including Devonshire Association and Dartmoor National Park Committee. Later interests centred on seashore and islands, especially Scilly, where retired in 1969. Spent spare time botanising with wife, tending alpine garden, playing bridge and doing the crossword. Liked wild rugged places: 'ecology, my trade, is also my escape from the irksome burden of being human'.

Hayward, Peter J. [Joseph] b. 1944. Marine biologist, author of *Seashore* (2004). Man of Kent, though born in Lichfield. Began career as scientific assistant in Natural History Museum, where introduced to his lifelong specialism, marine bryozoa or sea-mats. Read zoology with geology at Reading University. Thence to University of Wales, Swansea as research student, gaining PhD in population biology and taxonomy of sea-mats. Now senior lecturer at the university, and authority on bryozoa worldwide from Antarctic to coral reefs. Author with Professor John Ryland on four volumes on marine bryozoa in Linnaean Society Synopses series. Co-author of the popular *Collins Pocket Guide to the Seashore*, co-editor and contributor to the masterly *Handbook of the Marine Fauna of north-west Europe* (1995), as well as the Naturalists' Handbooks on seaweed and sandy shore habitats. Zoological editor of Journal of Natural History. Away from natural history, is a 'collector of unconsidered trifles'. enjoys gardening, cooking and 'the occasional bottle of wine'.

Hepburn, Ian [called 'Hep']. 1902-1974. Schoolmaster and botanist, author of *Flowers of the Coast* (1952). Chemistry master at Oundle School 1925-64, for most of the time as Housemaster, latterly as Second Master. An early leading amateur ecologist, publishing scientific papers on coastal vegetation in Cornwall and Norfolk (which he had known from boyhood) and on the Northamptonshire limestone. Served on Council of British Ecological Society and active as journal editor; member of several local natural history clubs and trusts in Northants and Cambridgeshire. A modest man, kind, courteous, patient, sardonic, popular with students and staff. Devoted teacher, fond of music (the music prize at Oundle is called the Hepburn Prize).

Hewer, H.R. [Humphrey Robert]. 1903-1974. Zoologist and university teacher, author of *British Seals* (1974). Graduate and later lecturer, assistant professor and reader in zoology at Imperial College London 1926-64; professor, University of London from 1964. Early work on colour change in salamanders and fish, and variation in burnet moths. Energetic teacher, introducing biological films to classes, including a famous pioneer film on gannets, made with Julian Huxley; active member of British Association. Mammal work began in wartime as chief rodent officer to Agriculture Ministry, an association he later maintained as chairman of pest control and farm animal welfare committees (CBE 1971). Secretary-general International Congress of Zoology (OBE 1959). Work on seals began 1951, combining laboratory techniques with island camps, studying living seals in natural surroundings.

Hoskins, W.G. [William George]. 1908-1992. Local historian, writer and university teacher, author of *The Common Lands of England and Wales* (1963 with Dudley Stamp). Virtual creator of landscape and local history as a respectable discipline, best known for his classic *The Making of the English Landscape* (1955) and programmes for BBC television in 1970s. A Devon man born and bred, gained doctorate in Devon landowning; thence Professor of English Local History (a unique department) at Leicester 1948-51, returning briefly 1965-68; Professor of Economic History at Oxford 1951-65. CBE 1971. Over 20 books, incl. *Devon and its people* (1954), *The Midland Peasant* (1957), *Local History in England* (1959) and *English Landscapes* (1973), all in his delightfully spicy and well-matured style. Active on Common Lands commission, Dartmoor Preservation Association and numerous local historical societies. A man of strong pleasures and strong prejudices; retired from university life 'to escape the growing lunacy of administrative chores'. Recreation: 'parochial explorations'.

Hosking, Eric. 1909-1991. Bird photographer and lecturer; photographic editor of the New Naturalist library 1943-91. *The* natural history photographer; started professionally in 1929 with ancient plate camera, used until 1947. First published collection *Friends at the Zoo* (1933) helped inspire children's zoo at Regent Park and swan photograph of the Swan Vesta matchbox. Regularly featured in weekly magazines from mid-1930s when treader of the lecture circuit. Pioneer of high-speed flash. Photographer on countless British, European and world birding expeditions, including those to Coto Donana, Bulgaria, Jordan and Pakistan in the 1950s and 1960s led by Guy Mountfort. OBE (1977). Single-minded: spent honeymoon on side of a Scottish mountain photographing eagles. Lost an eye to a tawny owl, hence title of autobiography: *An Eye for a Bird* (1970).

Huxley, Julian [Sorell]. 1887-1975. Ornithologist, writer and major influence on the new natural history; editor of the New Naturalist library 1943-75. Scion of a famously learned family, Oxford graduate in biology where lectured 1910-12, 1919-25. Well-known studies of courtship behaviour of grebes, divers and herons laid foundation for modern study of ethology; less well known were his laboratory studies on ontogeny with E.B. Ford and others. Secretary of Zoological Society 1935-42, much involved with London Zoo. Author of a stream of books and essays from 1930s on systematics, evolution and behaviour. Was then 'almost the only senior British zoologist who thought ecology and behaviour important, he made field natural history scientifically respectable again'. Among the most influential of applied scientists, chairman of Wildlife Conservation Special Committee 1945-47 and director UNESCO 1946-48. FRS 1938, knighted 1958. Stimulating, sagacious, sound judge, accomplished raconteur. A brilliant but inconstant man, usually sociable and humorous but occasionally paralysed by depression, inherited from his Arnold grandparents.

Imms, A.D. [Augustus Daniel]. 1880-1949. Entomologist and university teacher, author of *Insect Natural History* (1947). Cambridge graduate, cutting his teeth on systematics of midge larvae before leaving for India to teach. Returned to become reader in agricultural entomology, Manchester 1913-18; chief entomologist at Rothamsted 1918-31 and reader in entomology at Cambridge 1931-45. FRS 1929, Fellow of Downing College (1940). The authority on insect morphology and systematics, won lasting influence as author of the standard *General Textbook of Entomology* (1925). Lifelong asthmatic, whose somewhat dour, reserved manner was eventually 'warmed by the genial college atmosphere'. In all four phases of his career he was obliged to design new laboratories from scratch.

Ingram, David [Stanley] b. 1941. Plant pathologist, author of *Plant Diseases* (1999 with Noel Robertson). Educated at Yardley Grammar School, Birmingham, and University of Hull, where obtained PhD in plant pathology under Noel Robertson. Distinguished and busy academic career, alternating between Cambridge and Scotland. Lecturer in botany, later Reader in Plant Pathology, Cambridge University 1974-90; Regius Keeper, Royal Botanic Gardens, Edinburgh 1990-98 and now Master of St Catherine's College, Cambridge. Sits on all kinds of national and international committees and councils to do with horticulture, plant pathology and conservation, as well as editing several scientific journals. Part-time Professor of Horticulture at Royal Horticultural Society and President of British Society for Plant Pathology. OBE, FRSE (1993). Author and editor of academic books including *Encyclopaedia of Life Sciences* (1985), the nine-volume *Advances in Plant Pathology* (1982-93), *Molecular Tools for Screening Biodiversity* (1988) and *Science in the Garden* (2002). Also pens a science column in *The Herald*. Personable, relaxes listening to classical music and jazz, gardening and 'strolling around capital cities'.

Johnston, J. [James] **Laughton**. b. 1940. Conservationist and schoolmaster, author of *The Natural History of Shetland* (1980 with R.J. Berry). A Shetlander by birth; varied life as naval cadet, oilrig worker, poet, zoologist, primary headteacher. Served as NCC officer on Shetland 1968-75 and later as head warden on Rhum; now area officer for Perthshire for Scottish Natural Heritage.

Jones, J.W. [John William], called Jack. 1913-1983. Zoologist, author of *The Salmon* (1959). Welshman, graduated in zoology at Liverpool University, going on to complete PhD on 'studies on salmon'. After wartime service at a radar installation in Orkney, returned to Liverpool University zoology department as lecturer, later Reader in Fisheries Biology. Pioneer research on salmon spawning, later extended to trout and charr, involved long hours by riverside in winter filming salmon in a large observation tank, and resulted in a celebrated film. Other fish interests included eel fisheries and the gwyniad of Lake Bala. Expert on ageing fish by their scales. Established a field station at Lake Bala, and founded an influential school in freshwater fishery biology at the university. Founded coarse fisheries conferences at Liverpool, and, with Peter Tombleson, the Fisheries Society of the British Isles (1967), serving as its president, 1969-78. Helped launch Journal of Fish Biology (1969). OBE (1973) for services to fisheries. A stroke, followed by deteriorating health left him an invalid, and retired in 1979. 'Very Welsh', pugnacious, unassuming, loyal and generous to his circle, 'but would not suffer fools gladly'. Not an angler, but interested in music and painting.

Lack, Andrew b. 1953. Botanist and university teacher, author of *The Natural History of Pollination* (1996, with Michael Proctor and Peter Yeo). Son of ornithologist David Lack, and a keen naturalist from childhood. Graduated Aberdeen University, later gaining a doctorate on the pollination of knapweeds at Cambridge under Peter Yeo. Demonstrator and research fellow Swansea University 1979-86, now senior lecturer in ecology at Oxford Brookes University. Research interests include plant population genetics and the history and philosophy of biological science. Other publications include a tropical flora, *Illustrated flora of Dominica* (1998) and a textbook *Instant Notes in Plant Biology* (2001, with David Evans). Popular lecturer and 'enthusiastic identifier of shrivelled leaves'. Plays the violin, sings and enjoys long walks in good plant country.

Lockley, Ronald [Mathias]. 1903-2000. Writer, naturalist and island dweller, author of *Sea-Birds* (1954, with James Fisher). Born in Cardiff, son of a railway manager, educated at Cardiff High School, afterwards working with his sister on a chicken farm. A keen naturalist from boyhood, in 1926 he took a lease on Skokholm island and settled there, rebuilding the farmhouse from wreckage, studying birds, raising chinchilla rabbits, and practising self-sufficiency. He set up Britain's first bird observatory there in 1933, and was soon ringing more than 6,000 birds a year. With Julian Huxley he made one of the first professional nature films, *The Private Life of the Gannet* (1934), which won an Oscar, and later wrote definitive studies of *Shearwaters* (1942) and *Puffins* (1953). Worked for naval intelligence during the war, moved to Orielton, later bought the Field Studies Council. Wrote books on island life, travel and natural history, including a biography of Gilbert White and *The Private Life of the Rabbit* (1964), which inspired Richard Adams's novel, *Watership Down*. Helped plan Pembrokeshire National Park and found West Wales Field Society. Emigrated to New Zealand in 1977, where he continued to pursue his love of islands and birds. 'Mischievous and unsentimental,' well-preserved in old age, his face, said one admirer, had 'something of the serenity that comes from life in lonely places'.

Longfield, Cynthia [Evelyn]. 1896-1991. Naturalist and traveller, 'Madame Dragonfly', author of *Dragonflies* (1960 with Philip S. Corbet and N.W. Moore). Daughter of wealthy Anglo-Irish family with estates at Castlemary, Co Cork. Privately educated, self-taught naturalist living close by Natural History Museum where became lifelong Associate. Unmarried; one of the great natural history world travellers, starting with 1924 St George expedition to the Pacific Islands, and later taking in Mato Grosso, SE Asia, tropical and South Africa etc., collecting insects and plants. Member of local fire service during war, helped save Museum collections during the Blitz. Active member London Natural History Society and Royal Entomological Society; author of the standard *Dragonflies of the British Isles* (1937). Returned to Ireland after 'retirement', but continuing to travel into 1970s. Grandniece wrote a biography, *Madame Dragonfly* (1991).

Lousley, J.E. [Job Edward, called 'Ted']. 1907-1976. Botanist and banker, author of *Wild Flowers of Chalk and Limestone* (1950). The outstanding field botanist of his generation, by profession a departmental manager of Barclays Bank in London. Leading light of London Natural History Society and BSBI, serving latter in turn as Treasurer, Secretary and President. Expert on docks, compiler of Floras of London, Isles of Scilly and Surrey, amassed largest herbarium in private hands. Dedicated recorder, conservationist and seeker of rare flowers. Conservative, orderly and patient, could seem austere, though not in his writing. Frightened of cows.

Lowe, Frank A. [Aspinall]. 1904-1985. Ornithologist, author of *The Heron* (1954). Leading north-country naturalist, active member of Lancashire and Cheshire Fauna Society and wildlife correspondent for *Bolton Evening News* for 59 years (earning a Guinness record). Worked for family business of manufacturing chemists. Regional organiser for BTO crested grebe and heron enquiries, observed herons on Scarisbrook estates from tree-top hides. A modest man: 'I find it easier to write about herons than about Frank A. Lowe' – but a talented naturalist, photographer and writer.

Macan, T.T. [Thomas Townley, called 'Kit' or 'T.T.M.']. 1910-1985. Entomologist and field studies director, author of *Lakes in Lakes and Rivers* (1951 with E.B. Worthington). Brought up in SW England, educated at Wellington and Cambridge, graduating in biology. Member of 1933 oceanographical expedition to Indian Ocean, specialising in starfish. Joined Freshwater Biological Association 1935, becoming Deputy Director 1946 and remaining there until retirement in 1976, apart from war service as entomologist in Royal Army Medical Corps. At FBA, specialised first in corixid bugs, later in mayfly nymphs and gastropods, publishing series of keys and papers. Doctorate 1940 was mixture of bugs and starfish. Best known for long-term studies of streams and a moorland fishpond, continuing lake studies into his retirement from his yacht 'Sylvana' and private laboratory. Keen promoter of field studies, organised annual Easter field course at Windermere and scout camps. 'He rarely used the complex and highly expensive apparatus now common in research laboratories.' Wrote five books on freshwater biology, all used by students. Founding editor of journal, *Freshwater Biology*; leading roles in international association of limnology. Hon. lecturer Lancaster University and 1970s visiting professor at two American universities. Reserved but warmhearted; keen sailor; wife called him 'The Doctor'.

Maitland, Peter S. [Salisbury]. b. 1937. Zoologist and ecological consultant, author of *Freshwater Fishes* (1992 with Niall Campbell). Scottish freshwater ecologist, educated in Glasgow, where gained doctorate in freshwater ecology. Demonstrator, later lecturer at Glasgow University 1959-67, then joined Nature Conservancy (later ITE) 1967-86, specialising in freshwater habitats and fish biology, and in charge of IBP Project at Loch Leven. Senior lecturer St Andrews University from 1970s. Now consultant, working on conservation of salmonids and other rare fish. Past editor *Freshwater Biology*. Well-travelled in connection with freshwater research and probably the leading authority on wild British fish; author of the popular *Guide to the Freshwater Fish of Britain and Europe* (1977) and a textbook, *The Biology of Fresh Waters* [2nd ed. 1990].

Majerus, Michael [Eugene Nicolas]. b.1954. Geneticist and entomologist, author of *Ladybirds* (1994) and *Moths* (2002). Of Luxembourg and Swiss ancestry, educated at Merchant Taylors' School, graduated in botany and zoology at Royal Holloway College, London, going on to PhD in the control of larval colour in the Angle Shades moth. From 1980, Reader of Evolution at Cambridge University and Fellow of Clare College (1991). Active and enthusiastic communicator, prepared to talk to almost anyone prepared to listen. Pursuit of moths has taken him all over the world from arctic Lapland and alpine Canada to tropical Africa and Australia. Organises ladybird surveys, scientific consultant for wildlife films, enjoys teaching children about insects. Lives with Tina and three children in a house in a field and 'still likes it a lot'.

Manley, Gordon. 1902-1980. Meteorologist and university lecturer, author of *Climate and the British Scene* (1952). Pioneer of upland and polar weather studies. Leading member and sometime President of Royal Meteorological Society. Cambridge graduate, well-travelled in high latitudes in 1920s. During 1930s took long series of weather records from a hut on summit of Cross Fell 'personally experiencing the worst of the weather conditions of which he writes'. Geography lecturer at Cambridge and wartime officer of Cambridge University Air Squadron. Professor of Geography at Bedford College, London 1944-64. Recreation: 'travel among mountains'.

Markham, Roy. 1916-1979. Virologist, author of *Measles, Mumps and Mosaics* (1954 with Kenneth M. Smith). Cambridge graduate, joining the Plant Virus Station 1940 as assistant to K.M. Smith, where gained his PhD in 1944. Talented experimentalist, designing and testing own equipment, including 'the Markham Still'. Specialised in mosaic virus of turnips and in fundamental biochemical study of nucleic acids. A pioneer of use of electron microscope. FRS 1956. From 1967-79 Director of John Innes Institute, University of East Anglia. He brought a handyman's practical skills to laboratory equipment; more at home with research than administration. Outwardly a cheerful sociable red-faced man, but essentially private and self-effacing; left no papers.

Marren, Peter [Richard] b. 1950. Naturalist and writer, author of *The New Naturalists* (1995) and *Nature Conservation* (2002). Educated Loughborough College School, Exeter University and University College, London, where studied nature conservation. Worked for the Nature Conservancy Council 1977-91 in Scotland and England, as area officer and author-editor. More recently independent consultant, writer and journalist, writing for national newspapers, including obituaries of deceased naturalists for *The Independent.* Regular column in *British Wildlife* as 'Twitcher in the Swamp' and, formerly, *The Countryman.* Author of 14 books on topics ranging from wild flowers and woodlands to bibliography, military history and rural life. Not much of a naturalist compared with others in this series, but does his best; special interests include fungi, moths, and the history of natural history, including the lives of old as well as new naturalists.

Matthews, Leonard Harrison [called 'Leo' or L.H.M.]. 1901-1986. Zoologist and zoo director, author of *British Mammals* (1952) and *Mammals in the British Isles* (1982). 'Last of the great travelling naturalists', graduate of King's College, Cambridge, taking part in the *Discovery* expedition to southern oceans 1924-29 studying whales and elephant seals. Special lecturer, later research fellow Bristol University 1933-41, still travelling 1945-51. War-work on aircraft radar. Repute as the leading mammalogist led to appointment as scientific director of Zoology Society of London, in charge of London Zoo 1951-66. FRS 1954. After retirement wrote astonishing stream of books on whales, seals, penguins etc. *Wandering Albatross* is an account of his adventures in South Georgia and the South Seas. Never a 'lab hand', preferred to study animals in the field (or the sea) and helped popularise them through writings and broadcasts; founder member of Mammal Society. Great story teller, jovial and optimistic, 'when in detective mode and wearing his cape and carrying a swordstick, he always seemed ... an incarnation of Chesterton's *The Man who was Thursday*'.

Mellanby, Kenneth [called 'Ken']. 1908-1993. Entomologist, scientific administrator and writer, author of *Pesticides and Pollution* (1967), *The Mole* (1971) and *Farming and Wildlife* (1981), an editor of New Naturalist library 1971-86. Son of a professor, read natural sciences at London University and obtained doctorate at London School of Hygiene and Tropical Medicine (1933). Began career as a medical entomologist at the London School 1933-45, investigated causes and cure for scabies with help of a band of volunteers in Sheffield (*Human Guinea Pigs*, 1945): 'the man who saved us from "the itch".' Spent latter part of war in Burma and New Guinea investigating transmission and control of scrub typhus. Resumed academic career as Reader in Medical Entomology at London University, then as first principal of University College, Ibadan in Nigeria 1947-53, for which received CBE (1954). Head of entomology, Rothamsted 1955-61, then first director of Monks Wood experimental station, 1961-75, in both roles warning of unwanted side-effects of persistent pollutants, especially pesticides, and advocating biological methods of pest control. Mistrusted the doom-sayers of the Green movement, but forcefully argued for more sustainable farming: in his view most environmental problems stemmed from inefficiencies of livestock farming. If we ate less meat, Britain could both feed itself and have more wildlife. In gentler strain, studied moles and wrote a children's story about them (*Talpa: the Story of the Mole*). A man of 'sharp and capricious wit', also described as 'the extreme English moderate'. 'Calm, courteous and quizzical', listed Who's Who hobby as 'austere living'.

Mitchell, John b. 1934. Naturalist and environmental historian, author of *Loch Lomondside* (2001). A self-taught naturalist who 'came to fully understand the importance of detailed note-taking after joining Catterick Field Club as a National Service soldier', whose excursions were led by a certain Derek Ratcliffe. Worked as a professional jazz musician until 1966, when became warden of the then Nature Conservancy, with responsibility for Ben Lui, Glen Diomhan on Arran and the Loch Lomondside area until retirement in 1994. He also taught nature conservation and local history on further education courses at Glasgow and Strathclyde Universities, and his many published articles range from fern collecting, Loch Lomond's bygone 'strange beasts', folk lore, mine workings and their flora, yew trees and birds of prey. He played a crucial part in the establishment of Scotland's first National Park at Loch Lomond and The Trossachs.

Moore, [Henry] **Ian.** 1905-1976. Agricultural biologist and administrator, author of *Grass and Grasslands* (1966). Specialist in the utilisation of pasture grasses about which he wrote a dozen books. Reader in Crop Biology at Leeds University 1930-48; Principal at Seale Hayne Agricultural College, Newton Abbot 1948-71 and governor of Plant Breeding Institute, Cambridge. CBE 1964.

Moore, Norman [Winfred] b. 1923. Ecologist and conservationist, author of *Dragonflies* (1960 with Philip S. Corbet and Cynthia Longfield) and *Hedges* (1974 with E. Pollard and M.D. Hooper). Grew up in rural East Sussex, son of a doctor, and educated at Eton and Trinity College, Cambridge, completing his degree after the war. Wartime service in the Royal Artillery, where wounded and taken prisoner in 1944. From 1948-53, lecturer in zoology at Bristol University, where pioneering study of dragonfly territorial behaviour and ecology won him a PhD. From 1953 pursued career in scientific nature conservation, initially as the Nature Conservancy's regional officer in SW England, where carried out important studies of Dorset heaths and the buzzard. From 1960-74 Head of Toxic Chemicals and Wildlife Division at Monks Wood, where, as a sideline, he also studied hedges with his colleagues Hooper and Pollard. From 1974 to retirement in 1983 was the NCC's chief advisory officer, specialising in agriculture. During this time founded and chaired the Farming and Wildlife Advisory Groups (FWAG). Visiting Professor of Environmental Studies at Wye College, Kent. Founded and chaired Odonata Specialist Group of the IUCN, producing a world plan for dragonfly conservation. His 'professional autobiography' *The Bird of Time* (1987) was also an honest appraisal of the realties and difficulties of nature conservation in a time of rapid technological change. *Oaks, dragonflies and people* (2001) was based on his experiences of looking after a back garden nature reserve at home in Swavesey, Cambs, including perhaps the world's most famous dragonfly pond. One of the most influential figures in nature conservation over a half a century, he remains, in his eighties, graceful, courteous and wise.

Moss, Brian b. 1943. Freshwater ecologist, author of *The Broads* (2001). Born Stockport, Cheshire, read botany at Bristol University where he gained a PhD in 'the algal ecology of small pools'. Went to Malawi, and later Michigan, USA to study the impacts of nutrients on lake ecology. However best-known work is on the Broads, where he pioneered the idea of 'ecological engineering' by introducing water fleas to graze the plankton. Later studied effect of climate change on lakes, using temperature-controlled experimental tanks, and developed a means of classifying lakes related to their catchment. Lecturer, Bristol University 1972-79, Reader, University of East Anglia 1979-89, and now Professor of Botany and Head of Department at Liverpool University. Wrote a standard text book, *Ecology of Freshwaters.* Vice-President British Ecological Society and past-President of British Phycological Society, much involved not only in teaching and research but advising lake managers and policy makers.

Mountfort, Guy [Reginald]. 1905-2003. Ornithologist, conservationist and businessman, author of *The Hawfinch* (1957). One of the founders of international nature conservation, also well known as a travel book writer and co-author of the seminal *Field Guide to Birds of Britain and Europe* (1954) with his friend R.T. Peterson and P.D. Hollom. Born in London, son of a society portrait painter, left grammar school at 16 to lowly jobs – washing laboratory bottles and selling typewriters door to door. Found his niche in advertising, for Douglas Motorcycles, General Motors in Paris, and Proctor & Gamble before joining Mather & Crowther advertising agents as director. Wartime service in artillery and British Army Staff in Washington, reaching rank of Lt. Colonel, fitting in bird watching whenever possible. In 1950s and 60s, organised influential wildlife expeditions to Spain, Bulgaria and Jordan, writing his 'Portrait' books about them afterwards, as well as *The Vanishing Jungle* (1969) based on his environmental work in Pakistan. Helped found the World Wildlife Fund (1961), and was the architect of its crusade to save the tiger from extinction, which established a series of reserves across India. Hon. Sec, later president, British Ornithologist's Union. OBE (1970), WWF Gold Medal (1978). Autobiography: *Memories of Three Lives* (1991). Tall, pipe-smoking, energetic and diplomatic, a born organiser and leader, he was also a keen gardener and accomplished sketcher and photographer. He wrote *The Hawfinch* because he enjoyed the challenge of studying a truly elusive bird.

Murton, R.K. [Ronald Keir called 'Ron']. 1932-1978. Ornithologist, author of *The Wood Pigeon* (1965) and *Man and Birds* (1971). Graduate of University College, London, entering government service as MAFF ornithologist 1955-67. Brilliant field work on wood pigeon diet and feeding behaviour caused ministry to change their control methods. From 1968, he joined ITE Monks Wood as assistant to Kenneth Mellanby. Later work there became more lab-oriented, concerned with physiology and breeding cycles of birds of agricultural land. Hon Professor of Hull University. Entertaining lecturer and talented photographer; a brilliant scientist, cut off in his prime.

Neal, Ernest [Gordon]. 1911-1998. Naturalist and schoolmaster, author of *The Badger* (1948). Born Boxmoor, Hertfordshire, son of a Baptist minister, he was a keen naturalist from boyhood, educated at Taunton School and Chelsea Polytechnic where gained a London University degree in botany, zoology and chemistry. From 1936 taught biology at Rendcomb College, near Cirencester, and later at his old Taunton School where became head of science and housemaster before retirement in 1971. He became fascinated with badgers at Rendcomb, undertaking detailed study of the local setts with sixth formers, and later filming them. A founder, later chairman and president, of the Mammal Society in 1954, organised nationwide sett survey, and study of reproductive processes in badgers led to a PhD in 1960. In demand for radio programmes in 1950s and 60s, his other books included *Exploring Nature with a Camera* (1946), *Topsy and Turvy – My Two Otters* (1961), an account of otter cubs he reared, two books based on his travels in Africa, a textbook *Biology Today* (1975), four more books about badgers. Co-founder of Mammal Society (1954), campaigned for better protection for badgers and served on scientific committees investigating badgers and TB. Helped found Somerset Wildlife Trust. MBE (1976). Many travels in Africa from 1960s, including research on banded mongoose in Uganda. Memoirs: *The Badger Man* (1994).

Nethersole-Thompson, Desmond. 1908-1989. Ornithologist and local politician, author of *The Greenshank* (1951). One of the best field ornithologists, well known as the author of five classic monographs of Scottish birds and the popular *Highland Birds* (1971). Irish parentage, studied history at London School of Economics. Once a fanatical egg-collector, skills later put to productive use in studies of nest behaviour: 'To study birds you need a predatory hunger for the nest.' Abandoning teaching, moved to Speyside 1934, thereafter living on research grants and writing, dedicating life to bird study. Passionate socialist, local politician, standing twice for Parliament; represented parish on Inverness County Council, becoming vice-president District Councils Association of Scotland. A large, usually duffel-coated figure, not a society man but not without his measure of supporters and disciples.

Newton, Ian b. 1940. Ornithologist and applied scientist, author of *Finches* (1972). Leading expert on bird ecology and biogeography, specialising in finches, waterfowl and birds of prey, especially the sparrowhawk. Graduated Bristol University, gained doctorate in finch behaviour at Oxford, followed by research on bullfinch damage in orchards. Joined NERC 1967, initially studying population ecology of geese and finches, later the impact of pesticides on birds of prey. Director of Monks Wood field station, Head of Avian Ecology, NERC 1989-99. Past president British Ecological Society, British Ornithological Union. OBE (1999), FRS (1993), FRSE (1994), RSPB's Gold Medal and other awards. Other books include *The Sparrowhawk* (1986), *Population limitation in birds* (1998), *The Speciation and Biogeography of Birds* (2003). Hobby: cultivating apple trees.

Nicholson, E. [Edward] **Max.** 1904-2003. Ornithologist, civil servant, writer, director of the Nature Conservancy, original thinker, unrivalled founder of institutions; and, almost incidentally, author of *Birds and Men* (1951). Born at Kilternan, near Dublin, son of a photographer, educated at Sedbergh School and Hertford College, Oxford, where he read history and organised expeditions to Greenland and the Amazon. Polio, in childhood, left him with a mild speech impediment. After a short career as a journalist, joined the Political and Economic Planning think-tank, producing scores of policy studies influential in postwar Britain. In 1940, joined Ministry of Shipping, as head of convoy allocations, and attended many of the wartime and postwar conferences. Later, as head of Herbert Morrison's private office, helped steer the 1947 Town and Country Planning Act, the 1947 Agriculture Act and the 1949 Act that set up the National Parks and the Nature Conservancy. The latter he directed from 1952-66 and effectively established nature conservation on the national agenda. Along the way he wrote an important series of bird books, helped found the British Trust for Ornithology in 1933, and the WWF (1961), organised the 'Countryside in 1970' conferences, and was deeply involved in the IUCN, the International Biological Programme (IBP) and the International Commission on National Parks. Also found time to edit *British Birds* and the 9-volume *Birds of the Western Palearctic*. On retirement he wrote *The Environmental Revolution* (1970), his 'guide for the new masters of the world', and *The System* (1967), his denunciation of the ways of Whitehall. He founded Land Use Consultants and a series of pleasant 'jubilee walkways' through London. In his eighties, he set up an environmental think-tank, the New Renaissance Group, and continued to champion urban ecology, especially in London. Only modestly honoured by the state: CB 1948, CVO 1971; he is said to have refused a knighthood. Always very much his own man, he attributed his vision and lucidity of thought to sleeping through school lessons 'to avoid intellectual sterilisation'. He was by turns irascible, conciliatory, provocative and autocratic, 'An unusual mixture of Oxbridge committee man and radical'. A lifelong Londoner, he never lost his emotional love of birds, nor his conviction that ecology held most of the answers to the world's problems.

Two Nicholson-isms: 'Imagination is the stuff that dreams are made of'; 'To me, the gift of continuous perception is all the heaven I want'.

North, F.J. [Frederick John]. 1889-1968. Geologist and museum curator, author of *Snowdonia* (1949 with Bruce Campbell and Richenda Scott). Lifelong advocate and populariser of geology from his base as Keeper of Geology at National Museum of Wales 1914-59. Trained as a palaeontologist specialising in fossil brachiopods, but from 1920s wrote semi-popular books on slate, coal, ironstone and limestone, and interpreted subject broadly in public lectures, broadcasts and writings. Historian, cartographer and archaeologist; keen caver and hobby photographer. Founder member of British Association for History of Science.

Page, Christopher N. [Nigel]. b. 1942. Botanist and horticulturalist, author of *Ferns* (1988). From 1971 scientist in charge of ferns, fern allies and gymnosperms at Royal Botanic Garden, Edinburgh. Well-travelled since undergraduate years in quest of these ancient plants, with help of pilot's licence and own Land Rover. Broad research interests include horsetails, palaeo-botany, phytogeography and rehabilitation of historic fern gardens. Editor *Fern Gazette* 1974-85. Boundless energy and enthusiasm; fern polymath; lists interests as 'photography, painting, walking, writing and recording all I see'.

Pearsall, W.H. [William Harold]. 1891-1964. Ecologist and university professor, author of *Mountains and Moorlands* (1950) and *The Lake District* (posth. 1973 with Winifred Pennington). Possibly the greatest British plant ecologist, a northerner in love with the lakes and the hills from boyhood. Work on lake succession led to doctorate; reader in botany at Leeds 1922-38, followed by first professorial chair at Sheffield 1938-44, then Quain Professor at University College, London, 1944-57. Founding director of Freshwater Biological Association 1929-37; Charter member and Chairman of scientific policy at the Nature Conservancy 1949-63. FLS 1920; FRS 1940. Influential in science education, founding conservation diploma at UCL. At his best in the field, to be seen striding up mountains in all weathers. Tall, erect, fit, deaf (result of being knocked about in First World War), sometimes obstinate. His wife wrote that 'he never in all his life did nothing; he was either botanising, climbing, fishing, golfing, painting or writing'.

Pennington, Winifred [Anne] (Mrs T.G. Tutin). b. 1915. Ecologist, palaeobotanist and university teacher, author of *The Lake District* (1973 with W.H. Pearsall). Brought up in Lake District, disciple of Sir Harry Godwin and authority on post-glacial vegetation. Doctorate in pollen analysis at Reading University; lifelong associate and teacher at Leicester University where made Hon. Professor in 1980. Leading researcher at Freshwater Biological Association 1967-81 in the heart of the Lake District. *Festschrift* in her honour: Lake Sediments and Environmental History, eds E.Y. Haworth and J.G. Lund (1984). Author of *The History of British Vegetation* (1969). FRS 1979. Recreations: 'gardening and plain cooking'.

Perrins, Christopher M. [Miles]. b. 1935. Ornithologist and university teacher, author of *British Tits* (1979). Oxford graduate; lifelong association with Edward Grey Institute of Field Ornithology, where took over long-term studies of tits from 1957, becoming Director 1974 and Professor of Ornithology 1992. Fellow of Wolfson College. One of the leading contemporary ornithologists and active member of all the main bird societies. Active also in university life and administration as member of the board of faculty. LVO 1987. Author of six bird books, including *The Mute Swan* (1986 with M.E. Birkhead).

Pollard, Ernest. b. 1940. Ecologist, author of *Hedges* (1974, with M.D. Hooper and

N.W. Moore). Son of a Nottingham lace-maker, graduated in horticulture at Reading which led to a PhD on the ecology of insects in hedges and crops. Joined Nature Conservancy (later ITE) at Monks Wood in mid-60s, specialising in field-based studies of invertebrates, including Roman snail and white admiral butterfly. Ran the Butterfly Monitoring Scheme, introduced in 1976, and co-author of *Atlas of Butterflies in Britain and Ireland* (1984) and *Monitoring butterflies for ecology and conservation* (1993). Early retirement 1986, now farmer in Kent and ecology consultant.

Potter, Stephen [Meredith]. 1900-1969. Writer, radio producer and hobby philologist, author of *Pedigree: Words from Nature* (posth. 1973 with Laurens Sargent). Began varied career as university lecturer in English, publishing studies of Lawrence, Coleridge and an amusing account of English as taught. Later famous for his One-Upmanship books 1947-58. Joined BBC 1938 as writer-producer and director; also sometime magazine editor, theatre critic and book reviewer. Keen naturalist and companion of James Fisher on several bird expeditions and at the Savile Club. *Pedigree* is the fruit of a hobby, philology, and surprisingly serious in tone. One source describes him as 'an overgrown schoolboy .... tall and rangy with rough fair hair that stood on end .... he smoked, scattering ash wildly'. Obsessive games player. Title of autobiography: *Steps to Immaturity*.

Prime, Cecil, T. [Thomas]. 1909-1979. Botanist and schoolmaster, author of *Lords and Ladies* (1960). Biology teacher, later chief science master at Whitgift School, Croydon. Spare-time study of *Arum maculatum* in 1940s won him a doctorate at London University. A leading field botanist, vice-president of Linnaean Society and active member of BSBI; author of nine books on the British flora. A quiet man, well known and popular in botanical circles, mild-mannered but not a lover of fools, 'scornful of ostentation, hypocrisy and cant'.

Proctor, Michael [Charles Faraday]. b. 1929. Ecologist and university teacher, author of *The Pollination of Flowers* (1973 with Peter Yeo). Cambridge graduate, interested in pollination from early 1950s, gaining doctorate in rockrose research. Briefly with Nature Conservancy in Wales before joining staff at Exeter University where lecturer and later Reader in Plant Ecology 1956-86. First-rate photographer and ecological polymath, specialising variously in bryophytes, palaeobotany, phytosociology, whitebeams and the ecology of fens and bogs. Editor *Watsonia* 1959-71. Imbued with apparently limitless knowledge allied to an analytical mind; active on local conservation bodies and in BSBI. In the front rank as researcher and field demonstrator. As classroom lecturer perhaps took an over-optimistic view of the intelligence of his students.

Ramsbottom, John. 1884-1974. Mycologist and museum curator, author of *Mushrooms and Toadstools* (1953). Mancunian, joined Natural History Museum as assistant in 1910 and remained there until 1950, for last 20 years as Keeper of Botany. Awarded OBE 1919 for service in army medical corps. A sociable man, a leading and loyal member of numerous societies to do with fungi, horticulture, roses and bibliography. One of the foremost promoters and popularisers of mycology, much in demand for lectures, forays and at congresses. A cultured man, learned in fungal lore: 'he seemed to have read everything, met every living mycologist, and forgotten nothing'. Genial, teller of stories 'both proper and improper', notorious procrastinator. Dressed formally for the field 'in his dark city clothes as though he had just stepped from the Museum into the woods'.

Ratcliffe, Derek A. [Almey] b. 1929, author of *Lakeland* (2002) and an editor of New Naturalist series since 1993. Born in London, but moved to his adopted home town, Carlisle, aged nine, where attended the local grammar school. Obtained degree at Sheffield University, starting with zoology but switching to botany under the influence of Prof Roy Clapham. Studied hill vegetation in Snowdonia under one of the Nature Conservancy's first bursaries, resulting in a PhD. After National Service, joined Nature Conservancy staff, surveying mountain vegetation in Scotland (*Plant Communities of the Scottish Highlands*, jointly with Donald McVean, 1962). Helped organise national survey of the peregrine 1960-62, which revealed breeding failure attributed to pesticides. Later, examination of eggs in collections showed shell-thinning, connected with exposure to DDT. Later extended this survey work to golden eagle. Directed the first comprehensive evaluation of Britain's natural habitats, leading to *A Nature Conservation Review* (1977), which he edited, and to a large part wrote. From 1970, Deputy Director (Science) of Nature Conservancy (from 1973, the NCC), later Chief Scientist, in charge of scientific policy. He was the author and brain behind *Nature Conservation in Great Britain* (1984), a review of losses and gains which explicitly recognised the 'cultural' importance of nature. Since retirement has studied and photographed birds in Britain and Lapland. *In Search of Nature* (2000) is his memoir of days among the hills in Cumbria, the Borders and the North Pennines studying birds and plants. One of the few with a high reputation both in botany (vegetation, ferns, bryophytes) and zoology (birds), and one of the best field naturalists of modern times. Also a good photographer and writer. Modest, hardy and indefatigable, has the characteristic quiet intensity of a great observer. Books include *Birds of Mountain and Upland* (1990), *The Raven* (1997) and his masterpiece, *The Peregrine Falcon* (1980, 2nd expanded edition 1993).

Raven, John [Earle]. 1914-1980. Classics scholar, botanist and gardener, author of *Mountain Flowers* (1956 with Max Walters). Cambridge man, son of a great naturalist, classics graduate of Trinity College, Cambridge, lecturer and Tutor in ancient philosophy; Fellow of Trinity and later of Kings College, scholar of Greek pre-Socratic philosophy. Possessor of an almost mystical love of plants, especially mountain plants, which he sought, painted and grew. Tall, angular, nervous, usually good company, 'high spirits, discovery and good jokes went hand in hand'. Walk-on part as a 'pacifist puritan' in Simon Raven's *Shadows on the Grass.* Memoir, *John Raven by his friends* (1981) suggests a brilliant and popular man, inwardly plagued by nervous illness and self-doubt.

Robertson, Noel [Farnie]. 1923-1999. Teacher and plant pathologist, author of *Plant Diseases* (1999, with David Ingram). Born Dundalk, educated at Trinity Academy, Edinburgh University where he read botany and Trinity College, Cambridge and obtained a Diploma in Agricultural Sciences. Began career in Ghana studying swollen shoot disease of cocoa, which led to a PhD. Lecturer in botany at Cambridge 1948-59, Professor of Botany, University of Hull 1959-69, where he taught David Ingram, returning in 1969 as Professor of Agriculture at Edinburgh and Principal of East of Scotland College of Agriculture 1969-83. Retired 1983 and immediately went to Pakistan for six months to write that country's agriculture research plan. Spent retirement in the Borders 'struggling to cultivate narcissi'. FRSE (1969), CBE (1979). A man of parts: teacher, scientist, frustrated farmer, naturalist, beekeeper and family man. A Scottish Presbyterian and 'socialist with radical conservative tendencies', he was an influential, well-regarded but self-effacing teacher, 'constantly torn between love of research and teaching, and his desire to work for the good of his students'. The proofs for *Plant Diseases* arrived just a few days before his sudden death.

Rothschild, Miriam [Louisa] b. 1908. Zoologist, conservationist, free-thinker and *grande dame*, author of *Fleas, Flukes and Cuckoos* (1952 with Theresa Clay). Daughter of N. Charles Rothschild, pioneer conservationist and expert in fleas, following in his footsteps. Largely self-educated, enrolled in university courses in twin loves of zoology and English literature, but never took degree examinations ("you always wanted to hear somebody talk on Ruskin when it was time to dissect a sea urchin"). In quest of experience in parasitology, studied snails at Marine Biological Association, Plymouth, before taking on massive task of cataloguing father's collection of fleas (6 volumes 1953-83). In war joined the secret code-breaking team at Bletchley, also working for Min.of Ag. producing a novel form of chickenfeed from seaweed. Working with international scientists, studied physiology and anatomy of fleas and chemical defences of butterflies and moths. Addicted to causes: pioneered habitat creation using wild flower seed, much involved in conservation and animal welfare issues: 'All my life I have tilted against hopeless windmills'. Books include an *Atlas of Insect Tissue* (1985), a biography of her uncle Lord Walter Rothschild, the great collector, and *Butterfly cooing like a Dove* (1991), 'a crazy book about aire and angels'. CBE (1982) for services to taxonomy, FRS (1985), many other honours, DBE (2000). Irresistible blend of charm, eccentricity, deep learning, a fertile imagination and natural authority. Symbols: a head shawl and big white Wellingtons (she objects to leather). Recreation: 'watching butterflies'. Her Seven Wonders: the monarch butterfly, the tiger-moth ear-mite, the Jungfrau, the life-cycle of the parasitic work, *Helipegus*, the jump of the flea, carotenoid pigments, Jerusalem glimpsed in a sandstorm.

Russell, [Edward] **John.** 1872-1965. Soil chemist, research director and writer, author of *The World of Soil* (1957). Major influence on world agriculture as director Rothamsted 1912-43. Trained initially as inorganic chemist but at Wye Agricultural College 1901-12 turned to agricultural science and welfare of farm workers. Work on soil nutrients led to application of principles of soil science to food production in developing countries. FRS 1917; knighted 1922 for services to agriculture in Great War. Author of standard work, *Soil conditions and plant growth* (1912). After retirement came world lecture tours and books on soil fertility, world population and food supply, as well as a history of agricultural science. Great walker and traveller; an optimist, a good judge of character, approachable: remembered for his tea parties and 'at homes'. Small, quick movements; bright blue eyes with an 'expression of innocent candour'. Oldest New Naturalist author. Autobiography: *The land called me* (1956).

Salisbury, E.J [Edward James]. 1886-1978. Botanist and botanic garden director, author of *Weeds and Aliens* (1961). Distinguished British botanist, founder member of British Ecological Society and major influence on living plant biology and horticulture. His early research was on ecology of soil, coasts and woods; later on the reproductive biology of individual plants. Graduate of University College, London, successively lecturer, Reader, Fellow and Quain Professor there 1918-43. FRS 1933, CBE 1939, knighted 1946 for public services to agriculture and botany. Director of Kew Gardens 1943-56. Popular lecturer but at his best on field excursions where unmatched as a demonstrator with a story to tell about every plant.

Books include the influential *The Living Garden* (1935) and *Downs and Dunes* (1952). Great talker, his stories were 'always interesting, often witty .... though liable to be deemed over-lengthy by the impatient listener'. Seemed 'very pleased with E.J. Salisbury'.

Shorten, Monica [Ruth] [Mrs A.D. Vizoso]. 1923-1993. Zoologist, author of *Squirrels* (1954). Clergyman's daughter, Oxford graduate studying mammals under Charles Elton at Bureau of Animal Population in Oxford. Worked initially on ecology of rats and mice, thereafter on the grey squirrel on which she became the authority. From 1954, researcher at MAFF on squirrels and toxic chemicals. Author of *The World of the Grey Squirrel* (1973 with Prof. F. Barkalow). Studied woodcock for Game Conservancy, but scientific work cut short by family commitments and ill health. Drily humorous, a good critical scientist and a natural writer.

Simms, Eric. b. 1921. Writer, ornithologist and sound recordist, author of *Woodland Birds* (1971), *British Thrushes* (1978), *British Warblers* (1985), and *British Larks, Pipits and Wagtails* (1992). Always a field naturalist, also a history graduate, magistrate and bomber pilot and instructor during Second World War (DFC 1944). BBC wildlife sound recorder from 1950, and later director of wildlife films. Prolific broadcaster on radio and television; active in nature conservation, locally and nationally. 'Dollis Hill [London] is to him what Selborne was to Gilbert White'. Author of 20 books, including *The Public Life of the Street Pigeon* (1979) and an autobiography, *Birds of the Air* (1976). Attributes literary productivity to good memory and allowing his thoughts to flow.

Smith, Kenneth M. [Manley]. 1892-1981. Virologist and research director, author of *Measles, Mumps and Mosaics* (1954 with Roy Markham). The pioneer of both plant and insect virus research. Naturalist and butterfly-collector in his schooldays, graduated Royal College of Science; First World War invalid; became lecturer in agricultural entomology, Manchester University 1920-27. Joined Potato Virus Research Station, Cambridge [later Plant Virus Research Station] where awarded doctorate 1929 and appointed Director 1939-59. Pioneer, with Roy Markham, of use of electron microscope. Fellow of Downing College, FRS 1938, CBE 1956 for services to science. At age 70 went to USA as Visiting Professor and stayed for seven years 'among the most fruitful of my career'. The major advancer of knowledge of viruses, author of textbooks on agricultural botany and plant viruses, as well as clearly written semi-popular works. A quiet rather frail man, 'a familiar figure in Cambridge, cycling to work every day whatever the weather'. Tennis player, theatre goer, keen gardener and, during Second World War, smallholder.

Smith, Malcolm [Arthur]. 1875-1958. Physician and herpetologist, author of *The British Reptiles and Amphibians* (1951). Qualified doctor, medical officer to British Legation in Bangkok 1902-25, becoming Siamese court physician (related in *A Physician at the Court of Siam*, 1947). A great world herpetologist; his many collecting expeditions in India and Far East resulted in *Monograph of the Sea-snakes* (1926) and the reptile sections of *Fauna of British India* (1931-43). Founder and president of British Herpetological Society; associate of Natural History Museum from 1926; Fellow and zoological secretary of Linnaean Society 1938-49. 'A happy combination of field naturalist and museum man.' 'He had an ingenious way of capturing adders, picking them up with the side piece of his spectacles.'

Smith, Stuart. 1906-1963. Textile chemist and ornithologist, author of *The Yellow Wagtail* (1950). Mancunian with doctorate in textile chemistry, in charge of fibre research at British Cotton Industries research association and an authority on silk technology. Quick, lucid mind, experimental approach to ornithology allied to expertise as a bird photographer led to *Birds Fighting* (1955) as well as the influential *How to Study Birds* (1945). Leading light of BTO, founder of Manchester Ornithological Society.

Stamp, L. [Laurence] **Dudley.** 1898-1966. Geographer, writer and university teacher, editor of New Naturalist series 1943-66 and author of *Britain's Structure and Scenery* (1946), *Man and the Land* (1955), *The Common Lands of England and Wales* (1963 with W.G. Hoskins) and *Nature Conservation in Britain* (posth. 1969). Best-known geographer of his generation through authorship of text books on geography and development of land-use studies. Ill-health as child, and educated largely at home. Graduate of King's College, London, Reader in economic geography 1926-45 and later Professor of Geography at London School of Economics 1945-58, during which time he collected several degrees in geography and geology. Organiser of Land Utilisation Survey 1931-35. Influential member of Scott Committee on rural land utilisation 1941, Royal Commission on Common Land 1955-58 and in development of town and country planning. CBE 1946, knighted for services to land-use 1965. Chairman England Committee of the Nature Conservancy 1958-66. Leading member International Geographical Union. Big man, with zest for life and willingness 'to help lame dogs over stiles'.

Stearn, William T. [Thomas], called 'Willie'. 1911-2001. Botanist, bibliophile and taxonomist, author of *The Art of Botanical Illustration* (1950, with Wilfred Blunt; new edition outside series 1994). Renowned botanical scholar, born in Cambridge and became assistant, later librarian to RHS Lindley Library 1933-52, with a break in the war years where served in the RAF Medical Corps. In 1952 joined Botany Dept of Natural History Museum, in charge of the general herbarium. 1975-83 visiting professor at botany dept, Reading University. President at different times of Linnaean Society, Ray Society and Society for the History of Natural History. His main work was on horticultural taxonomy, with a stream of monographs or articles on onions, lilies, epimediums, peonies and other plants, and erudite studies of the history of botany. He was author of *Botanical Latin* (four editions, 1966-92) and a scholarly 176-page introduction to Linnaeus' *Species plantarum* (1976). Expert on the flora of Greece, and much involved in British field botany on BSBI mapping schemes and as the referee for *Allium*. His knowledge was encyclopaedic, in breadth and depth, and in many ways he was, like Linnaeus, 'the compleat naturalist'. Many awards and honours including CBE. A Quaker, his recreations were 'gardening and talking'.

Steers, J.A. [James Alfred]. 1899-1987. Geomorphologist, author of *The Sea Coast* (1953). Cambridge geographer, graduate and later Fellow of St Catherine's College. Professor of Geography at Cambridge 1949-66 and the authority on the moulding of coastlines. Member of expeditions to coral reefs in Australia 1928, 1936 and Caribbean 1939; but concerned above all with surveys of British coast from a classic study of *Scolt Head Island* (1934) to the standard work *The Coastline of England and Wales* (1946). Much involved with coastal protection and a keen exponent of coastline preservation on Nature Conservancy, National Parks Commission and National Trust's Enterprise Neptune. CBE 1973. A reserved, meticulous man, brusque at times, but generally effective at all that he did. Lectures so packed with facts 'they required close attention'. A practical geographer, not a theorist. Stamp collector and railway buff.

Summers-Smith, [J.] **Denis.** b. 1920. Engineer and ornithologist, author of *The House Sparrow* (1963). Graduated in metallurgy at Glasgow University; infantry officer during Second World War, after which gained PhD in physics at Reading University. From 1953 based in Stockton, Durham as development engineer for ICI at Tees-side, retiring in 1980. Took up study of sparrows while living at Highclere in late 1940s, attracted 'by their accessibility and as successful animals'. Now the world authority on the group with two more monographs published. Active field ornithologist, representing BTO in Tees-side area.

Summerhayes, V.S. [Victor Samuel]. 1897-1974. Botanist and herbarium curator, author of *Wild Orchids of Britain* (1951). Doyen of orchids, in charge of orchid herbarium at Kew for 39 years 1924-64, expert on African and Polynesian, as well as British, Orchidaceae and established Kew as a world centre on the family. OBE (1950). In 1920s, one of the pioneer ecologists, taking part in Oxford Spitsbergen expedition 1921, on which he wrote papers with C.S. Elton. Active member British Ecological Society, treasurer 1938-57. Slight in build, 'quick and incisive in speech and manner'. A kindly man with a few amusing oddities: he wore a 'disgusting' hat when botanising and pronounced species 'svecies'.

Swinnerton, H.H. [Henry Hurd]. 1875-1966. Geologist and university teacher, author of *Fossils* (1960). Son of Wesleyan minister; graduate, lecturer and the first Professor of Geology at University College Nottingham 1912-46. Leading promoter and teacher of geology, introducing field studies and visual aids in classroom. Reached wide audience through authorship of popular books, *The lands beyond the Bible story, The earth beneath us* etc. Lifetime active member of Geological Society. Specialised in fossil reptiles and fish; promoter of local studies in East Midlands. CBE 1950. Age didn't weary him: he wrote the New Naturalist book when in his 80s.

Thompson, Harry V. [Vassie?]. 1918-2003. Zoologist and biological consultant, author of *The Rabbit* (1956, with Alastair N. Worden). Graduated at London University and became a researcher at the Bureau of Animal Population in Oxford, where Charles Elton was a great influence. As a conscientious objector, he spent the war years in Scotland on forestry work, later joined the Ministry of Agriculture, based at Tolworth. Much involved in myxomatosis and its aftermath, and studying rabbit behaviour both in Britain and Australia; later, as a consultant, interested in the long-running controversy about TB and badgers. Active member of the Mammal Society. Founder and first director of Tangley Place research laboratory. President, Universities Federation for Animal Welfare, 1985-95. Tall, always well dressed, he loved reading and could recite large chunks of poetry by heart. His *European Rabbit-history and biology of a successful coloniser* (1994), written with New Zealand-based Carolyn M. King, examined the extraordinary success of rabbits in widely differing parts of the world.

Tinbergen, Niko [short for Nikolaas]. 1907-1988. Animal behaviourist, university teacher and Nobel Prize winner, author of *The Herring Gull's World* (1953). Famous co-founder (with Konrad Lorenz) of animal behaviour studies. Born in The Hague, keen ornithologist from early youth, watching gulls and studying behaviour of hunting wasps for 32-page doctorate thesis. After 14-month 'honeymoon' in Greenland, in 1933 became instructor in zoology department at Leiden, becoming Professor there in 1947. Chose stickleback as his demonstration animal. Met Konrad Lorenz 1936, and founded science of ethnology with classic studies of greylag goose and herring gull. Joined zoology dept. in Oxford 1949, founding research unit studying gull behaviour; Fellow of Merton College; Professor of Animal Behaviour 1966-74. Contributor to German, Dutch, British and American journals. FRS (1962); awarded joint Nobel Prize for Physiology and Medicine 1973 (with Konrad Lorenz and Karl von Fische). Author of numerous books on social behaviour and instinct; books for laymen include *Curious Naturalists* (1959). Late in career, worked on cure for autism. Great enthusiast and teacher, 'a combination of obstinacy and flexibility'; approach always down-to-earth and commonsensical, 'devoting equal attention to observation and experiment'. Elder brother also won a Nobel Prize (for economics). Scientific Biography: *Niko's Nature. A life of Niko Tinbergen and his Science of Animal Behaviour* by Hans Kruuk (2003).

Tubbs, Colin R. [Rodney]. 1937-1997. Naturalist and conservationist, author of *The New Forest* (1986, new edition outside series 2001). The son of a Portsmouth coal merchant, left school at 16, yet became an energetic and successful conservationist in his home county of Hampshire, as well as the foremost authority on the New Forest and The Solent. At first intending to follow a career in forestry, he impressed the Nature Conservancy enough to land the important post of its Hampshire officer, which he retained for 33 years until retirement in 1993. Fighting the Conservancy's corner at numerous public inquiries and commissions, he won a reputation for thorough scientific preparation, often based on his own research and observations, as well as exceptional lucidity and dedication. Among numerous scientific papers on birds, especially waders, and ecology, he wrote a pioneering ecological history of *The New Forest* (1968), a monograph on *The Buzzard* (1974) and a detailed study of *The Solent* (1999), published posthumously, in the New Naturalist tradition of scientific rigour married to readability. Influenced by the common grazings of the Forest, he was a key player in the European Forum on Nature Conservation and Pastoralism, advocating sustainable, low-intensity farming methods. Affable, conscientious, always refreshingly down-to-earth, his early death from cancer was a sad loss.

Turrill, W.B. [William Bertram]. 1890-1961. Botanist and herbarium curator, author of *British Plant Life* (1948). Keen naturalist from youth, largely self-taught. Joined Kew as assistant 1909 and remained there until retirement, from 1946-57 as Keeper of herbarium and library. Influential in changing plant systematics from herbarium-based study to one of living plants, incorporating modern studies in ecology, cytology, chemistry and genetics. Author of the standard work on the Balkan flora (1929). Founder of Systematics Association, editor *Botanical Magazine* 1948-61, keen gardener and bibliophile specialising in pamphlets. To British botanists best known for his meticulous studies of knapweeds (1954) and bladder campions (1957) with E.M. Marsden-Jones, published by Ray Society. OBE 1953, FRS 1958.

Vesey-Fitzgerald, Brian [Percy Seymour]. 1900-1981. Writer, editor and professional countryman, author of *British Game* (1946). Prolific author of books on wildlife, pets, country sports, gardening etc. 1937-69. Editor of *The Field* 1938-46; veteran broadcaster (*Field Fare* 1940-5; *There and Back* 1947-9) and newspaper and magazine columnist. Numerous interests included fairgrounds, gypsies and boxing; sympathies embraced poachers as well as gamekeepers.

Walters, S. [Stuart] **Max.** b. 1920. Botanist and botanic garden director, editor of the series from 1981 to date, and author of *Wild Flowers* (1954 with John Gilmour), *Mountain Flowers* (1956 with John Raven) and *Wild and Garden Plants* (1993). Cambridge botanist, graduate of St John's and past Fellow of Kings College. Lecturer in botany and curator of botany school herbarium 1948-73, director of University Botanic Garden 1973-83. Leading experimental taxonomist, working mainly with British flora; active field and local botanist: co-author of *Atlas of the British Flora* (1962) and editor of *Flora Europaea* (1964-80). VMH 1984. One of the best-built bridges between scientific botany and horticulture; a lover of plants in all their aspects and guises; and genial company.

Webb, Nigel [Rodney]. b. 1942. Animal ecologist, author of *Heathlands* (1986). Graduate of University of Wales, specialising in soil fauna of Danish heaths. In 1967 joined Nature Conservancy (now ITE) at Furzebrook research station specialising in heathland ecology, especially invertebrates, and conservation management; now a leading European authority on heaths. Active member local conservation trusts and ecological societies.

Williams, C.B. [Carrington Bonser, but always called 'C.B.']. 1889-1981. Agricultural entomologist, author of *Insect Migration* (1958). Son of Liverpudlian bank manager, keen collector and breeder of butterflies and moths in youth. Graduated in life sciences and agriculture at Cambridge; studied thrips at John Innes Institute, thence to West Indies to study pests of sugar cane where became interested in insect migration. This took him on to colonial entomological service in Africa studying locusts; a continuous flow of scientific papers 1917-70 including the landmark (but dull) *Migration of Butterflies* (1930). Joined staff at Rothamsted 1932-55 as head of entomology under Sir John Russell, specialising in influence of weather on insect activity and numbers; more field naturalist than lab man, developed long-running insect light-trapping programme. One of the first fully numerate ecologists, developing predictive mathematical models and formulae outlined in *Patterns in the balance of nature* (1964). FRS 1954. On retirement, moved to Kincraig in Inverness-shire, where promptly embarked on an intensive study of local insects. Popular with colleagues; fond of mathematical puzzles and paradoxes in the manner and humour of Lewis Carroll. Once overheard in a Trinidad jungle pulling the legs off a giant centipede, murmuring 'She loves me, she loves me not.'

Wooldridge, S.W. [Sidney William]. 1900-1963. Geomorphologist and promoter of field studies, author of *The Weald* (1953 with Frederick Goldring). Founder of British geomorphology, graduate of King's College, London, where successively lecturer, reader and Professor of Geography 1921-63. Initially a geologist, made his name in 1930s studying river development and erosion chronology in SE England. Wrote the standard work, *The physical basis of geography* (1937 with R.S. Morgan). Active promoter of field centres helping to found the first one at Juniper Hall. A home-based geographer, much involved in human settlement geography and planning for gravel extraction; happiest leading field excursions in The Weald. CBE 1954, FRS 1959. Congregationalist with ready command of biblical phrases; good teacher with characteristic 'pungency of phrase', golfer and cricket umpire; sang baritone leads in Gilbert and Sullivan.

Worden, Alastair N. [Norman]. 1916-1987. Zoologist, biochemist and consultant, author of *The Rabbit* (1956 with Harry V. Thompson). Qualified in clinical medicine at Cambridge, becoming Professor of Animal Health at University College of Wales, aged 28, where directed research on rabbits and their control. Established and chaired contract consultancy the Huntingdon Research Centre 1950-78, concerned with animal nutrition and health. Late in life became interested in toxicology, studied for new doctorate at Addenbrookes and appointed Professor of Toxicology, Bath 1973-85. Member of dozens of medical and agricultural societies and committees; first chairman Mammal Society and editor of *Journal of Animal Behaviour* 1950-65. 'He lived his life at top speed, his words tumbling over each other to keep pace with his thoughts.' Not a field naturalist, but president of his local naturalists society for 20 years.

Worthington, E. [Edgar] **Barton.** b. 1905. Ecologist and science administrator, author of *Life in Lakes and Rivers* (1951 with T.T. Macan). Cambridge graduate, career alternating between Britain and Africa. Took part in African lakes expeditions 1927-31 and African research survey 1934-37. Secretary and first full-time director of Freshwater Biological Association 1937-46; returned to Africa in late 1940s as science and development advisor; thence deputy scientific director for Nature Conservancy 1957-65 and scientific director International Biological Programme 1964-74. Active in water biology and international nature conservation, concerned with environmental impacts of drainage and irrigation. Author of several books, including *The Ecological Century* (1983). CBE 1967. Still travelling in his late eighties.

Wragge-Morley, Derek. 1920-1969. Biologist and scientific journalist, author of *Ants* (1953). Precocious young scientist, carried out original research on ants in his teens and read two papers for the Berlin International Congress of Entomology in Berlin 1938; youngest ever Fellow of Linnaean Society (1942). Graduate in natural sciences at Cambridge. War-time work on insect pests for Ministry of Agriculture; Macaulay Fellow for genetics research, Edinburgh (1946-49); invitation lecturer at Institute of Social Anthropology, Oxford, working on genetics and social behaviour of animals and insects. From 1950, career in science journalism and consultancy. First science correspondent for Financial Times 1952-61; scientific and technical adviser for Hambros Bank 1962-69. A promising youthful talent despite ill-health which left him semi-invalid; great enthusiast for ants and insect behaviour and a pioneer in the popularisation of science. The critical failure of *Ants* was a bitter disappointment, but, contrary to rumour, he got over it.

Yeo, Peter. b. 1929. Botanist and librarian, author of *The Pollination of Flowers* (1973 with Michael Proctor). Cambridge botanist, Fellow of Wolfson College, specialising in taxonomy and pollination while also librarian at University Botanic Garden 1953-93. Exceptional among botanists in being equally expert on insects (and birds). Well travelled in search of flowers. Books include *Solitary Wasps* (1983 with Sally Corbet) and *Hardy Geraniums* (1985).

Yonge, C. [Charles] **Maurice.** 1899-1986. Marine biologist and university teacher, author of *The Sea Shore* (1949) and *The Oyster* (1960). Promoter and populariser of marine science of world repute. Son of Yorkshire headmaster, graduating in zoology at Edinburgh after dabbling in history, forestry and journalism. Lifelong interest in form and function of marine invertebrates, specialising in bivalves. Studied oyster physiology at Plymouth Laboratory in 1920s. Organised and led Cambridge Great Barrier Reef expedition 1927-29, afterwards writing the first major study of coral reefs, *A Year on the Great Barrier Reef* (1930). Professor of Zoology, Bristol University 1932-44 and Glasgow University 1944-64, continuing research into his 80s. Advised on establishment of marine stations all over world; chairman Scottish Marine Biological Association, member of all kinds of advisory bodies on science and fisheries. FRS 1954, CBE 1957, knighted 1967. Commanding teacher despite stammer, writer of excellent popular books on marine life. Eternal boyish enthusiasm, devotee of curio shops, reserved but approachable: 'the Master'.

Appendix 2

An Updated New Naturalist Bibliography and Print-runs

The number of copies sold includes book club editions published in the UK, but not foreign editions or translations, the paperback (unless otherwise stated) or the Bloomsbury reprints. 'Editions' usually differ from reprints or 'impressions' in that they incorporate more substantial revisions involving a degree or resetting. In practice, though, some of the new editions incorporated little, if any, revision. I have in each case followed the sequence printed on the half-title of each book, which we must accept as definitive, even if it is occasionally inaccurate. The foreign editions and translations listed may not be complete. I would be interested to learn of any others that have come to the reader's attention.

	Title, author and number in the series	**Editions**	**Hardback copies sold in Great Britain**
1.	*Butterflies* by E.B. Ford	First edition **1945**, Second edition **1946**, Third edition **1957** reprinted 1962, 1967, 1971, Fourth edition (without colour plates) **1977** reprinted 1977, o/p 1983 Readers Union 1953, Fontana paperback 1975; Bloomsbury Books reprint 1989.	53,600 (1st edition 20,000)
2.	*British Game* by Brian Vesey-Fitzgerald	First edition **1946** reprinted 1946. Readers Union 1953, 1959, o/p 1971.	40,000 (1st edition 20,000)
3.	*London's Natural History* by R.S.R. Fitter	First edition **1945**, reprinted 1946. Readers Union 1953, 1959, o/p 1965. Collins paperback reprint 1984. Bloomsbury Books reprint 1989.	40,000 (1st edition 20,000)
4.	*Britain's Structure and Scenery* by L. Dudley Stamp	First edition **1946**, Second edition **1947**, Third edition **1949**, Fourth edition **1955**, Fifth edition **1960**, Sixth edition **1967** reprinted 1970, o/p 1975. Fontana paperback 1960, Collins paperback reprint 1984.	62,000 (1st edition 20,000)

5. *Wild Flowers* by John Gilmour and Max Walters	First edition **1954**, Second edition **1955**, Third edition **1962**, Fourth edition **1969**, Fifth edition **1973** reprinted 1978, o/p 1984. Fontana paperback 1972, Collins paperback reprint 1985, Bloomsbury Books reprint 1989.	23,450 (1st edition 7,000)
6a. *Natural History in the Highlands and Islands* by F. Fraser Darling	First edition **1947**, o/p 1954.	c. 25,000
6b. *The Highlands and Islands* by F. Fraser Darling and J. Morton Boyd	First edition **1964** reprinted 1969, Second edition **1973** reprinted 1977, o/p 1983. Fontana paperback 1969. Bloomsbury Books reprint 1989.	16,500 (1st edition 8,000)
7. *Mushrooms and Toadstools* by John Ramsbottom	First edition **1953**, 2nd impression 1954, 3rd 1959, 4th 1963, 5th 1969, 6th 1972 reprinted 1977, o/p 1983. Bloomsbury Books reprint 1989.	24,000 (1st edition 8,000)
8. *Insect Natural History* by A.D. Imms	First edition **1947**, Second edition **1956**, Third edition **1971**, reprinted 1973, o/p 1983. Fontana paperback 1973, Bloomsbury Books reprint 1989.	40,200 (1st edition 30,000)
9. *A Country Parish* by A.W. Boyd	First edition **1951** o/p 1959.	7,000
10. *British Plant Life* by W.B. Turrill	First edition **1948**, Second edition **1958**, Third edition **1962** reprinted 1971, o/p 1980. Readers Union 1959, Bloomsbury Books reprint 1989.	33,000 (1st edition 20,000)
11. *Mountains and Moorlands* by W.H. Pearsall	First edition **1950** reprinted 1960, 1965, Revised edition (with Winifred Pennington) **1971** reprinted 1977, o/p 1985. Fontana paperback 1968, Bloomsbury Books reprint 1989.	26,300 (1st edition 15,000)
12. *The Sea Shore* by C.M. Yonge	First edition **1949** reprinted 1958, 1961, Second edition **1966** reprinted 1971, 1976, o/p 1980. Readers Union 1959, Fontana paperback 1963, Bloomsbury Books reprint 1989.	43,000 (1st edition c. 25,000)
13. *Snowdonia* by F.J. North, Bruce Campbell and Richenda Scott	First edition **1949**, o/p 1959.	18,000

14. *The Art of Botanical Illustration* by Wilfrid Blunt with the assistance of William T. Stearn	First edition **1950**, Second edition **1951**, Third edition **1955**, Fourth edition **1967**, Fifth edition **1971** (referred to on the half-title as reprints). o/p 1976. US ed. by Scribners 1951, translated into Japanese 1985. New enlarged edition **1994** by Antique Collectors' Club.	24,500 (1st edition 12,000)
15. *Life in Lakes and Rivers* by T.T. Macan and E.B. Worthington	First edition **1951** reprinted 1959, 1962, Second edition **1968**, Third edition **1974**. o/p 1984. Readers Union 1959. Fontana paperback 1972, Bloomsbury Books reprint 1990.	26,450 (1st edition 12,000)
16. *Wild Flowers of Chalk and Limestone* by J.E. Lousley	First edition **1950**, Second edition **1969** reprinted 1971, 1976, o/p 1985. Bloomsbury Books reprint 1989.	20,000 (1st edition 12,500)
17. *Birds and Men* by E.M. Nicholson	First edition **1951**, o/p 1965. Readers Union 1959. Bloomsbury Books reprint 1989.	16,000
18. *A Natural History of Man in Britain* by H.J. Fleure	First edition **1951** reprinted 1959. Revised edition **1970** (with Margaret Davies). o/p 1981. Fontana paperback 1971, Bloomsbury Books reprint 1990.	18,500 (1st edition 12,500)
19. *Wild Orchids of Britain* by V.S. Summerhayes	First edition **1951**, Second edition **1968** reprinted 1969, 1976, o/p 1982. Collins paperback reprint 1985.	19,500 (1st edition 11,500)
20. *The British Amphibians and Reptiles* by Malcolm Smith	First edition **1951**, Second edition **1954**, Third edition **1964**, Fourth edition **1969**, Fifth edition **1972**, o/p 1979.	18,000 (1st edition 6,500)
21. *British Mammals* by L. Harrison Matthews	First edition **1952** reprinted 1960, 1963, Second edition **1968** reprinted 1972, o/p 1975. Readers' Union 1960, Bloomsbury Books reprint 1989.	27,000 (1st edition 12,500)
22. *Climate and the British Scene* by Gordon Manley	First edition **1952**, 2nd impression 1953, 3rd 1955, 4th 1962, 5th 1971, o/p 1980. Fontana paperback 1962.	19,900 (1st edition 7,500)
23. *An Angler's Entomology* by J.R. Harris	First edition **1952**, Second edition **1956** reprinted 1966, 1970, 1973, 1977, o/p 1983. US Editions A. S. Barnes (1966) and The Countryman Press (undated). Bloomsbury Books reprint 1989.	19,000 (1st edition 7,500)

24. *Flowers of the Coast* by Ian Hepburn	First edition **1952** reprinted 1962, 1966, 1972, o/p 1978.	12,500 (1st edition 6,000)
25. *The Sea Coast* by J.A. Steers	First edition **1953**, Second edition **1954**, Third edition **1962**, Fourth edition **1969** reprinted 1972, o/p 1983.	18,000 (1st edition 8,000)
26. *The Weald* by S.W. Wooldridge and Frederick Goldring	First edition **1953**, 2nd impression 1960, 3rd 1962, 4th 1966, 5th 1972, o/p 1984.	15,000 (1st edition 6,000)
27. *Dartmoor* by L.A. Harvey and D. St Leger Gordon	First edition **1953**, Second edition **1962** reprinted 1963, 1970. Revised edition **1977**, o/p 1983. Fontana paperback 1973.	14,500 (1st edition 6,500)
28. *Sea-Birds* by James Fisher and R.M. Lockley	First edition **1954**, o/p 1962. US edition (Houghton Mifflin) 1954. Bloomsbury Books reprint 1989.	10,000
29. *The World of the Honeybee* by Colin G. Butler	First edition **1954** reprinted 1958. Revised edition **1962** reprinted 1967, 1971. Revised edition **1974** reprinted 1976, 1977, o/p 1985. US edition (Macmillan) 1954, (Taplinger) 1974.	23,000 (1st edition 7,500)
30. *Moths* by E.B. Ford	First edition **1954**, Second edition **1967**, Third edition **1972** reprinted 1976, o/p 1981.	14,500 (1st edition 8,500)
31. *Man and the Land* by L. Dudley Stamp	First edition **1955**, Second edition **1964**, Third edition **1969** reprinted 1973, o/p 1983.	18,500 (1st edition 10,000)
32. *Trees, Woods and Man* by H.L. Edlin	First edition **1956**, Second edition **1966**, Third (revised) edition **1970** reprinted 1972, 1978, o/p 1983.	16,950 (1st edition 8,000)
33. *Mountain Flowers* by John Raven and Max Walters	First edition **1956** reprinted 1965, 1971, o/p 1980. Collins paperback reprint 1984.	12,000 (1st edition 7,000)
34. *The Open Sea: The World of Plankton* by Alister Hardy	First edition **1956**, reprinted with revisions 1958, reprinted 1962, 1964. Second edition **1970** reprinted 1971, o/p 1979. US edition (single volume, Houghton Mifflin) 1956, Readers Union 1959, Fontana paperback 1963.	26,500 (1st edition 10,000)
35. *The World of Soil* by Sir E. John Russell	First edition **1957**, Second edition **1959**, Third edition **1963**, Fourth edition **1967**, Fifth edition **1971** reprinted 1975, o/p 1980. Readers Union 1959, Fontana paperback 1961.	15,350 (1st edition 8,000)

36. *Insect Migration* by C.B. Williams	First edition **1958**, Second edition **1965** reprinted 1971, o/p 1980. German edition 1961.	9,800 (1st edition 6,000)
37. *The Open Sea: Fish and Fisheries* by Sir Alister Hardy	First edition **1959**, Second edition **1964**, Third edition **1970** reprinted 1971, 1976, o/p 1983. US edition, see *World of Plankton*.	16,000 (1st edition 8,000)
38. *The World of Spiders* by W.S. Bristowe	First edition **1958**, Revised edition **1971** reprinted 1971, 1976, o/p 1978. US edition (Taplinger) 1971, Readers Union 1976.	13,150 (1st edition 6,500)
39. *The Folklore of Birds* by Edward A. Armstrong	First edition **1958**, o/p 1962. Second enlarged edition printed in US by Dover Publications 1970.	6,000
40. *Bumblebees* by John B. Free and Colin G. Butler	First edition **1959** reprinted 1968, o/p 1974.	6,500 1st edition 5,000)
41. *Dragonflies* by Philip S. Corbet, Cynthia Longfield and N.W. Moore	First edition **1960**, o/p 1972. Collins paperback reprint 1985.	5,000
42. *Fossils* by H.H. Swinnerton	First edition **1960** reprinted 1962, 1970, 1973, o/p 1978. Bloomsbury Books reprint 1989.	17,000 (1st edition 5,000)
43. *Weeds and Aliens* by Sir Edward Salisbury	First edition **1961**, Second edition **1964**, o/p 1972.	8,000 (1st edition 5,000)
44. *The Peak District* by K.C. Edwards	First edition **1962** reprinted 1962, 1964, 1970. Second edition **1974**, o/p 1979. Fontana paperback 1973, Bloomsbury Books reprint 1989.	14,500 (1st edition 5,000)
45. *The Common Lands of England and Wales* by W.G. Hoskins and L. Dudley Stamp	First edition **1963** reprinted 1964, o/p 1979.	9,000 (1st edition 5,000)
46. *The Broads* by A.E. Ellis	First edition **1965**, o/p 1974.	8,500
47. *The Snowdonia National Park* by W.M. Condry	First edition **1966**, Second edition **1967** reprinted 1976, o/p 1981. Fontana paperback 1969.	15,500 (1st edition 7,500)
48. *Grass and Grasslands* by Ian Moore	First edition **1966**, o/p 1981.	6,500
49. *Nature Conservation in Britain* by Sir Dudley Stamp	First edition **1969** reprinted 1970, Second edition **1974**, o/p 1982.	13,500 (1st edition 8,000)

50. *Pesticides and Pollution* by Kenneth Mellanby	First edition **1967**, Second edition **1970** reprinted 1971, o/p 1981. Fontana paperback 1970.	15,500 (1st edition 6,500)
51. *Man and Birds* by R.K. Murton	First edition **1971** reprinted 1973, 1977, o/p 1981. US edition (Taplinger) 1973, Readers Union 1973.	9,500 (1st edition 6,000)
52. *Woodland Birds* by Eric Simms	First edition **1971** reprinted 1976, o/p 1982. Readers Union 1976, Bloomsbury Books reprint 1989.	11,000 (1st edition 8,000)
53. *The Lake District* by W.H. Pearsall and Winifred Pennington	First edition **1973** reprinted 1977, o/p 1982. Readers Union 1977.	9,000 (1st edition 6,000)
54. *The Pollination of Flowers* by Michael Proctor and Peter Yeo	First edition **1973** reprinted 1975, 1979, o/p 1985. US edition (Taplinger) 1973.	8,000 (1st edition 4,000)
55. *Finches* by Ian Newton	First edition **1972** reprinted 1975, 1976, 1979, o/p 1981. US edition 1972, Readers Union 1972. Collins paperback reprint 1985.	15,000 (1st edition 5,000)
56. *Pedigree: Words from Nature* by Stephen Potter and Laurens Sargent	First edition **1973**, o/p 1980. US edition (Taplinger) 1973.	4,000
57. *British Seals* by H.R. Hewer	First edition **1974**, o/p 1980. US edition (Taplinger) 1974.	3,500
58. *Hedges* by E. Pollard, M.D. Hooper and N.W. Moore	First edition **1974** reprinted 1975, 1977, 1979, o/p 1985. US edition 1974, Readers Union 1979.	11,500 (1st edition 4,500)
59. *Ants* by M.V. Brian	First edition **1977**, o/p 1981. Readers Union 1977.	4,000
60. *British Birds of Prey* by Leslie Brown	First edition **1976** reprinted 1976, 1979, 1982, o/p 1986. Readers Union 1976-82. Bloomsbury Books reprint 1989.	20,000 (1st edition 6,000)
61. *Inheritance and Natural History* by R.J. Berry	First edition **1977**, o/p 1985. US edition (Taplinger) 1977, Readers Union 1977, Bloomsbury Books reprint 1989.	6,000
62. *British Tits* by Christopher Perrins	First edition **1979** reprinted 1980, o/p 1985, part remaindered 1983. Book Club edition 1979.	14,250 (1st edition 4,000)

63. *British Thrushes* by Eric Simms	First edition **1978**, o/p 1986, part remaindered 1983. Book Club 1978.	14,000 (1st edition 6,000)
64. *The Natural History of Shetland* by R.J. Berry and J.L. Johnston	First edition **1980** reprinted 1986. Revised as *A Naturalist's Shetland* by Johnstone alone, published by Poyser (1999).	6,000 (1st edition 4,000)
65. *Waders* by W.G. Hale	First edition **1980** reprinted 1981, 1982, o/p 1985.	6,200 (1st edition 4,000)
66. *The Natural History of Wales* by William M. Condry	First edition **1981** reprinted 1982, o/p 1986. Bloomsbury Books reprint 1989.	4,000
67. *Farming and Wildlife* by Kenneth Mellanby	First edition **1981** reprinted 1983, o/p 1986.	3,500 1st edition 2,500)
68. *Mammals in the British Isles* by L. Harrison Matthews	First edition **1982**, part remaindered 1985. Book Club 1982.	3,000
69. *Reptiles and Amphibians in Britain* by Deryk Fraser	First edition **1983**. Book Club 1983, Bloomsbury Books reprint 1989.	3,500
70. *The Natural History of Orkney* by R.J. Berry	First hardback edition, first state **1985**, second state 1985, paperback 1985. Book Club (paperback) 1985. Revised as *Orkney Nature*, published by Poyser (2000).	Hardback 1,250 (1st state 725) Paperback 4,000
71. *British Warblers* by Eric Simms	First hardback edition, first stage **1985**, second state 1985, paperback 1985. Book Club (pbk) 1985.	Hardback 1,250 (1st state 725) Paperback 9,600
72. *Heathlands* by Nigel Webb	First hardback edition **1986,** reprinted 1986, paperback 1986. Book Club 1986.	Hardback 3,000 (1st edition 1,000) Paperback 2,000
73. *The New Forest* by Colin R. Tubbs	First hardback edition **1986**, paperback 1986, reprinted 1988. Book Club 1986. Revised and reprinted by New Forest Centenary Trust (2002).	Hardback 1,800 Paperback 3,650
74. *Ferns* by Christopher N. Page	First hardback edition **1988**, paperback 1988, Book Club 1988.	Hardback 1,600 Paperback 2,550
75. *British Freshwater Fish* by Peter S. Maitland and R. Niall Campbell	First hardback edition **1992**, paperback 1992.	Hardback 1,500 Paperback 1,500

76. *The Hebrides* by J. Morton Boyd and Ian L. Boyd	First hardback edition **1990**, paperback 1990. Revised in 3 volumes as *The Hebrides*: 1) *A natural tapestry*, 2) *A mosaic of islands*, 3) *A habitable land*?, published by Birlinn Ltd (1996).	Hardback 1,500 Paperback 2,000
77. *The Soil* by B.N.K. Davis, N. Walker, D.F. Ball and A.H. Fitter	First hardback edition **1992**, paperback 1992.	Hardback 1,500 Paperback 2,000
78. *British Larks, Pipits and Wagtails* by Eric Simms	First hardback edition **1992**, paperback 1992.	Hardback 1,600 Paperback 3,000
79. *Caves and Cave Life* by Philip Chapman	First hardback edition **1993**, paperback 1992.	Hardback 1,500 Paperback 2,000
80. *Wild and Garden Plants* by Max Walters	First hardback edition **1993**, paperback 1993.	Hardback 1,500 Paperback 3,000
81. *Ladybirds* by Michael E. N. Majerus	First hardback edition **1994**, paperback 1994.	Hardback 1,500 Paperback 2,500
82. *The New Naturalists* by Peter Marren	First hardback edition **1995**, reprinted 1996, paperback 1995.	Hardback 2,000 (1st edition 1,500) Paperback 1,750
83. *The Natural History of Pollination* by Michael Proctor, Peter Yeo and Andrew Lack	First hardback edition **1996**, reprinted 1996, paperback 1996. US edition (Timber Press)	Hardback 2,000 (1st edition 1,500) Paperback 2,000 US 2,500
84. *A Natural History of Ireland* by David Cabot	First hardback edition **1999** reprinted 1999, paperback 1996.	Hardback 3,120 (1st edition 2,100) Paperback 3,000
85. *Plant Disease* by David Ingram and Noel Robertson	First hardback edition **1999**, paperback 1999.	Hardback 2,000 Paperback 2,000
86. *Lichens* by Oliver Gilbert	First hardback edition **2000**, paperback 2000.	Hardback 2,500 Paperback 2,000
87. *Amphibians and Reptiles* by Trevor Beebee and Richard Griffiths	First hardback edition **2000**, paperback 2000	Hardback 3,000 Paperback 2,000
88. *Loch Lomondside* by John Mitchell	First hardback edition **2001**, paperback 2001	Hardback 2,500 Paperback 2,000
89. *The Broads* by Brian Moss	First hardback edition **2001**, paperback 2001	Hardback 2,500 Paperback 2,000
90. *Moths* by Michael Majerus	First hardback edition **2002**, paperback 2002	Hardback 3,000 Paperback 2,000
91. *Nature Conservation* by Peter Marren	First hardback edition **2002**, paperback 2002	Hardback 3,000 Paperback 2,000
92. *Lakeland* by Derek Ratcliffe	First hardback edition **2002**, paperback 2002	Hardback 3,000 Paperback 2,000

93. *British Bats* by John Altringham	First hardback edition **2003**, paperback 2003	Hardback 3,000 Paperback 2,000
94. *Seashore* by Peter J. Hayward	First hardback edition **2004**, paperback 2004	Hardback 3,000 Paperback 2,000
95 *Northumberland* by Angus Lunn	First hardback edition **2004**, paperback 2004	Hardback 3,000 Paperback 2,000

MONOGRAPHS AND SPECIAL VOLUMES

	Title, author and number in the series	**Editions**	**Hardback copies sold in Great Britain**
M1.	*The Badger* by Ernest Neal	First edition **1948**, Second edition **1962**, Third edition **1969** reprinted 1971, Fourth edition **1975**, Fifth edition **1976**, o/p 1977. Paperback (Pelican) edition 1958. Translated into German and Japanese (1974).	19,100 (1st edition 11,000)
M2.	*The Redstart* by John Buxton	First edition **1950**, o/p 1974. Swedish edition 1951.	6,000
M3.	*The Wren* by Edward A. Armstrong	First edition **1955**, o/p 1964.	3,000
M4.	*The Yellow Wagtail* by Stuart Smith	First edition **1950**, o/p 1975.	6,000
M5.	*The Greenshank* by Desmond Nethersole-Thompson	First edition **1951**, o/p 1966.	4,000
M6.	*The Fulmar* by James Fisher	First edition **1952**, o/p 1958. Collins paperback reprint 1984.	3,000
M7.	*Fleas, Flukes and Cuckoos* by Miriam Rothschild and Theresa Clay	First edition **1952** reprinted 1952, 1957, o/p 1964. Readers Union edition 1953, US edition (Philosophical Library) 1952. Paperback editions (Pelican) 1957, (Arrow Books) 1960.	8,000 (1st edition 2,000)
M8.	*Ants* by Derek Wragge Morley	First edition **1953**, o/p 1954.	3,900
M9.	*The Herring Gull's World* by Niko Tinbergen	First edition **1953** reprinted 1960, 1963, 1971, 1976, o/p 1981. US hardback and paperback editions (Praeger and Basic Books) 1960-71. Translated into German (1955), Italian, Swedish and Japanese (1974).	12,750 (1st edition 5,000)

M10.	*Mumps, Measles and Mosaics* by Kenneth M. Smith and Roy Markham	First edition **1954**, o/p 1956. US edition (Praegar) 1954.	3,000
M11.	*The Heron* by Frank A. Lowe	First edition **1954**, o/p 1962.	3,000
M12.	*Squirrels* by Monica Shorten	First edition **1954**, o/p 1971.	6,000
M13.	*The Rabbit* by Harry V. Thompson and Alastair N. Worden	First edition **1956**, o/p 1968.	5,000
M14.	*Birds of the London Area since 1900* by Richard Homes and the London Natural History Society	First edition **1957**, o/p 1958. Revised edition 1963 by Rupert Hart-Davis.	2,500
M15.	*The Hawfinch* by Guy Mountfort	First edition **1957**, o/p 1971.	2,750
M16.	*The Salmon* by J.W. Jones	First edition **1959**, 2nd impression 1961, 3rd 1968, 4th 1972, o/p 1978. US edition (Harpers) 1961.	9,250 (1st edition 4,000)
M17.	*Lords and Ladies* by C.T. Prime	First edition **1960**, o/p 1969. Reprinted by BSBI 1981.	2,400
M18.	*Oysters* by C.M. Yonge	First edition **1960** reprinted 1966, o/p 1972. US edition (Macmillan) 1960.	5,500 (1st edition 4,000)
M19.	*The House Sparrow* by D. Summers-Smith	First edition **1963** reprinted 1967, 1976, o/p 1976.	6,000 (1st edition 4,000)
M20.	*The Wood-Pigeon* by R.K. Murton	First edition **1965**, o/p 1979.	6,000
M21.	*The Trout* by W.E. Frost and M.E. Brown	First edition **1967** reprinted 1972, 1975, o/p 1979. Fontana paperback 1971.	8,300 (1st edition 6,000)
M22.	*The Mole* by Kenneth Mellanby	First edition **1971** reprinted 1971, 1974, o/p 1980. US edition (Taplinger) 1973, Readers Union 1973.	6,350

Appendix 3

A New Collector's Guide

The New Naturalist library is probably the most avidly collected series of natural history books in English, as the high asking prices for the scarcer ones indicate only too well. The library has most if not all of the ingredients of collectability. The books are well produced in uniform format. They have strikingly attractive dust jackets designed by important artists. They were, collectively and often individually, what Crispin Fisher once described as 'landmarks in contemporary study'. While there are enough of them to go round, they are just scarce enough to engage the hunting instinct, making the assembly of a complete collection difficult but attainable. Last, and probably least in collecting terms, most of the titles are still a good read. All they lack is the bloom of antiquity to achieve the book collector's dream *tout court.*

Most New Naturalist collectors are, I would guess, naturalists themselves – at least those I know all are. The series has more appeal to the naturalist who likes books than the book-lover who likes nature. I feel on less certain ground in suggesting that their greatest appeal is to a particular generation, that is, those of us who grew up with these books from the 1940s through to the mid-1970s. There are probably not a few people with jobs in the biological sciences or nature conservation who could claim that the New Naturalist library changed their lives. As an A-level biology student, *Butterflies* was my favourite book, though I never became very interested in genetics; as an undergraduate I knew chunks of J.E. Lousley's *Wild Flowers of Chalk and Limestone* almost by heart and used it to plan field trips; it was *Mountain Flowers* that led me to the botanical wonders of Ben Lawers and Caenlochan and fostered an interest in biogeography. I took pleasure and pride as the late Nature Conservancy Council's conservation person in Aberdeenshire in my custodianship of most of Chapter Six! New Naturalist titles appear regularly as top ten choices in the 'natural classic' guest column in *BBC Wildlife*. Among the recent ones have been *The Badger, The Open Sea, British Amphibians and Reptiles* (twice) and *The Sea Shore*. Unless we were abnormally precocious or unusually wealthy, the New Naturalists were probably not our first natural history books, but from teenage onwards they probably took the lead in terms of opening our eyes to ecology and what Fraser Darling called the 'wholeness of nature'. In those days good bookshops would stock 30 or 40 titles, often side by side on the same gleaming shelf. During the past 20 years they have been less forcefully brought to our attention and there is now, one must suppose, a new generation of naturalists to whom this series is not the seminal influence that it was. Even the word naturalist seems to have gone out of fashion. We are the old or middle-aged naturalists now. The new generation are 'environmentalists', 'conservationists' or merely 'green'. *Plus ca change.*

Well-preserved copies of the older New Naturalist titles are already part of the Rare and Fine Books market, though few of them are in fact rare. Shop-fresh second-hand copies are still surprisingly plentiful but are always in demand. Grubby, worn or jacketless copies, on the other hand, will continue to repine on the booksellers' shelves as they do today, until someone takes pity and offers a pound or two to take them away. That is in the nature of the game, of course, but it is particularly the case with this series. The reason is clear: one torn or stained copy can ruin the appearance of the entire shelf. You will never be happy with it. Ten years ago, I thought I was above all that, and that my collection was for use not ostentation, as Gibbon once said of the harem of a Roman emperor. Today, though, I notice that those now threadbare books I bought in my teens have all been replaced. But this obsession with condition is purely a matter of personal taste. As the author of *Modern First Editions*, Joseph Connolly put it, if condition truly isn't important to you, then quite simply it isn't important. And you will be able to knock together a set of New Naturalist 'greenbacks' in no time at all and blow the savings on a month in the West Indies.

Condition

New Naturalist books are well made and relatively hard wearing. You rarely find the pages sagging from their bindings, or boards that are split and falling away from the spines, or plates falling out. It is also clear that in so many cases these books have been looked after. Modern first editions from the 1940s and 1950s in well-preserved dust jackets are usually scarce, but that is not the case with the New Naturalist library. Moreover, you sometimes come across 30-year-old copies in truly remarkable condition, as fresh as the day they were bought. I have a copy of *Weeds and Aliens* that shows no sign of ever having been opened, let alone read. Its brilliant colours suggest that it was not kept on open book shelves either, nor in a box in some damp garage. Where has it been all this time? The snag with pristine copies like this, of course, is that the book is a bit *too* good: *I* don't like to use it either, and, since I'm rather fond of *Weeds and Aliens*, I suppose I'm in the market for a more dog-eared copy that I can take into the fields and woods and press flowers in. Collecting is a form of madness, I agree.

One reason why New Naturalists tend to be better preserved than some of their contemporaries is that they were not intended for use as field identification guides, so many of them escaped the horrors of the rucksack or the oven-like heat of the car back-seat. Nor were they in general bought for young children to decorate with crayons and sticky substances. I have the sneaking suspicion that a few people having bought them found them too difficult, and so back they went on the top shelf, permanently (this, of course, applies to some titles more than others. Old copies of *The Sea Shore* and *Wild Flowers of Chalk and Limestone* are *always* well used). Some books may not be as old as they seem. When you find that miraculously well-preserved copy of *British Game* or *The Redstart*, be wary of jumping to conclusions about its age. True, the stated printing date is 1946 and 1950 respectively, but these titles, and several others, were cut and bound up according to demand, and there is evidence that printed quires and wrappers were lying around in the vast Collins printing factory well into the 1960s. You can find jackets of *The Yellow Wagtail* which were *printed* in the late 1960s (as proven by the advertisements they carry), though the rest of the book was printed nearly 20 years earlier. Today, this seems an odd way of going about things, a combination of thrift and optimism very much at odds with the present-day book publishing scene.

The green buckram casing of the New Naturalist books has in general withstood the test of time, as have the cotton bindings. The cloth does not stand up well to prolonged sunbathing, however, and jacketless copies of the older books are often faded. The most vulnerable part of the binding, the gilt lettering stamped on the spine, can fade away altogether (or, mysteriously, certain letters fade and others don't). According to Ron Freethy, the New Naturalist board discussed the latter point at length after a reviewer had drawn attention to it, but there was evidently nothing to be done about it.

The most easily damaged part of the book is, of course, the dust jacket. Some people evidently threw away the jackets after purchase (E.B. Ford seems to have been one of them – *see* p. 93), a practice much more common a couple of generations ago than today. New Naturalist jackets were from the start intended to be an important part of the book's design, so much so that Dudley Stamp once suggested they be printed as a frontispiece on the American editions. They were printed on thick paper suitable for lithography, and hence are fairly resistant to chipping and tearing. But they are prone, as every collector knows, to fading, browning and spotting, and not being laminated, soot and grime seems to stick to them. Rust spots and toned paper are to some extent defects of age. It is unreasonable to expect a book to remain as fresh 50 years on than as it was on the day it was sold. Nevertheless, if one could plot the ratio of well-preserved copies against time, I believe it would be not a straight line but a curve. It is much more difficult to find pristine jackets on books printed before 1960 than after 1970, more so than the difference of ten years would explain. One obvious reason is that most New Naturalist dust jackets – reprints as well as new ones – were between 1972 and 1985 sealed inside plastic 'duraseal' wrappers to prolong their life in the shop. Mint copies of any title published after 1972 are therefore easy to find. I suspect that another reason is that, while most houses today are centrally heated at a fairly constant temperature, those in the 1940s and 1950s were usually heated by coal fires or rather inefficient electric fires and radiators. Books would be subjected to cold and damp to a greater degree than today, especially if they were kept in a library or some other room not in regular use. Prolonged exposure to cold, damp conditions result in the book equivalent of pneumonia, the unsightly yellow-brown blotches known as 'damp-stains'. Conversely heat and drying, as for example when the book is shelved too close to a radiator, are said to be responsible for the condition known as 'foxing' in which the jacket, endpapers and page edges develop a rash of red-brown freckles. A certain degree of foxing seems almost inevitable with the older books, and it shows up more obviously on the paler-coloured jackets. It is hard to find a jacket of the 1949 *Snowdonia*, for example, without a few blotches and spots. A small amount of foxing does not detract unduly from the book's collectability (though it should, of course, reduce the price), but damp-staining is a serious minus, since it means that the paper has started to rot. Books stored for long periods in damp conditions also start to smell unpleasantly. This, I think, is particularly unfortunate with this series since the pages of fresh copies have a distinctive and rather pleasant smell, with a touch of vanilla in it, rather like hay drying in the sun. Not the most recent titles though, which assault the nasal senses with the power of plastic melting in battery acid.

The most common jacket defects are fading and browning, both of which are caused by sunlight. The inks used to print the jackets are also notoriously prone to fading, especially the red pigment, which can disappear altogether, and to a lesser extent the green, which tends to turn blue. Those with red-based colours, like *The Badger*, *The World of Spiders* or *The Hawfinch*, are often faded on the spine. Equally sensitive are jackets like *Britain's Structure and Scenery*, *Butterflies*, *The Fulmar* and

British Mammals, where sunburn gradually turns the whites and pale greys to an unlovely sepia. There may be copies of *The Fulmar* in which the bird on the spine is still lily-white and not the colour of coffee, but I have not seen any (but I have seen the proof of the jacket and can confirm that when new the bird was as white as snow). In terms of fading, the worst of the lot may be the orange jacket of *The World of the Honeybee* which, indeed, behaves very much like honey, turning an unpleasant opaque shade of yellow if you leave it standing in the sun.

Most collectors do not like inscriptions of any kind unless they are by the author himself, or someone closely associated with that particular book – say one of the photographers, or a person acknowledged in print by the author. The book should resemble the condition in which it was bought new as closely as possible. There seems to be an unwritten rule that the more interesting the inscription, the more disgusting the condition of the book, or at least that is my experience. It is not at all unusual to find signed copies of some of the later titles, the reason being that at least one enterprising bookseller arranged to have his order author-signed (I wonder more sellers haven't thought of it). On the other hand, a *Butterflies* signed by E.B. Ford, or an *Insect Natural History* by A.D. Imms would have rarity value indeed, though it might not affect the price much.

The degree of defectiveness that one is willing to tolerate is entirely a personal matter, though there are titles like *The Fulmar* where mint, unblemished copies may simply not exist. Some collectors insist on Fine copies of the more recent titles, but for the sake of pragmatism will accept minor tears and spotting on earlier first editions. A few, however, are very, very choosy. Dr Tim Oldham, a natural history bookseller, recalled one couple who would prowl book fairs all over the country in search of the perfect copy. 'Madam would sit whilst her husband ran around bringing her examples from various dealers, and these would be carefully examined. They developed their own terminology over the years especially for these books, and the affair took on the air of a courtship ritual. Alas, they have now completed their collection which I hear is housed behind locked glass in a dark room with the spines turned inward.' (Perhaps one should forebear to mention that keeping books behind glass is not a good idea – they grow breathless, and start to smell of drains.)

Editions

I list the complete printing history of each title, as far as I am able, in Appendix 2. The revisions that went into each new edition were usually quite minor because Collins could not afford to reset the text each time. In some cases a replacement section was added, often in smaller print, such as the chapter on insect flight in the later editions of *Insect Natural History*. Sometimes there were a few less, or a few different, plates but the difference would be barely discernible. The most significant exceptions were the books that were revised for the Fontana paperbacks in the 1960s and then used as the basis for a new hardback edition, at some loss of printing quality.

The text apart, there are two technological changes that the New Naturalist collector should be aware of. The first was the change-over in half-tone printing from gravure (i.e. printed on engraved rotary cylinders) to letterpress (i.e. printed on blocks) around 1960. The half-tones in the older books have a 'period' feel, like contemporary magazine illustrations, while the letterpress plates, printed on glossier paper, have a much more up-to-date look. The changeover was made for reasons of economy; letterpress blocks last longer, and are cheaper to make, than

Table 11. Commonly encountered defects of second-hand New Naturalists in roughly descending order of undesirability and a few compensating merits

Cloth cover dented or bumped (I hope you sued the Post Office).
Owner's neat inscription or modest bookplate.
Owner's hideous inscription or stamp, or 'happy birthday young Timmy from Auntie Moggs and leave the bugs outside next time etc. etc.", all in purple ink.
Jacket chipped but not torn.
Jacket stuck inside adhesive film, under which the inks have started to sweat.
A few tears, the odd chunk missing, spine faded, edges and endpapers sunned and other signs of a busy life opposite a south-facing window.
Torn jacket crudely repaired with sellotape which has discoloured, curled up and died, leaving a dirty yellow reminder to invest in some proper cloth tape next time.
Jacket and edges dirty or badly foxed, pages have lost their original crispness and grown limp, like fortnight-old lettuce.
Jacket damp-stained and discoloured, evidently kept on the window sill, then in the cellar.
Jacket evidently chewed by children or pets, cloth badly faded, gilt lettering nearly gone.
Book has obviously spent some time in the toaster, browned, spotted and lustreless.
Ex-library copy. The jacket may be passable but inside limp, toned pages, rubber stamps, dog-eared corners and, Good God, I hope that's jam.
You surely don't want to descend any further?

Desirable attributes
A nice neat signature by the author.
Very desirable attributes
A nice neat ditto, but addressed to a person mentioned in acknowledgements – known as an 'association copy'.
Dreamable attributes
'To my old friend Billy Collins with thanks for everything from James Fisher.'
The author's own copy, in fabulous condition and full of small, neat marginal notes in pencil.
NB: The last two may not exist.

engraved cylinders. The plates survived the change quite well, gaining in contrast while losing a little sharpness here and there.

Changes in technology also produced visible changes in print quality. Until about 1977, the text was printed by letterpress using the beautiful Baskerville lettering pioneered by Collins in the 1930s. The first book to be printed by the modern method of film-set was *Inheritance* (1977), while all titles since *The Natural History of Orkney* (1985) have used a new, smaller typeface set by contractors using desk-top methods. Some of the titles reprinted in the 1970s were printed by off-set lithography instead of letterpress, and, again, the loss of quality shows in the lettering. However much we regret the changes, without the advent of less labour-intensive ways of printing books it is very unlikely that the series would have survived at all. The changes in jacket production from letterpress to off-set lithography have been discussed in Chapter 4. The difference is apparent when you compare the two. Jackets printed by letterpress are on thicker, creamier paper, the colours are softer and more clearly differentiated. Offset litho, coupled with the use of more transparent pre-mixed inks, gives a 'harder' finish, the colours are deeper and the overall effect is slightly flatter. I have the impression that the offset litho jackets wear better than the letterpress ones, probably because with the former the inks used were pre-mixed and react less to light. The difference between the two sets of jackets is rather similar to that between a limited print and a mass-produced colour reproduction. The forcefulness of the design does not suffer, but there is some loss of printing quality.

New Naturalist hunting

Acquiring a collection of New Naturalist books is a mercifully straightforward business, unencumbered by the special boxed presentation copies of *editions de luxe* on handmade paper. No one seems much interested in the Book Club or American editions, nor the paperbacks. The Bloomsbury reprints are handy for filling the gaps on the shelf, but no true New Naturalist collector will touch them. A straightforward collection of the hardback titles in the best condition the purchaser can afford is the almost universal aim. Some collectors stopped at a particular point – often at No. 50 (1967) or No. 70 (1985). Today, the monographs are at least as sought after as the main titles, though 20 years ago booksellers had to practically give them away. The New Naturalist journal is worthy of honorary NN status, or would be if it matched the other books better. And many regard New Naturalist No. 6 as two separate books, *The Highlands and Islands* being a substantially rewritten version of its predecessor *Natural History in the Highlands and Islands* with reduced colour plates.

To judge from book fairs and booksellers catalogues, by far the most desirable New Naturalists are first editions. There is not, however, the enormous disparity in price between first editions and later ones that you find in other parts of the book market. I would say that, on average, the first edition is worth about 33% more than a reprint, or perhaps as much as 50% for some of the scarcer or more recent books like *Weeds and Aliens* or *Man and Birds*. First editions of early titles like *Butterflies* and *Britain's Structure and Scenery* are difficult to find in fresh condition, even though 20,000 of each were printed. From the point of view of sheer usefulness, one might have considered that the later editions would be more in demand, especially the minority of titles that were revised substantially like *Life in Lakes and Rivers* and *A Natural History of Man in Britain*. Unfortunately production standards were slipping by the late 1970s and in some of the latest reprints the number of colour plates were drastically pruned or, in one notorious case (*Butterflies*), omitted altogether. Which is preferable, the 1977 *Dartmoor* with an updated chapter but less than half the original colour plates, or the badly out-of-date 1953 first edition? Much will depend on what you are buying the book for. If, however, you hope one day to sell the collection at considerably more than you paid for it, you would be wise to stick to first editions in Fine condition. Protect the jackets in cellophane, store the books on open shelves behind a dark curtain and let no one near them. Which, I think, would be a shame.

A word of warning about dust jackets. As the books become more expensive, fraudulent jackets are appearing on the market. There are two main types. One is the colour-photocopied dust jacket, until recently easy to spot as a crude travesty of the real thing. But modern laser-guided copying techniques produce a much more accurate reproduction of the original jacket. They still look suspicious, however: the paper will probably be chalk white without that subtle patina of age that even the best preserved genuine jackets tend to have. But the latter can be easy to miss when the fake jacket is clad in cellophane. The other fraudulent practice is to wrap a jacketless first edition in a later dust jacket. It is usually obvious when this has happened, since the titles listed on the back will not correspond with the date of the book. These things are sent to try us. In my experience they are still quite rare, and no reputable bookseller would dream of doing such things (though they are sometimes caught out by it themselves). But bear it in mind at that car boot sale or bargain mail order.

Where do you find old New Naturalists for sale? Many second-hand bookshops will stock at least a few, and the better ones may have a shelf-full or more. The more there are, the more expensive they are likely to be. Here you will find *British Games, London Natural Histories* and *Sea Shores* a-plenty in varying states of hygiene. A second source is booksellers catalogues. A few years ago, many booksellers specialising in natural history, gardening or country sports used to offer a good range of titles. Though some still do, I have the feeling that the supply may be drying up. The reason is that collectable copies are fetching higher prices, and in this spiral of buying and selling the bookseller has to pass this on to his or her customers. I know of two or three booksellers who no longer stock many New Naturalist titles because the price of the latter has grown too high for their particular market. This trend becomes worse when the books have been purchased by a middle man so that there might be two lots of 25% to 33% profit margins to pass onto the customer. In these circumstances, the New Naturalist market may be moving out of the province of people who actually use the books to the book connoisseur who is willing to pay much higher prices for them. One should not overstate this trend, however. While one might have to pay £100 plus for Fine copies of the scarcest titles, most of them are still available for £30 or less, especially if you take your time. In other words, many out-of-print titles are still available for no more than the price of the recent hardbacks (considerably less in some cases). One hears stories of someone buying *The Fulmar* or *British Warblers* for £300. The stories may be true, but whoever it was was wasting his money and setting a bad example.

The best way of all to find New Naturalists in plenty is to attend one or more of the PBFA (Provincial Booksellers Fairs Association) book fairs held regularly in central London and market towns across the country. One can find details of these in the magazine *Book and Magazine Collector* (whose advertisements are another source of New Naturalists, though not usually a very productive one). The best venue for that rare title you are searching for is the natural history book fair held at Kew Gardens each September. In 1992, I saw nearly all the New Naturalist titles there, often in superb condition and (in some cases) at reasonable prices. There used to be several booksellers who carted a complete set around from fair to fair, but at the time of writing they seem to have gone to ground. Still, the second-hand book world is an ever-changing kaleidoscope of tiny businesses, and doubtless we shall see more complete sets gleaming on their special shelves at collecting-idiots only prices.

Which are the rarest hardback titles? That is easy; there are two:- *The Natural History of Orkney* and *British Warblers*. Both were published in 1985 at the time when the series was switching from casebound to paperback production, and only 725 copies of each were printed in hardback for libraries and 'the collectors' (see

Chapter 10). By the time the publishers realised that they had grievously underestimated the market, it was too late to print more. Some 500 paperbacks of each title were stripped and rebound as hardbacks, but since these were slightly smaller in size and have other minor differences, no one seemed to want them. To make matters worse, there was some crafty speculation in *British Warblers*, and many people missed their chance of buying one. It is still not too difficult to find mint copies of the 'first edition, first state' but the asking prices are always well in excess of £100. The rebound copies of 'first edition, second state' are on offer at about half that. Not surprisingly then, this is the commonest gap on the shelves.

Modern book collecting is a mad looking-glass world in which the least successful titles become the most sought after. How we sigh now for *A Country Parish*, the first main series title to go out of print (undeservedly, for it is an attractively written and beautifully illustrated book). Particular titles seem to swing in and out of fashion. Twenty years ago, the most sought-after title was *Sea-Birds*, and that book has actually fallen in price since then. Ten years ago, it was *Dragonflies*, following a sudden burst of enthusiasm for these insects reflected in the founding of a British Dragonfly Society. Today, it might be the monographs, sought by people who have completed their main series titles and don't want to stop there.

One title that is always sought after and expensive despite a fairly long and successful life in the shops is *The Art of Botanical Illustration*. This is one of the few that sells readily *without* a jacket. The reason is the demand from the art and gardening market for this classic textbook, both here and in the United States. It will be interesting to see if the recent publication of a new enlarged edition of '*The Art of*' reduces the demand for the original. This book also scores very highly in terms of beauty and readability. For the same reason, *The World of Spiders* and *The Open Sea* books are always in great demand, and command higher prices than one might otherwise have expected. Certain others like *Weeds and Aliens* and *The Pollination of Flowers*, are standard works on their subject, so that demand for them is continuous and by no means confined to collectors.

I have noticed a marked upward trend in the prices of the New Naturalist titles published between 1972 and 1983, which probably reflects growing scarcity. The print-run of these books was much smaller on average than their predecessors, and all are now out of print. Even *British Tits* and *British Thrushes*, which were a book club choice and so were printed in large numbers, are not the common or garden birds they were a few years ago. I do not want to encourage speculation by picking out individual titles, but I would quietly suggest that if No. 67 *Farming and Wildlife* and No. 68 *Mammals in the British Isles* are still gaps on your shelves, do snap them up as soon as you can because their print-run was miserably small. My other 'top tip' would be to persevere with the series and buy each new title in hardback as they appear. The current print-run of 1,500 hardback copies guarantees that one day they will be one with *A Country Parish* and *Pedigree: Words from Nature* unless people lose interest altogether.

Collecting in 2005

'Look at a collection of New Naturalists in their beautiful dust wrappers!' cries a bookseller of Tintern and Hay-on-Wye. 'Sensual and alluring? – absolutely!' 'Touch the green buckram bindings, the top quality paper. Savour the visual and tactile sensations.'

One does enjoy a bit of enthusiasm. When I was writing *The New Naturalists*, the series was bought and collected mainly by natural-historians. Today that may still be true, but there is a growing market for the books by bibliophiles and book collectors. The fact that titles immediately double in value the moment they go out of print (which is soon enough) also means there is inevitably some speculation in the secondhand book trade. The demand for mint, shop-fresh copies of recent titles has resulted in a steeply declining graph for books in less gleaming condition. In particular, a faded spine will knock a significant lump off the price, and for some titles that could be upwards of a hundred pounds.

Collecting fashion has also swung in favour of first editions – a sure sign of the ascendancy of the book collector over the naturalist. This does not affect recent titles, which are nearly all *de facto* first editions, but it means that earlier books in the series may be worth twice as much (or more) than reprints in comparable condition. For the user, as opposed to the collector, the opposite is true – we should want the *latest*, not the earliest version. At least we will get them at, relatively speaking, a bargain price.

In 1995, the most expensive titles by far were the hardback *British Warblers* and *Natural History of Orkney* which, as we all know, were disastrously under-printed. What I completely failed to predict was the rise in price of even more recent titles, especially those from volumes 72 to 81 inclusive – *Heathlands* to *Ladybirds*. Only 1,500 or so hardbacks of each were printed at this time of high pessimism, and they all sold out within months. If you didn't buy them right away, you missed the boat. *The Hebrides*, I remember, sold out so quickly that some unfortunates who had ordered it from their local bookshop went away empty-handed and, obviously, not rejoicing. Titles from No. 93 onwards were printed in hardback in slightly larger numbers – around 3,000 for the more recent ones – but even these tend to go out of stock within a year, and always within two years.

Unfortunately, as every collector will know, often to his cost, these now very pricey books have dust jackets printed in non-fast colours. It is hard to find a copy of *Freshwater Fish* whose strong red title band has not already faded on the spine to a duller shade of magenta or pink. Or a *Wild and Garden Plants* on whose paling yellow band the tiny title lettering has not virtually disappeared. Or a *New Forest* on which the tawny fallow buck has not faded to an unrealistic yellow, amid equally surreal blue tree-trunks. There is not, it would seem, a lot we can do about it unless we store the books in a darkened vault, though the colours will last much longer when kept well away from direct sunlight.

A short collector's guide to the New Naturalists

In the original edition of *The New Naturalists* I drew up a guide to the value of each book based on a system of stars. Slightly facetiously, and because this is not an exact science, I compared the cheapest category, a single star, with the price of a good but unostentatious bottle of wine, that is to say, about £10 to £15. Only the rarest and most sought-after books received four stars, which wasn't categorised explicitly, but was certainly meant to mean upwards of £100. The trouble with this, or perhaps any other method of valuing, is that books do not have a fixed price (not even in shops, thanks to the collapse of the Net Book Agreement). It all depends on condition, and on the market. At auction, including online sites such as ebay, a book is worth whatever the highest bidder feels like paying for it. Every book has its day. Thirty years ago, the New Naturalist with the highest price tag (about £50) might have been *Sea-Birds* or *A Country Parish*. Ten years ago, *Orkney* and *Warblers* were the first titles to break the £500 tape, at least on a 'good' day. Today certain other titles are catching up, notably *The Hebrides*, with *Freshwater*

Fishes, *Ferns* and *Ladybirds* not far behind. But it all depends on condition. A perfect *Freshwater Fishes* – and perfect for a book published in 1992 means really perfect, shop-fresh, unfaded, preferably unopened – might make £400, perhaps even more if you had two reckless and over-excited bidders. One with an even mildly sunned spine would be worth less than half that. Perfect copies, you see, are already rare, while slightly blemished ones relatively commonplace (commonplace, that is, in a fairly rare book). *Butterflies*, the first title in the series, is a common book, but a shop-fresh first edition in a like jacket would be about as rare as a gold tooth in the jaws of a duck. A really nice 'first' might set you back £50, even £75, while for another copy which has long lost the bloom of youth, you might pay £10. And a copy without a jacket would be the sort of book you pluck for a pound or two at a flea market or a market stall ('tell you what, guv, three for £1.50'). So this revised star-guide does not represent absolute value but a sliding scale of values from mint (or as near to mint as age allows) to what is politely known as a reading copy. Today, first editions are worth up to twice the amount of a reprint. For the scarcest titles, of course, there is only one hardback edition. The New Naturalist 'law of reversed success' applies: the most sought-after titles today were the least sought-after at the time.

One star means that you should be able to find a perfectly nice copy for not much more than £10-15, a stunning one for perhaps £20-30. At the cheap end, technically 'good' copies are available for a fiver, and bad ones should be given away.

Two stars are a bit pricier, say £20-30 for a reprint, and up to £50 for a really tip-top first edition. There is a steep downward slide towards blemished copies which no one really wants, unless you are more interested in the contents of the book than in collecting it. It still happens.

A three star book is a bit elusive, and you'll be lucky to find a nice one without paying a specialist's high prices (though that happens too). We are probably looking at £50-£100 here. In other words three stars is worth a bit more in 2005 than it did in 1995. Time was when only three or four books in the whole series would have earned three stars, and we could have blown the rest on high living. Ho hum.

Four stars means that the book has become rare. Really rare. Join the queue to pay more, often much more, than £100. A row of titles in gleaming mint jackets from No. 69 to No. 82 – the unlucky 13 – taking up about fifteen inches of shelf space would now cost you – what? – £5,000? I don't really want to think about it.

The following estimates are culled from recent book catalogues and online bookshops, ebay auctions and advice from Robert Burrow, who runs a bookshop and 'The New Naturalist Book Club' in Jersey. Thanks to Bob for all his help here.

Butterflies. A common title, but early ones are maddeningly difficult to find in fresh condition. They really *used* their books back in the 1940s. There are two forms of dust jacket: the first three editions were produced by letterpress but from the mid-1960s, they used a screen that turned white into a grubby shade of grey. Beware of the colourless 1977 edition. Rating: * + * for a first edition.

British Game. Printed in huge quantities and often surprisingly well preserved. That is because although the book was published in 1946, copies were being bound up from printed sheets long after that. I think this may be the commonest title in the series, which is odd because it is also probably one of the least read. Rating: *

London's Natural History. Also quite common though usually showing its age. Look for the rare copy with clean edges, tight, springy binding and a clean little tufted duck at the bottom. Rating: *

Britain's Structure and Scenery. The best seller of the series but surprisingly difficult to find a first edition without the usual spots and stains. From 1967 the colour plates were reduced from 48 to 32. From the mid-60s, the jackets were printed by offset lithography using a screen that gives a newsprint-like effect; the earlier ones were printed by letterpress with cleaner colours. Rating: * + * for a first edition.

Wild Flowers. The first of the dark jackets, effectively hiding spots, freckles, stains etc. Easy to find in Fine condition. Minor differences (including plates) between editions. Rating: *

Natural History of Highlands and Islands. Not uncommon and often well preserved. All are first editions. Worth buying for the colour plates. Rating: *

The Highlands and Islands. Though the jacket is the same as the preceding title, this is really a separate book. 1964 is the 'first thus'. Plenty around still. Technically a better book than *Natural History*, though only 4 colour plates compared with 32. Rating: *

Mushrooms and Toadstools. One of the best New Naturalists, and plenty of them about, though a fresh first may be elusive. The jacket of the 1977 reprint was photographed, then litho-printed, a disgusting sight. Rating * + * for a first edition.

Insect Natural History. Still a common title, though early editions are usually sunned and/or frayed. The later ones in duraseal wrappers are invariably in near-Fine condition. The 1971 edition contained a new chapter on insect flight set in a smaller typeface. Rating: *

A Country Parish. A scarce book, out of print since 1959 and hard to find in VG, let alone Fine, condition. Jackets are usually foxed, and often sunned and stained. But 7,000 were printed, so there *ought* to be enough copies to go round. I wonder where they are. Rating: ****

British Plant Life. Plenty around in varying degrees of wear and tear. The first editions usually sport interesting collections of spots and stains. The jackets of later ones were printed in wishy-washy colours which reduces the effect of the design. Rating: *

Mountains and Moorlands. Plenty of these, too. The 1971 edition is partly rewritten by Winifred Pennington though the plates are the same. Rating: *

The Sea Shore. Another of the commoner titles, with an enormous first edition. The pink shells on the spine tend to fade, and a good bright first edition might be elusive. Rating: * + * for a first edition.

Snowdonia. With 18,000 printed, this is not a scarce title, but you might have to hunt for a tight, bright copy – this was a bulky book and strained its bindings; there always seems to be some foxing, which is allowable, and sometimes damp-staining as well, which is not. Two states exist, one with authors' names on the spine of the book (but not the wrapper), the other without. The latter is probably earlier. Rating: **

The Art of a Botanical Illustration. Not scarce, but always in demand in almost any condition. At the moment though, it is decidedly *** in Very Good to Fine condition. Sun-tanned jackets abound, and it doesn't suit them.

Life in Lakes and Rivers. A lot of updating to successive editions, especially the 1974 one, blown up from Fontana plates. Freshwater ecologists will want the later, book collectors the earlier, edition. Another common title. Rating: *

Wild Flowers of Chalk and Limestone. Always a popular title for use on botanical excursions and hence often sunned from lying on car seats. The first edition is difficult to find in Fine condition. Check the 1969 edition for loose pages. Rating: **

Birds and Men. Though long out of print, it was printed in large numbers and is still quite a common title. Look for a nice clean one, and inspect the reverse for grime. They are all first editions. Rating: * + * for a clean copy.

A Natural History of Man in Britain. The second edition (1970) was a substantial revision bringing in Margaret Davies as co-author. Nice second editions easy to find, fresh firsts scarcer. Rating: * (*)

Wild Orchids of Britain. A popular book, always in demand. The second (1968) edition has an updating introduction and a new set of distribution maps. Rating: **

British Amphibians and Reptiles. A much-revered herpetological classic, and attractively illustrated, so always in demand. The jacket spine is prone to fading. This book was continually updated through five editions. Rating: **, may be an extra star for a fine first.

British Mammals. Like other pale covers, this one quickly shows signs of wear, especially from sunning, and, like other big books in the series the binding may show signs of strain. Dated, but still a classic. Rating: * (*)

Climate and the British Scene. Quite easy to find, and attractive. The switch to off-set did not suit the jacket design. The post-1962 editions are often in collectable condition; not so the earlier ones. Rating: * (*)

An Angler's Entomology. The printer made a dog's breakfast of the later editions of this title, mixing letterpress and litho with heedless abandon. The first edition escaped that fate. Rating: **, with an extra star for a fine first.

Flowers of the Coast. Distinctly scarcer than most of its contemporaries and with a particularly spotting-prone title, so I will award it two stars ** + maybe an extra * for a Fine first edition.

The Sea Coast. Another middling title, not common, not rare. The Third Edition included a new appendix about the 1953 floods. Rating: **

The Weald. Only the first edition has the full set of excellent Goldring plates, the highlight of this rather dull book. Demand from residential field courses kept it in print. Rating: **

Dartmoor. Only the first edition has all the colour plates and the 1977 reprint had hardly any. It did, on the other hand, carry a new chapter by the author. Your choice. Rating: * + * for first a edition.

Sea-Birds. At one time the most sought-after book in the series, but rather less so now as it is deemed to be out of date. Not as rare as it is cracked up to be: 10,000 were printed, plus another 2,500 in the States. They are all first editions. The colour band on the spine tends to fade. Rating: ***

World of the Honeybee. The size shrank with successive editions and the latest one which lacks the original prefaces was Lilliputian. The orange jacket turns yellow like lightning, especially on the spine. Keep well out of the sun. Rating: * + * for a first edition.

Moths. This is one of the titles where first editions are offered at twice the price of reprints. The dark wrapper ages well. Minor differences over three editions. Rating: **

Man and the Land. Only the first edition has the full set of colour plates. Successively revised over three editions. ** for first.

Trees, Woods and Man. Only the first edition has the full set of colour plates. Their number had shrunk from 24 to 12 for the last (1978) one. The 1978 ed. had a new preface lamenting the loss of the elm. Rating: *

Mountain Flowers. Another popular botanical classic, in demand for excursions which might explain the scarcity of Fine copies (mine went with me to Loch Tay, where I spilt a cup of coffee over it). Rating: **

Open Sea. World of Plankton. Always in demand for its beauty and readability, and as a classic by a famous biologist. The print-run was larger than its companion *Fish and Fisheries*, but the latter was in print longer. So honours are about even at **, or £70 for a Fine first, reprints about half that.

The World of Soil. Dark cover, but with red in it so prone to fading. Still quite easy at * + * for a first edition.

Insect Migration. A limited print-run on a specialised subject, and becoming scarcer. Rating: **

Open Sea. Fish and Fisheries. Always in demand for Hardy's watercolours and pleasant writing. Rating: **, as with *World of Plankton.*

The World of Spiders. A beautiful book, always in great demand. The title band on the jacket is hard to find unfaded, especially in the first edition. Rating **(*)

The Folklore of Birds. One of the scarcest titles in the series, on sale for only four years. On the other hand, those copies that do appear are often in excellent condition. Their price indicates that this is unfortunately a **** book.

Bumblebees. Also scarce, with a modest print-run. Like *Honeybee*, this title was published in slightly smaller format than the other titles: it seems they couldn't decide whether this was a main series title or a monograph and so compromised. Now a rare book, £150+ for a Fine first, half that for the reprint. ***(*)

Dragonflies. A scarce and famous book, always in great demand by dragonfly enthusiasts. Often in nice condition when found, but I fear that the days when you could discover one at your local jumble sale are over. Rating: ****

Fossils. Not difficult, though the dark jacket tends to fade. The first edition was the same as *Dragonflies* and *Bumblebees*, and so must be quite scarce. The book is badly out of date. Rating: * + * for a first edition.

Weeds and Aliens. Another sought after title, out of print since 1972. The first edition fetches double the price of the small reprint, for some daft reason. Rating: ** (*)

The Peak District. Still fairly common, the small first edition commanding at least twice the price of later reprints. Rating: * (*)

The Common Lands of England and Wales. In printing terms, almost as scarce as *Weeds and Aliens*, but you would never know it from the price. Rating: ***, dock a star for the reprint.

The Broads. One of the more elusive titles in only one edition. The supposedly modish laminated jacket looks old fashioned now, and is often chipped and torn (and also cracked, like crazy paving). Currently around £150 for a Fine copy. Rating: ***

Snowdonia National Park. Still quite common and easy to find in perfect condition: readers seem to have battered their *Broads* but safeguarded their *Snowdonias*. Rating: * + * for a first edition.

Grass and Grasslands. Scarce in print-run terms, but it didn't sell and remained in print throughout the 1970s, and so is not hard to find. Rating: **

Nature Conservation in Britain. Quite a common title, but note that only the first edition has the important foreword by Dudley Stamp. The second edition (1974) was updated, but the text figures are inferior. The red no-entry sign fades to orange. As most nature sanctuaries now 'welcome' visitors, maybe the red should change to green. Rating: *

Pesticides and Pollution. Everyone bought the Fontana paperback, and the hardback in its dramatic wrapper is less easily found than one might expect for such a well-known book. From now on, New Naturalist titles are easy to find in near-perfect condition. *Pesticides* was the last without duraseal wrappers or laminated jacket. Rating: **

Man and Birds. Another title where the first edition fetches almost twice as much as the reprint. So: *** for first and ** for reprint.

Woodland Birds. Same remarks apply. *** and **

The Lake District. Like many of the 1970s titles, this one is scarcer than it was, and the first edition is now definitely a two-star, if not a three-star, book, so **.

The Pollination of Flowers. A big book, still in use by university students and in demand from botanists. *** for the first edition, knock off a star for the reprint.

Finches. One of the better-selling 1970s titles, and still in demand due to its high reputation and pretty jacket. Again, a premium for the modest first edition, thus * (*).

Pedigree. Scarce, though frequently on display as a bookseller's 'prize' copy and usually in Fine condition (was this title subject to some speculation?). Only one edition. I would award it three stars rather than four – many book fairs will have it. Rating ***

British Seals. Don't tell anyone, but this title is actually scarcer than *Pedigree.* Until recently *Seals* was underpriced, but not any more. Very much a case of buy now while stocks last. No one seems to want the almost identical US version published by Taplinger. Rating: ** and rising.

Hedges. Plenty of copies about still, snug inside their nasty durasealed jackets, which afflict most of the titles published between 1974 and 1986 (a pity they don't come off, isn't it?). Again, while reprints are as common as sparrows, first editions are more like song thrushes: not as common as they used to be. Rating: *(*)

Ants. A scarce title, on sale for only a few years. Always seems to be in mint condition when found, may be because it was one tough read. It probably scrapes into the three star category now. Rating: ***

British Birds of Prey. The best-seller of the 1970s and still common, the now scarce first edition being much more desirable than the reprints. The book club edition has cheaper (darker green) bindings and no price on the front flap, though it does have 'Collins' on the spine. Both are issued in protective plastic film. The last reprints were issued in rather thin, wishy-washy jackets. *** for a first, ** for reprints.

Inheritance. Becoming harder to find and the price is rising. The first film-set title. Rating: **

British Tits. A large club edition (dark bindings, no price) and partial remaindering ensures plenty of copies about. ** for first, * for reprint and book club.

British Thrushes. A large club edition and partial remaindering made this one of the commonest of the recent titles in the late 1980s, most of which seem to have been sold by now. The red spines turn orange in sunlight. Rating: **

Natural History of Shetland. Silly prices are being charged for the first edition in its matt durasealed wrapper, which is not all that scarce, while no one seems to want the much cheaper 1986 reprint in its laminated jacket. *** and **

Waders. Still a fair number of copies around; specialist bird-book sellers usually stock it, ditto *Tits, Thrushes* and *Finches.* Do keep that horrible pink title band out of the sun. The first edition has crept into two stars, so *(*).

Natural History of Wales. Already quite scarce and moving up in price (about £65 for a fine copy in 2004). **

Farming and Wildlife. This title had a small print-run, and was on sale for only five years. Not surprisingly, it has become quite scarce. First-edition fetishists check: there was a reprint in 1983. **(*)

Mammals of the British Isles. Like all titles post-1981 destined for a short life in print, though I think the NHBS stocked it until 1991. Not as rare as the books on either side of it, for some reason. The weasel on the spine fades easily. No reprints of this one, though there was a Book Club edition with cheaper binding. **

Reptiles and Amphibians. Though the print-run was actually greater than the first editions of the previous two, it is now a much rarer book, and had, I remember, only a passing presence in the bookshops. Beware Book Club edition with cheaper binding. ****

Natural History of Orkney. The original hardback is probably the most elusive of all the mainstream New Naturalists, though the print-run tied with that of *British Warblers.* A copy was sold for £1,000 in 1998! The unsatisfactory 'second state', restricted to 500 copies, is worth less than half the original, but that is still a lot. Even the paperback is scarce. Beware 'fake' hardbacks built up from paperbacks, with laser-printed jackets. **** (both states). Pbk: **

British Warblers. Ties with *Orkney* as the rarest hardback title in the series, thanks to the unfortunate decision in 1985 to switch to paperback production, restricting the hardback edition to 750 copies for libraries and NN collectors. The 'second state' (paperbacks turned into hardbacks) of 500 copies is worth less than half the original hardback. The paperback is much commoner than the Orkney one. **** (both states). Pbk: *

Heathlands. The scarce first edition is worth much more than the reprint (which was available into the 1990s from the NHBS) or the large Book Club edition (which has a similar jacket but without the NN oval, and with cheaper bindings). The paperbacks of this and subsequent titles are not often collected, though they grow elusive, once the remaindered copies have passed away. The first edition is now around the £150 mark, so nudging into ****. ** for the reprint.

The New Forest. The hardback is scarce, and the spine vulnerable to colour-change. Like nearly all subsequent titles, the hardback sold out very quickly, while the paperback lingered for several years. Former on sale in late 1990s for £200-250, so: ****.

Ferns. The hardback in its delicious jacket has gone up in price quite horribly. The reason: only 1,600 books, twice that many people wanting it, and only a few copies on the market at any time. Beware inferior Book Club editions with dark binding. My copy was accidentally shot-blasted by a duck that had strayed into the study and then panicked: minor damage, which is to say, about £300 worth. Fine copies on sale in 2001 for £350-500. Shocking, isn't it? ****

British Freshwater Fish. Like *Ferns,* this title has become remarkably scarce and sought-after. It is also one of those most sensitive to sunlight, courtesy of its bright red title band. The hardback NNs 74-81 were restricted to 1,500 or 1,600 copies, the smallest hardback editions in the series apart from Orkney/Warblers. Expect to pay £400+ for a mint copy, half that for one with still bright but faded colours. ****

The Hebrides. Currently the most expensive of all the New Naturalist main series with the usual exception of first state Orkney/Warblers. I don't know why that is, since the hardback edition was apparently the same as other titles published in the early 90s. In 1998, I saw one on sale for £800! Even the paperback is rare. ****

The Soil. Another title with a light-sensitive jacket, in which the rich brown soil debris gradually turns to bleached-yellow desert over the years. There wasn't the rush for this more academic title that there was for *Hebrides,* and the remaindered paperback was on sale for a few pounds in the mid-90s. Even so, with only 1,500 hardbacks printed, it was bound to go up in price eventually, and it has. Currently around £150-200 for a mint copy. ****

British Larks, Pipits and Wagtails. The short-lived hardback is now scarce, the paperback much less so (indeed my local bookshop had a copy on sale in 2002). Prices for a mint hardback now nudging £200-250, though demand is fickle - sometimes they sell, and sometimes they don't. ****

Caves and Cave Life. This one took a while to creep up in price. Up to about 2001, you could find a hardback for under £100; some sellers are now charging £200+. ****

Wild and Garden Plants. Bob Burrow, who sells New Naturalists in Jersey, tells me this has seems to have become one of the scarcest titles. There is probably some law of chaos or randomness at work here; one month it will be *Wild and Garden Plants* that has gone to ground, in another *Ladybirds*, or *Ferns*. The hardback was snapped up, as usual. The paperback lingered for two years, before being ruthlessly remaindered. Recent hardback prices around £200+. Always look at the light-sensitive spine first. ****

Ladybirds. The first title to hit £30, but the big, chunky hardback was snapped up regardless, and was o/s by early 1996 after a shelf life of 18 months. The remaining paperback stock was bought by a specialist entomological bookseller in 2000. Another scarce and expensive book: available at around £100-150 in 1998, prices have risen to £350 or more for a mint copy. ****

The New Naturalists. The first edition of this title, of which the book you are reading is the second, went out of print within a few months, but was reprinted in 1996. I think all of them had a big white sticker attached to announce the 50th birthday of the series. You are advised to keep them on. The orange books on the spine are, alas, apt to turn yellow. The paperback is scarce, and, as paperbacks go, in demand. Prices I've seen quoted recently are in the £200 range for the first edition, £100 for the reprint and £50-70 for the paperback. Don't come to me for copies; I flogged or gave them all away years ago. ****, one might say.

The Natural History of Pollination. There are two 'states' of this book. The true first edition has a yellow title band, the words 'New Naturalist library' on the half-title, and the editors and aims of the NN series listed as usual on the title page. These are missing on the 'second state', which also has a deeper yellow title band and a price of £35 instead of £30. Both say 'First published 1996', and so it is not immediately obvious that the second state is technically a reprint. The first state is, of course, worth more than the second, though three times as many were printed. There is also an American edition, published by Timber Press Inc. Going up slowly in price, the former is currently about £100, the latter about £60. *** (**)

Ireland. The true first edition hardback sold out within eight months of publication. About 900 further copies were bound up from stock, and, as far as I know, there is no reliable way of telling 'first state' from second. Until one can, they are in practice all one edition, and an unusually generous one. 3,000 hardback copies of *Ireland* were printed, twice as many as its immediate predecessors. Even so, the book was o/s by February 2000, and it now commands a secondhand price of about £80-100. This is likely to rise. ** for now.

Plant Disease. First published in October 1999, the hardback sold 'surprisingly well' and was o/s by July the following year, though not reprinted. The paperback was a slow seller, and the last 500 were remaindered in June 2001. With only 2,000 hardback copies printed, this is set to become the rarest title since 1995. Buy now, whilst it's only **.

Lichens. Another fast-selling title, which helped to convince the publishers that they could now sell 2,500 to 3,000 hardbacks of any NN title, even at £35. It had sold out by the end of 2001. ** for now.

Amphibians and Reptiles. The hardback edition was increased at the last minute to 3,000 'to meet expected demand', a print-run shared by NNs 90-94. Two-thirds were sold in the first two months, and the book is now o/p, though you could still find it on sale in 2004. ** for now.

Loch Lomondside. The hardback print-run of *Loch Lomondside*, *Lichens* and *The Broads* was 2,500, but whether this will come to affect the secondhand price it is too soon to say. As usual, the hardback sold well, and was o/p by the end of 2002, while the remaining paperback stock had to be remaindered. ** for now.

The Broads. Published in November 2001, there were only 180 copies left in stock four months later. The hardback was o/p by 2004. ** for now, but I've a feeling that this one will rise and rise.

Moths. Hardback print-runs of 3,000 are by now the norm, ensuring a slightly longer life in print. Within four months, 2,000 hardbacks had been sold, but only 700 paperbacks. Will collectors be paying ten times their current price in ten years time? Don't ask me, I don't know nothing. **, as new.

Nature Conservation. This title, my second New Naturalist, sold as well as the others, but I had hoped for better things. The paperback, I see, is already o/s by spring 2004, when the hardback had nearly sold out, and I don't suppose it will be reprinted – ****! Oh – it has two stars.

Lakeland. Published October 2002, and still in print late 2004.

Bats. Published March 2003, and still in print late 2004.

Seashore. Published April 2004 and in print.

Northumberland. In print.

Monographs

Up to about 1980, the monographs would sit on the second-hand bookshelves like birds on a branch, while people snapped up copies of *Sea-Birds* and *Dragonflies* for thirty times the original sales price. Not all New Naturalist fans collect the monographs, but many do and for most titles there are fewer copies to go round. They are offered at prices that range from bargains to near-swindles, and I urge you to shop around for the commoner ones. The scarcer single-edition titles will always be expensive, for demand probably exceeds the supply of collectable copies. Consider a title like *Hawfinch* with a print-run of about 3,000. Assume the 'standard' wastage of 25% for books of that vintage, and that perhaps 500 copies went to public libraries and institutions. Suppose further, that of the remainder about half were bashed-up or lost their jackets, and we are left with fewer than 800 copies in collectable condition. And I reckon that there are over 1,000 New Naturalist collectors. Enough said?

The Badger. With no fewer than six editions, this grand old classic still graces many second-hand shelves. Most first editions show their age, however, with that infernal salmon-pink spine fading away to grey. Rating: * + * for a mint first edition.

The Redstart. Although there was only a single edition dated 1950, *The Redstart* (or the '*Restart*' as it is often listed) and *Yellow Wagtail* were obtainable throughout the 1960s in fresh jackets, hence the surprisingly fresh state of preservation of many copies. Rating: **

The Wren. A scarce book, but the price varies according to condition, Fine Wrens being at a premium. Currently averaging around £200, I saw one presumably tip-top copy on sale for £425. Decidedly ****

The Yellow Wagtail. See *Redstart.* Some copies of *Wagtail* have sooty smudges on the jacket, presumably a printing fault. This book seems marginally commoner than *Redstart,* though the print-run was about the same after 1,000 *Wagtails* had been pulped. Rating: **

The Greenshank. Another scarce book, and usually sunned. Like all Nethersole-Thompson's bird monographs, it is in demand from a large ornithological market. Rating: *** Prices currently in excess of £100 for Fine copies.

The Fulmar. Widely regarded as a book of legendary scarcity, though probably no more so than *Greenshank, Heron* or *Hawfinch.* The spine is nearly always sunned, and the strain of 500 pages often tells on the binding and jacket. Like the other scarce monographs, you should try shopping around for *Fulmars.* A presentation copy from James Fisher was on offer for £150 recently, but you should manage to knock at least a third off that for a copy you could live with. Rating: ****

Fleas, Flukes and Cuckoos. The first edition is scarce, the two reprints less so. Hence a large disparity in quoted prices. ***(**)

Ants. Despite its short life, this is by no means the most difficult of the monographs. The surprise is not that it is scarce, but that we find it at all. ***.

The Herring Gull's World. In demand as a classic study of animal behaviour. Rather scarce as a first edition, commoner as a reprint. The later reprints shamelessly omitted the foreword by Konrad Lorenz. Avoid the last, inferior reprint of 1976. Rating: *(*)

Mumps, Measles and Mosaics. Currently the rarest of the monographs, possibly because of its dull jacket. A Fine copy was sold recently for £300. ****

The Heron. Scarce, especially in its striking dust jacket, and only one edition. In recent years it has edged from *** to ****.

Squirrels. Not among the scarcer monographs, but still in short supply and a nice little book, too. Rating: ***

The Rabbit. Scarcer than *Squirrels,* and possibly more *** than **, though much depends on condition.

Birds of the London Area. Scarce, but never a high flier. I had never seen a copy in its dust jacket until 1992 when it seemed suddenly to be everywhere, though no one seemed to want it. It's now scarce again. ***

The Hawfinch. Scarce, sought after, having reached the pound per page point. Many dustwrappers are slightly smaller than the book. The reddish bird on the spine fades easily. I once saw a copy in a plain green sugar-paper jacket. Did they run out of Ellis jackets and wrap the last few in plain ones? Rating: ****

The Salmon. One of the commoner monographs, especially as reprints, printed in larger numbers for the angling market. A first is worth more than a reprint: *** compared to **.

Lords and Ladies. Scarce, and often ex-library. This is one of the rarest titles in the series, and now in the same high flying company as *Heron, Hawfinch et al.* It is a smashing book, but the slow sales didn't encourage more botanical titles. Rating: ****

Oysters. Also rather scarce, but it sold well enough to be reprinted once. Beautifully written. Rating: *** (and ** for the reprint).

The House Sparrow. Also rather scarce, despite two reprints, and sought after by birders. Rating: *** (and ** for the reprints).

The Wood Pigeon. The print-run would suggest that *Wood Pigeons* are as scarce as *Sparrows* and *Rabbits*, but one seems to see more of them about and for the moment I will leave it with **.

The Trout. The commonest monograph after *Badger* and *Herring Gull*, intended for the large fisherman's market. Reprints abound, but the first edition is a little more elusive. *** for the first, ** for the reprints.

The Mole. Quite a few *Moles* about, often in immaculate condition. Make sure it is the home-produced Collins one and not the US or Reader's Union variety (on the other hand, if that doesn't matter to you, ignore this advice). Rating: **

Appendix 4

The Australian Naturalist Library

A series of six titles about Australian wildlife was published by Collins in Sydney, Australia, between 1974 and 1981. They were obviously akin to the New Naturalist series, with an ecological approach marrying up-to-the-minute science with natural history and land-use. Physically, too, they were similar, bound in octavo, cloth boards, with gilt lettering, and with text figures and colour plates. However the titles were not individually numbered, and the dust jackets of all but the last were more like the natural history books published by Poyser, being off-white and bearing a coloured drawing below the title. A logo above the Collins imprint on the spine proclaimed the books to be members of The Australian Naturalist Library. The last volume, *A Coral Island*, differed in having blue, instead of brown, boards and a wrap-around photographic jacket.

The books are quite common secondhand, although unsurprisingly found more often in Australia and the United States than in Britain (though I'm told that some years ago a bookshop in Hay-on-Wye had a huge remainder stock of one of the titles, *Living Insects)*. For several titles there was a Book Club edition published in laminated pictorial boards with a like jacket in 1975, and some titles also had a revised quarto paperback edition in 1982-84. *Bird Life* and *Living Insects* were printed in the United States by Taplinger of Marlborough, New Jersey. All in all, the series seems to have been fairly successful, if short-lived. The books are available plentifully on online book services such as Abebooks.com, averaging between £5-15, depending on condition. In February 2004, the following number of copies were on sale on 'Abebooks': Fish (36), Insects (21), Bird Life (23), Spiders (5), Frogs (14), Coral Island (13).

The six titles are as follows:

J.M. Thomson (1974) *Fish of the Ocean and Shore*. 208 pp, 16 pp colour and 4 pp b/w plates by Margaret Senior. ISBN 0002114364. Book Club edn, Sidney 1975. A background to recreational (sport) and commercial fishing in Australian waters.

R.D. Hughes (1974) *Living Insects*. 304 pp, colour and b/w plates, and 69 text figures. ISBN 0800849299. Book Club edn, Sidney, 1974. USA edn 1975. A survey of insect life in Australia.

I. Rowley (1974) *Bird Life*. 284 pp, colour and b/w plates, text illustrations and tables. ISBN 0002164361. Book Club edn, Sidney, 1975. USA edn. 1975. New paperback edition, 1982. An introduction to Australian birds, including their ecology, with life-histories of selected species.

Barbara York Main (1976) *Spiders of Australia*. 296 pp, 44 colour illustrations, 12 b/w, text figures. New paperback edition, 1984.

Michael J. Tyler (1976) *Frogs*. 256 pp, 20 pp colour and b/w plates, text drawings. ISBN 000214429. Reprinted in paperback, 1976. New paperback edition, 1982.

Covers the frogs of Australia, New Guinea and nearby islands. The author had also written volumes about the frogs of individual Australian states.

Harold Heatwole (1981) *A Coral Island. The story of One Tree Island and its reef.* 200 pp, 16 colour plates, text illustrations. Intensive study of a small island near the Tropic of Capricorn at the southern tip of the Great Barrier Reef.

Notes

Chapter 2

1. 'The coming of peace reminded many people of the oldest and most enduring of simple pleasures, the countryside. For the benefit of the countrysiders, so long cut off by barbed wire and defence regulations from their haunts, the list was extended, though for some years it had already been noted for works devoted to natural history, horticulture and ornithology. Now, a very distinguished venture in this field was launched in the form of *The New Naturalist Series.*

   This ambitious collection had gone through the chrysalis stage of planning during a grim period of the war. It was first discussed in 1942 by W.A.R. Collins, Julian Huxley the biologist, James Fisher the ornithologist, and a director of Adprint Ltd. The project they discussed was the present chairman's own. For a long time he had pondered a new series of illustrated nature books which would be not merely a popular addition to the literature of natural history, but a series of definitive texts judged by the strictest scientific standards. It was therefore resolved to conduct a complete new survey of Britain's natural history with at least fifty titles, each written by a specialist and all edited by a committee of four – James Fisher, Julian Huxley, John Gilmour the botanist and Dudley Stamp the geologist; Eric Hosking was to be photographic editor.'

   From *The House of Collins* (1952). Fanfare: 1945 and after.

2. The co-option of 'a botanist and a geologist' to the editorial board took place in early 1943. The following letter from William Collins to Julian Huxley and James Fisher sets out the responsibilities of the editors and the relatively generous terms they were offered. This could be said to be the letter that launched 100 books!

   '6th January 1943

   Dear Professor Huxley and Mr Fisher,

   Messrs Wm Collins Sons & Co Ltd, and Messrs Adprint Ltd have jointly decided to ask you, together with a botanist and a geologist, to prepare a survey on *Natural History Books* on the lines discussed at several meetings.

   It is understood that you will give in this survey a very detailed synopsis of the text of every book, even indicating the individual chapters, and that you will also provisionally specify between 32 and 48 pictures per book.

   We further understand that you will give us by the end of February 1943 an interim report covering ten of such books which we in our mutual judgement consider the most suitable for launching a series of such variety and extent.

   A fee of £1,000, covering all expenses of the four editors in the preparation of these reports has been agreed upon, and the four editors have consented to allow their names to appear in the books of this series in view of the preparatory work, even if any of them is unable to supervise the production of the books in detail.

It was also agreed that, during the preparation of, or after the approval of the report[,] the editors would make themselves responsible for finding authors to undertake the writing of the books in the series at an inclusive royalty for author and editor of:-

6% up to 5,000 copies
7½% from 5,000 to 10,000
9% thereafter

In the event of any of the editors being unable to continue on the board, the remainder will make themselves responsible for finding an alternative editor of similar standing.

Yours faithfully
[W.A.R. Collins]
Director
Wm. Collins Sons & Co Ltd.'

3. As the above letter makes clear, the editors were originally charged with pairing ten titles and authors. The shortlist would almost certainly have included the first four titles of the series, *Butterflies*, *British Game*, *London's Natural History* and *Britain's Structure and Scenery*. Judging from Collins' circular from 1946, the others were probably *Insect Natural History*, *Wild Flowers*, *Mushrooms and Toadstools*, *Natural History in the Highlands and Islands*, *The Sea Shore*, and either *Climate and the British Scene* or the never-published *Fish*. Other titles in preparation by 1946 were *Mountains and Moorland*, *Fishing and Angling* (Eric Taverner), *A Country Parish* and *Moths*. Each title went on sale for 16 shillings.

Chapter 3

4. A lengthy correspondence between Collins & Sons Ltd and Adprint survives in a contemporary box-file. Adprint expressed the wish to retire from the New Naturalist series in 1950, since it was 'no longer an economic proposition' for them. An ambiguity over original colour slides had soured relations: Adprint maintained that Collins was under contract to pay £10 for each slide selected by the editorial board, whilst William Collins claimed he was bound to pay only for those actually used. The end of the Collins-Adprint partnership was one reason why books published after 1950 had fewer colour plates; the other was increased production costs. One of the losses was the beautiful maps and diagrams prepared for the books from the author's sketches by Adprint.

Chapter 5

5. A detailed portrait of E.B. Ford and his scientific career by Judith Hooper, an American journalist, was published in 2002 as *Of Moths and Men: An evolutionary tale* by Fourth Estate, London. In her vivid portrayal of Ford and his Oxford contem-

poraries, the author was helped by three New Naturalist authors, R.J. Berry, Michael Majerus and Dame Miriam Rothschild. The personalities, rivalries and closet politics behind the science is well-dissected and as readable as a novel. Her underlying thesis that the Oxford evolutionists were a bunch of eccentric amateur naturalists masquerading as world scientists and cooking their results is based on numerous misunderstandings. For a corrective account of Ford and Kettlewell's work on melanism, see Michael Majerus's *Moths*, especially chapter 9.

I asked Michael Majerus to comment on the criticisms of E.B. Ford and Bernard Kettlewell's work on the Peppered Moth in Judith Hooper's book. This is his reply.

Of Moths and Journalists
by Michael Majerus

The rise and fall of the melanic Peppered Moth is one of the most famous examples of Darwinian evolution in action. The black form, *carbonaria*, first recorded in Manchester in 1848, increased rapidly in industrial areas, becoming the predominant form in just fifty years. It remained at high frequencies in areas affected by industrial pollution until after the Clean Air Acts in the 1950s. Since then, *carbonaria* has been declining. J.W. Tutt first proposed an explanation of the rise of *carbonaria* in 1896. He postulated that on trees denuded of lichens by sulphur dioxide and blackened by soot, birds found the black moths harder to find than the pale ones, so more black moths survived to reproduce. In unpolluted regions, where the pale form had the better camouflage, it remained predominant. In the 1950s, Bernard Kettlewell, under the mentorship of E.B. Ford, tested and confined this hypothesis with his classical bird predation and mark-release-recapture experiments, in both polluted and unpolluted woodlands.

Over the next 40 years, other workers studied the finer facets of the case: observing the precise behaviour of the moth, making and testing computer-model predictions of the decline in the frequencies of forms, and refining Kettlewell's experiments. These studies have added elements to the story, such as the per-generation dispersal-distance of the moths. Or changed minor details of the story, like the fact that the Peppered Moth rests on lateral branches and twigs in the canopy more commonly than on tree trunks. But the story of the melanic moth remained the example *par excellence* of Darwinian evolution in action. However, in the last few years the Peppered Moth's reputation, and with it that of Kettlewell, Ford and others who have worked on the case, have become unfairly tarnished.

In my New Naturalist book *Moths*, I defended melanism in the Peppered Moth as an example of evolution in action. However, in the same year, Judith Hooper published her book *Of Moths and Men*, where she attacks Tutt's hypothesis, undermines Kettlewell's work on the moth with veiled accusations of bullying by Ford and fraud by Kettlewell, and assassinates the characters of both these two scientists, and British naturalists generally. As a British naturalist, I am grateful to Peter Marren for the opportunity to defend myself, Ford, Kettlewell and other British naturalists.

So, what of Hooper's book? On the characters of Ford and Kettlewell I have nothing to say. I knew neither of them, so cannot comment on their personalities, although I have long respected the focus, drive and innovation of Ford's work in the field of ecological genetics, and Kettlewell's knowledge and abilities as a

Lepidopterist and naturalist. Rather, I will comment briefly on the factual accuracy of Hooper's book, on the accusations of fraud and say a word in defence of British naturalists, new or otherwise.

Hooper's book is strewn with factual errors, and she demonstrates repeatedly that she understands little of evolution, and even less about the moth that is her subject. Let us take the very first sentence of chapter 1. It reads, "To begin at the beginning, the Lepidoptera are divided into two orders: butterflies (*Rhopalocera*) and moths (*Heterocera*)". But they are not. Moths and butterflies all belong to a single order, the Lepidoptera, with the butterflies comprising a small group of superfamilies within the order. So, fact one is wrong, as are many of the other 'facts' in Hooper's book. But Hooper's book does not deal only in fact.

Much of *Of Moths and Men* is conjecture, opinion or interpretation. Let's consider one of the suggestions of data-fudging. Hooper makes large of the fact that Kettlewell increased the number of marked moths he released during his experiments. She ties the consequent increase in recapture rates to a letter from Ford to Kettlewell dated 1st July, in which Ford wrote: "It is disappointing that the recoveries are not better ... However, I do not doubt that the results will be very worthwhile ....". Hooper translates this passage as: "Now I do hope you will get hold of yourself and deliver up some decent numbers". Yet, if you read the data, it is clear that Kettlewell's recapture rate increased on the morning of 1st July, as a result of increased releases on 30th June, and that could not possibly have been in response to Ford's letter. Perhaps Hooper, when she read Kettlewell's table of results, failed to understand that day follows dawn, and night follows dusk, so moths recaptured on a particular date are *not* the result of moths released on that date.

Finally, what of Hooper's opinion of British naturalists, who are described on the flysheet as "eccentric"? In her first chapter, Hooper assassinates the character of 'moth men', who, she says, have the "stunted social skills of the more monomaniacal computer hackers, going about with misbuttoned shirts and uncombed hair, spouting taxonomic Latin". She cites Ted Sargeant, one of Kettlewell's strongest critics, who considers moth collectors to be weirder than butterfly collectors. According to Hooper, Sargeant is awed by moth enthusiasts who "can go up to a streetlight and start naming these things ..... It's an extraordinary talent". But it isn't extraordinary. Countless children across the world can recognize hundreds of different Pokemon characters, and provide details of their characteristics, their evolutionary potential and their powers in contest. How is this different to a 12-year-old who can recognize several hundred species of moth, knows when they fly, and what their larvae feed on? Calling out names to a group around a moth trap, the names I use are in English, not Latin, for I learnt them, out of interest and fascination, when I was a child, and the English names were easier.

As a child, I learnt much about moths and butterflies, their behaviour, ecology, hereditary patterns and evolution from Ford's two New Naturalists books, *Butterflies* and *Moths*. I was flattered when HarperCollins asked me to write a new book for the series on Moths, but I knew that our knowledge and understanding had advanced sufficiently in the forty-plus years since Ford's *Moths* to make the project worthwhile.

Biology advances unevenly. A novel insight, or new technique, can lead to a great leap forward in understanding, followed by smaller advances as hypotheses are examined and refined, added to or amended, and empirical evidence is gathered in support. The additions and amendments to detail do not invalidate previous work: they simply improve it. Thus, while Albert Einstein's view of the universe was more correct than Sir Isaac Newton's, we still revere the former, as we do Einstein, despite the refinements to his ideas made by Stephen Hawking. This is as it should be. No doubt our current understanding of the processes of biological evolution will be amended and added to in the future. But this does not mean that our current perceptions are wholly wrong; they are just, as yet, incomplete.

Similarly, increased understanding of the natural history of the Peppered Moth since the 1950s does not refute Kettlewell's work, it validates it. One or two details have changed, but the tale is still largely the same. The rise and fall of the melanic Peppered Moth still stands as one of the best examples of Darwinian evolution in action.

6. Contemporary reviews of *Butterflies*:

   'I have read nearly every book published about British butterflies from 1720 onwards, and I declare without hesitation that Dr Ford's brilliant treatise at once steps into the position of an indispensable and classic work' – Compton MacKenzie.

   '[The colour plates] are almost beyond praise for beauty and accuracy' – Arthur Tansley.

   At least one elderly entomologist was less impressed. '…Having twice looked over and tried to read the vol. – I had to give it up. It is to me a mixy-maxy-queer hoch-poch. It jumps over all 'obstacles' and keeps one turning over the pages to see where you are at – ? – It leads to nowhere – … You have been 'Had' – It is a catch penny for the younger generation and at 16/-?? no Sir.

   – Alexander M. Stewart. Letter to a young friend, quoted in *Butterflies of Scotland* (1980) by George Thomson.

Chapter 8

7. An internal minute preserved among the early New Naturalist papers reveals that 'the original idea was to have 6 or 8 good selling monographs'. *The Fulmar*, *The Wren* and *Fleas, Flukes and Cuckoos* were referred to as 'super-monographs' because of their larger size. It was calculated that Collins would need to sell five to six thousand copies of *The Fulmar* before it went into profit.

Chapter 10

8. The series came closer to ending in 1984 than I had realised when writing *The New Naturalists*. Both Michael Walter, who returned as natural-history editor for a few years after the death of Billy Collins in 1977, and Adrian House, who deputised for Collins at New Naturalist meetings, were convinced by falling sales that the New Naturalists had had their day. An internal minute by Adrian House dated 27 September 1984 and marked 'strictly confidential' addressed the future of the series. 'The series is the foundation stone of our serious Natural History publishing,' he wrote. In its early years, the books won prestige and exceptional sales because of their quality both of the contents and of their production ('especially the

jackets and colour plates'). The reputation of the series remained high, but sales had plummeted while costs, in the inflationary Seventies, had soared. The colour plates had been thrown overboard to reduce costs 'and furthermore the original design has come to look distinctly dated'. Other causes for falling sales were competition, 'the general hard-back recession', increased specialisation of the natural sciences and lack of funds for promotion.

Another 'possible factor' was the Editorial Board whose average age was now 68: 'No member has the distinction of Julian Huxley, the gift for communication of James Fisher, or the pictorial energy of the young Eric Hosking'. Some of the recent and forward titles 'may be too specialised'. And whereas in the past authors could write on small advances, relying on royalties from healthy sales to follow, that consideration no longer applied. 'It is unlikely that the most articulate, distinguished, energetic, promising or ambitious authors will devote time to writing for the series'. Finally, diminishing sales produced diminishing expectations. And titles were not receiving as much individual attention as in the past.

'The inherited policy was to retain the series for its prestige,' House went on, 'but to soft pedal its activities (*sic*) as being of doubtful commercial success.' 'We now think we should only continue it if we can produce books which are both commercially sound and important to learning.' Future books should have either popular potential 'with an allowance for promotion expenditure and deserving full-blooded selling'. Or 'specialist value' 'economically produced as trade paperbacks'. Titles which are too specialist to be commercial or too long overdue 'should be dropped from the programme'.

An alternative course of action would be to close down the series altogether. This 'would not necessarily be ignoble, as it would be equivalent of dropping in 1960 a series which had originated in 1920'. However, 'the principal argument against (it) is that it might give the impression that we had lost our fundamental concern with Natural History publishing'.

This last consideration seems to have been the clincher. The future titles were neither more nor less specialised than before, but House may have underestimated the conservatism of book-buyers, as well as the distaste among the literate for 'dumbing down'. After a series wobble in the mid-eighties the series carried on much as before, that is, with an elderly Board, little or no promotion, and derisory levels of advance. Perhaps we new naturalists are a peculiarly stubborn breed.

9. The refinding of *Aspicilia*

by Oliver Gilbert

This essay was one of twelve 'interludes' intended to follow each chapter in NN *Lichens*, to lighten the text and give a better sense of the experience of lichen-hunting. Sadly, the editors rejected them. See what *you* think. Here we go:

'Setting off into low cloud and driving rain to search for a remote lochan that may, or may not, contain a rare lichen, is not much incentive when the alternative is to hog the fire in one of Scotland's best hotels. However, the weather front that was to bring these conditions was still 3 hours away so we headed up the stalkers' path into the hills. Soon we were fording streams and tussling with peat hags as the short, well-drained, turf of the Durness limestone gave way to Cambrian quartzite bog. The day's objective was to refind *Aspicilia melanospis*, the first and only British record of which had been made by Peter James and Dougal Swinscow in 1958. I knew the name of the tiny lochan where it had been found, but the grid reference was for an adjacent square and there was reputedly limestone in the catchment, which was not the case with the lochan that was our goal. A phone call to London requesting more information had been met with an answer phone message. It might have been prudent to delay setting out, but for sometime this 'lost lichen' had been shining like a lamp in my mind drawing me northwards.

Eventually we breasted a rise and in front of us lay a pool of brown peaty water. It was still in the grip of winter and, surrounded by snow patches, it looked like a thousand other water bodies in the Highlands – hardly worth a second glance. The only stones projecting from the water were at the outflow. I knelt to examine these, and each was completely covered with a thick crust of the *Aspicilia.* This beautiful, silvery-grey species is to strongly lobate that it has recently been transferred to a new genus *Lobothallia.* As I triumphantly raised my face from near water level it was hit by a blast of icy rain, the sky grew dark, and I realised my world had been reduced to a few feet by thick cloud. Waves started to break over the rocks; it was as if the Gods of the mountain were angry that their secret had been discovered. The next twenty minutes were desperate as, in the gathering storm, I wrote notes with frozen fingers, back turned to the wind, and hopelessly tried to observe lichens through five layers of water, one on the specimen, others on both sides of my hand-lens and spectacles. The *Aspicilia* was growing in a semi-inundated zone with *Dermatocarpon luridum.* Local eutrophication was indicated by the presence of *Physcia caesia,* though whether this was associated with red deer, sheep or birds was not clear.

A cursory inspection of the rest of the lochan, and others in the vicinity, failed to find any more of the lichen but by now white horses were breaking on the shore and conditions were becoming impossible. The objective achieved, we started the return journey, but far from being at peace, my mind was active with a question that would not go away "What had drawn that pair to such a remote and seemingly unpromising spot 37 years ago?" I was determined to find out.'

Postscript

This article is reproduced, with minor changes, from The New Naturalists Book Club newsletter no. 21, January 2004, with the permission of Bob Burrow and the author, John Sykes. The Book Club is run by Mr Robert Burrow at 'Books and Things', First Tower, St Helier, Jersey JE2 3LN or telephone 01534 759949, email burrow@itl.net. Club membership costs £5.00 a year, and members receive three or four newsletters a year, with snippets about the series, as well as members' wants and sales advertisements.

The Big Bang Theory

by John Sykes

It was one of those padded November afternoons, cold if you stood still, warm if you took exercise. Heavy, grey clouds lumbered low overhead, and it had seemed like dusk all day: a day for sitting by an open fire with a glass of Hine Antique, flicking desultorily through *Grass and Grasslands*, and dreaming nostalgically of those idyllic honeymoon days in Swanage reciting chapters of *Waders* alternately with one's beloved, as the hot-water bottle struggled to retain its feet.

At 14 St James's Place the lights too were struggling, struggling to penetrate the smoky atmosphere of the Board Room. Four men were variously engaged in waiting for something to happen. On a large polished table lay scattered a few books in their arresting dust-jackets – *The Natural History of Computers* by Smith and Markham, *Grottoes and Grotto Life*, by Alexander Pope, with its foreword by H.H. Swinnerton, who in his youth had been acquainted with the author himself, Blofeld's *Lives of the Saints*, and the classic *Pond Life* by Compton and Edrich.

Of the four men, Huxley was intent on finalizing his Christmas order from his Great Universal Stores mail-order catalogue, while two of the others, Gilmour and Fisher, both had their pipes lit, and were each trying to outdo the other in imitations of the exhaust beat of a Gresley V2-class steam-locomotive. Spasmodically, they discussed the relative merits of Bactrian and Dromedary camel dung in Allbright and Munchausen's Patent Levantine Shag.

"You know, Fisher," said Gilmour, "Thesiger tells me that for the best flavour the stuff should strictly be kept in a fresh goat's bladder for at least three months."

Before Fisher could reply, a sudden flash of light in the corner of the room caused them to turn round. It was the fourth man, Hosking. There he was, in his tinted goggles, testing grades of magnesium for his next photographic expedition. "Oh bugger!" he cried.

The fifth member of the team was on duty elsewhere. Outside the main entrance Commissionaire Stamp (it was his turn that day) was feeling rather pleased with himself. He had, that very morning, finished writing his fifth book in two months, *Tectonic Plates for Lower Form Pupils*, and, resplendent in his green buckram uniform and Ellis jacket, was eagerly awaiting the arrival of Billy Collins so that the board meeting could begin. Stamp was rather enjoying himself, for on his transistor radio, which he had placed by the door, the disc-jockey Shakin' Ed Salisbury was broadcasting a lively selection of Eric Simms' Greatest Hits, a particular favourite of Stamp's. In fact, he was still tapping his feet to Simms' infectious rhythms, when he noticed the familiar tall figure of Collins approaching along the street. But Collins, who generally came alone, was accompanied by a blonde, mysterious and furry lady, who lit up the darkening London street as the awakening sun dispels the dank vapours of the night.

When they reached the entrance steps, Collins was all smiles. "Ah Stamp, so good to see you. Tea and buns in half an hour, old chap?" He slipped Stamp a couple of bob as the latter opened the door for the Glamorous One and Collins himself.

* * *

"He's late again," said Fisher, interrupting an uneasy quiet that had settled in the room.

"Trying to persuade more of his authors to stop retraining as insurance salesmen," came Gilmour's reply.

Suddenly the top sound of high-heeled shoes reverberated from along the corridor. "Collins?!" exclaimed a startled Fisher and Gilmour together. But it was not Billy Collins who strode first through the door. No, surely, it was one of the Shining Ones, an incandescence of pink and silver femininity. Open-mouthed, as if pulled by strings, the editors rose. Hosking, caught on the hop in his goggles, fumbled to remove them as Collins closed the door behind him.

"Good afternoon, gentlemen. Today I have departed from our usual routine. May I introduce Miss Barbara Cartland?" After an electric hiatus it was Gilmour who grasped the measure of the situation. "THE Barbara Cartland? The Doyenne of Romantic Fiction, the Epitome of the Feminine Force? This is indeed a great honour." "A great honour," mumbled the others in humble imitation.

Miss Cartland smiled. "And you must be Mr Gilmour. I have heard so much about you in the best circles." Collins had meanwhile hung up Miss Cartland's fur coat on the Guinness Toucan hat-stand and invited her to sit down.

"Gentlemen," he pronounced, "You all know we have been experiencing a spot of bother finding an author for "Land Molluscs". And then Dodd disappointed me. He refused to deal with anything but the Knotty Ash subspecies. How lucky, therefore, that last week, behind the W.I. stall at Hemel Hampstead market, I encountered Miss Cartland here. I mentioned my predicament, and at once she offered to jump into the breach. Isn't that so, Miss Cartland?"

"Why yes, Billy darling. To put it bluntly, gentlemen, you need me. My books sell by the million, and you have print-runs of – what? – three or four thousand? It is clear," – and here she smiled enigmatically at Gilmour – "you need a boost, and that means a new approach. I don't think even Billy here can see all the possibilities. You lack the human touch – too many facts, facts, facts, and not enough Truth. Cut out the facts, and let the romance flow. Let me show you what I mean."

Whereupon Miss Cartland took up her voluminous pink-chiffon handbag, opened the gold leopard clasps, and, from among the essentials of a practising Romantic novelist – pots of mascara, pointing trowels and a full set of Harris brushes – drew out what appeared to be a most sumptuous box of chocolates.

"Here is your future, gentlemen. Bound by Bayntun in full pink contemporary morocco, stamped on the front in gilt with hearts and cherubs blowing celestial trumpets: this book will sell. It is, I'm afraid, goodbye to buckram."

Her voice was exultant. Turning to the spine, she read out the crisp gold title lettering: *"Slugs and Snails"* by Barbara Cartland. Then, deftly undoing the charming pink bow with which the volume was loosely tied, she opened it and read the dedication: "To Billy, with a thousand kisses".

Collins blushed; but, unable to contain his pleasure, exclaimed: "Why, Babsy, I mean Miss Cartland, it's stunning! But I only asked for a specimen chapter and a list of contents!".

"With me, Billy darling, you get the whole thing," she drawled languorously. "It never takes me more than a week to finish a book."

"Even Stamp's not that quick," whispered Hosking to a dazed Fisher.

Miss Cartland handed the volume to Collins, who turned the book over in his hands to catch the angle of the light. He was about to open it, when there was a knock at the door. A clattering of cups, saucers and spoons heralded the appearance of Stamp with the refreshments. "Assam today," he announced, "with a little Bactrian for Fisher." As he put the tray down, he noticed The Shining One and stared.

"Ah, Stamp. Splendid!" said Collins. "This is Miss Barbara Cartland, our famous novelist. Miss Cartland, this is Stamp.

"Oh, but I haven't a spare c-cup," stuttered Stamp.

"Don't worry, old chap," came Gilmour's calming voice. "Miss Cartland can have mine." With that, Gilmour, transfigured, gazed serenely at Miss Cartland, as if contemplating an early Italian Madonna.

Stamp poured. "Sugar, Miss Cartland?"

"Thank you."

"Bun, Miss Cartland?"

"Why certainly, Stamp! I like a good confection."

"When we've had our buns," resumed Collins, "I shall ask our guest to read us an extract from her wonderful book."

It was when Fisher turned away from the others to wipe some cream off his glasses that he looked up and noticed there seemed to be something slightly different about the grand portrait behind him on the wall. The painting, by an unknown artist, of Gilberto Bianco di Venderenato, the great Italian diarist and *basso profundo*, had had to be cleaned the previous year after Gilmour, conjecturing a resemblance to Friedrich Nietzsche, had pencilled in a most opulent moustache. To Fisher now, the features of the face appeared to have hardened to an expression of utter disapproval. Putting it down to the dazzle from The Shining One's tiara, he resumed his position as Miss Cartland began her reading.

* * *

When Miss Cartland had finished and replaced the book in her bag, the Board stood up as one and cheered. For three solid minutes they cheered. "This is it," shouted Huxley. "This is the future. Miss Cartland, you have made an ageing man very happy. You are sublime!"

You could not, at that instant, have wiped the victorious smile from Billy Collins' face.

However, in this moment of intoxicating fervour and ecstatic anticipation, there occurred the catastrophic incident that was destined to rob natural history publishing of its most significant development in modern times.

In the confusion that inevitably followed the explosion, the details have, regrettably, become somewhat blurred. It has, nevertheless, been established that Hosking, as he withdrew from his pocket a handkerchief to dry the tears of joy steaming from his remaining eye, neglected to remember that this same handkerchief was holding one of his more virulent grades of magnesium. Inadvertently, he spilled the said substance on to Fisher's smouldering pipe, which the latter had placed on the table so as to be better able to voice his adulation. One can only be appalled at the consequences of this perilous juxtaposition. Fisher's Patent Levantine Shag, notorious in smoking circles for its fearsome unpredictability, must have reacted with the magnesium with such deafening ferocity that Miss Cartland's handbag, cosmetically inflammable as it was, itself was ignited and exploded with horrifying suddenness.

Fortunately, the Board was given just enough time to duck under the table, which received the full force of the blast but afforded sufficient defence against the ensuing shrapnel from Miss Cartland's exploding accoutrement. Lipstick, mascara, nail-varnish flew everywhere. It was mayhem. Miss Cartland herself, unaccustomed to such behaviour in a boardroom of such standing, was taken completely by surprise. Though shaken and somewhat coloured by the blast, damage to her person was superficial, due in the main to her unparalleled understanding of cosmetic protection. This, and the exemplary presence of mind of the ever-practical Stamp, who administered on-the-spot treatment in the shape of a good washing-down with cold tea and application of several poultices fabricated *ex-tempore* from unconsumed cream buns, undoubtedly saved Miss Cartland from lasting physical damage.

When the dust had settled, the worst aspect of the catastrophe came to light: the pink manuscript was no more, save for a few shreds of ribbon on the ceiling and a noticeable smell of burnt sugar pervading the room.

Eventually Miss Cartland gathered herself together sufficiently to address the Board. It is unnecessary to add that she was beside herself.

"....and puppy dogs' tails!" she snorted. "Never again!"

"Stamp," sighed Collins, "I think you'd better escort Miss Cartland and call a cab."

And that was the last the Board ever saw of Miss Barbara Cartland, the Queen of Romantic Fiction.

* * *

It may be of interest to some of our readers to know, that years later, after John Gilmour passed away, among his effects was found a tantalising glimpse of what might-have-been. Inside his own copy of "Fossils" there was discovered a pretty pink envelope with a still discernible fragrance of lavender, enclosing two pieces of charred typewritten script. Considering the nature and condition of the material, Gilmour must, it is supposed, have come across it in the debris following the disastrous meeting with Miss Cartland. He had kept and cherished them all these years! I am sure you will agree with me that they constitute two of the most intriguing fragments in the whole of literature and are, in my considered opinion, comparable in quality to the wisps of beauty contained in the few fragments of Sappho's Odes that have come down to us.

I append the pieces here. In the first, we are blessed indeed to have the page of contents itself.

TABLE OF CONTENTS

The second fragment, from page 381, gives us an unmistakable vision of the depth of knowledge and passion that Miss Cartland brought to bear on her subject.

COURTSHIP AND MARRIAGE

...dismissed her devoted retainers in their gold lamé tunics and slithered over to where the Marquis was waiting.

"We are alone at last," she said softly, as a frisson of expectancy rippled through her body. She knew from Archie's trembling antennae that he felt the same.

"Oh, Serena, my own dear slug, how happy I am that the tumult of the day is over."

"Yes, the nuptial banquet laid on by His Royal Highness the Prince of Snails was exquisite. The hostas, especially, were divine!"

The red sun had set, but still diffused the thin lines of cloud with pink, apricot and gold. Soon Leo and Virgo would be visible in the southern sky. Serena Slug could not have been happier.

"Look," said the Marquis, pointing. "As far as the eye can see is the Molluscia Estate, and now it is all yours to share! Over there is where my mother laid me as an egg, and yonder the Little Gem lettuces that were the delight of my youth." Suddenly he shuddered: he had seen the spot where the hateful Harry Hedgehog had dined on his poor brothers and sisters, one damp summer morning.

Archie kissed her. Serena, the new Marchioness of Molluscia, felt an intense awareness of life in its fullness invade her whole being, the yearning ache of mollusc desire surging through her body like an uncontrollable fire, passion

> churning inside her. At last she would give herself to her mollusc master, Archie, the Fifth Marquis of Molluscia. As she lay enraptured, their eyes met.
>
> "Oh Serena," he gasped.
>
> "Oh Archie," she groaned.
>
> Then, slowly and masterfully, Archie raised his....

Here, unfortunately, the fragment breaks off, but what instantly impresses is the dynamic style, which New Naturalist titles have ever since been endeavouring to recapture, as yet unsuccessfully in my view. It is indeed true that certain passages in "British Warblers" and "The Soil" remind one strongly of Miss Cartland's heady style and voluptuous language, but it must be deplored that nothing since has been totally free from a stolid factuality and a reprehensible absence of romantic interest.

Yes, with what abject misery we must lament the destruction of that Great Pink Book, an achievement that, had it withstood the assault of magnesium and Levantine Shag, would surely have raised the reputation of its author to a level equal with, if not indeed higher than, the very greatest of the Ancients.

And so I end this chronicle with a plea: a plea for a stylish writer with a love of Romance and a naturalist's bent to step forward and help re-shape our beloved series. Is there anybody out there?

General Index

Text illustrations of the subject are indicated by italics.